Protein-Ligand Interactions

Protein-Ligand Interactions

Proceedings of a Symposium held at the University of Konstanz, West Germany, September 2–6, 1974

Edited by
Horst Sund and Gideon Blauer

Walter de Gruyter · Berlin · New York 1975

Editors:

Gideon Blauer, Dr. rer. nat.
Professor of Biochemistry, Department of Biological Chemistry,
The Hebrew University, Jerusalem, Israel

Horst Sund, Dr. rer. nat.
Professor of Biochemistry, Fachbereich Biologie,
Universität Konstanz, Konstanz, West Germany

Cover design: R. Hübler. – Printing: Druckerei Gerike, Berlin. – Binding: Mikolai, Berlin. –
Printed in Germany
ISBN 3 11 004881 7

Preface

The present volume contains the proceedings of a symposium on "Protein-Ligand Interactions" held at the University of Konstanz, West Germany, on September 2–6, 1974.

Protein-ligand interactions are of great biological significance and cover a wide spectrum of systems. They range from respiration to the action of receptors and repressors and to the subtle influence of electrolytes on proteins. The last decade has witnessed great progress in a molecular approach to biopolymer-small molecule interactions, which includes the application of thermodynamics, quantum mechanics and modern kinetics to protein-ligand interactions, particularly to enzymatic reactions.

Symposia or congresses usually are devoted either to a large variety of subjects which are only loosely connected, or they are highly specialized in a narrow field. The basic idea of the present Symposium has been to assemble both biologically and physico-chemically oriented scientists in order to discuss current topics in the wide field of protein-ligand interactions. It has been the organizers' hope to promote an interdisciplinary interaction and stimulation of ideas between scientists looking at specific problems from different viewpoints. It is for this reason that the discussions at this Symposium have had no time limit placed on them, as far as this is possible.

Photoprints of the submitted typescripts were circulated among the participants before the Symposium in order to facilitate the progress of the proceedings and discussions. The discussions were not directly recorded. The participants were requested to provide a written report of what they considered worth including in the publication. This information is presented in this volume.

As usual for symposia, there was no refereeing of the papers submitted, the contents of which are solely the authors' responsibility.

As a natural consequence of the aims of this meeting, the number of participants had to be severely limited. A large number of outstanding workers in the fields discussed, unfortunately, could not be heard directly. We hope that the reading of these proceedings will, nevertheless, provide some stimulus and inspiration for future research also for those who did not attend the meeting.

Our appreciation is extended to all participants who have cooperated so effectively in order to promote this Symposium and publication of its proceedings. Special thanks are due to the Stiftung Volkswagenwerk, whose generous financial support has made this Symposium possible. The support given by the University of Konstanz, where the meeting took place, by the Gesellschaft der Freunde und Förderer der Universität Konstanz and by the Byk Gulden Lomberg Chemische Fabrik GmbH, Konstanz, is gratefully acknowledged.

We would like to acknowledge the cooperation of the staff of the Verlag Walter de Gruyter, Berlin. We are also very grateful for Mrs. S. Lau's devoted assistance in all matters concerning the Symposium.

In conclusion, the editors would like to express their hope that meetings of a similar kind on various biological and biophysical topics of interest will be held in the future and will be instrumental in the promotion of new ideas and approaches in the natural sciences.

Konstanz, September 8, 1974

H. Sund
G. Blauer

Contents

List of Contributors

E. Antonini, Istituto di Chimica, Facoltà di Medicina e Chirurgia, Università di Roma, Roma, Italy

M.D. Barkley, Department of Chemistry, University of California, San Diego, California, USA

F.J. Barrantes, Max-Planck-Institut für Biophysikalische Chemie, Göttingen, West Germany

K. Beyreuther, Institut für Genetik, Universität Köln, Köln, West Germany

G. Blauer, Department of Biological Chemistry, The Hebrew University, Jerusalem, Israel

J. Bode, Gesellschaft für Molekularbiologische Forschung mbH, Braunschweig-Stöckheim, West Germany

W. Boos, Fachbereich Biologie, Universität Konstanz, Konstanz, West Germany

S. Bourgeois, The Salk Institute for Biological Studies, San Diego, California, USA

J. Brahms, Centre National de la Recherche Scientifique, Institut de Biologie Moléculaire de la Faculté des Sciences, Paris, France

M. Brunori, Istituto di Chimica Biologica, Università di Roma, Roma, Italy

M. Buehner, Forschergruppe Röntgenstrukturanalyse, Physiologisch-Chemisches Institut, Universität Würzburg, Würzburg, West Germany

J.P. Changeux, Neurobiologie Moléculaire, Institut Pasteur, Paris, France

E. Chiancone, Istituto di Chimica, Facoltà di Medicina e Chirurgia, Università di Roma, Roma, Italy

J.B. Cohen, Neurobiologie Moléculaire, Institut Pasteur, Paris, France

A. Colosimo, Istituto di Chimica Biologica, Università di Roma, Roma, Italy

C. Coulondre, Département de Biologie moléculaire, Université de Genève, Genève, Switzerland

J. Deutsch, Church Laboratory of Chemical Biology, California Institute of Technology, Pasadena, California, USA

H. Dieter, Fachbereich Biologie, Universität Konstanz, Konstanz, West Germany

E.E. Di Iorio, Friedrich Miescher-Institut, Basel, Switzerland

T. Fanning, Institut für Genetik, Universität Köln, Köln, West Germany

J.G. Files, Biological Laboratories, Harvard University, Cambridge, Massachusetts, USA

J. Flossdorf, Gesellschaft für Molekularbiologische Forschung mbH, Braunschweig-Stöckheim, West Germany

J.-M. Frère, Service de Microbiologie, Faculté de Médecine, Institut de Botanique, Université de Liège, Liège, Belgium

D. Ganem, Biological Laboratories, Harvard University, Cambridge, Massachusetts, USA

N. Geisler, Institut für Genetik, Universität Köln, Köln, West Germany

D. Gho, Institut für Genetik, Universität Köln, Köln, West Germany

J.-M. Ghuysen, Service de Microbiologie, Faculté de Médecine, Institut de Botanique, Université de Liège, Liège, Belgium

W. Gilbert, Department of Biochemistry and Molecular Biology, Biological Laboratories, Harvard University, Cambridge, Massachusetts, USA

R.S. Goody, Abteilung Biophysik, Max-Planck-Institut für Medizinische Forschung, Heidelberg, West Germany

A. Gordon, Department of Neurology, School of Medicine, University of California, San Francisco, California, USA

J. Gralla, Department of Biochemistry and Molecular Biology, Biological Laboratories, Harvard University, Cambridge, Massachusetts, USA

C.A. Gross, Institute of Molecular Biology and Department of Chemistry, University of Oregon, Eugene, Oregon, USA

H.D. Heilmann, Abteilung für Chemie, Ruhr-Universität Bochum, Bochum, West Germany

E.J.M. Helmreich, Physiologisch-Chemisches Institut, Universität Würzburg, Würzburg, West Germany

P. Hemmerich, Fachbereich Biologie, Universität Konstanz, Konstanz, West Germany
M.P. Heyn, Biozentrum, Universität Basel, Basel, Switzerland
H.J. Hinz, Fachbereich Biologie, Universität Regensburg, Regensburg, West Germany
P.H. von Hippel, Institute of Molecular Biology, University of Oregon, Eugene, Oregon, USA
F. Hucho, Fachbereich Biologie, Universität Konstanz, Konstanz, West Germany
R. Jaenicke, Fachbereich Biologie, Universität Regensburg, Regensburg, West Germany
A. Jobe, The Salk Institute for Biological Studies, San Diego, California, USA
J. Kania, Institut für Genetik, Universität Köln, Köln, West Germany
A. Karlin, Department of Neurology, College of Physicians and Surgeons, Columbia University, New York, New York, USA
P. Kathmann, Institut für Genetik, Universität Köln, Köln, West Germany
M. Kempfle, Physiologisch-Chemisches Institut, Universität Bonn, Bonn, West Germany
K. Kirschner, Biozentrum, Universität Basel, Basel, Switzerland
R. Knippers, Fachbereich Biologie, Universität Konstanz, Konstanz, West Germany
R. Koberstein, Fachbereich Biologie, Universität Konstanz, Konstanz, West Germany
M.-R. Kula, Gesellschaft für Molekularbiologische Forschung mbH, Braunschweig-Stöckheim, West Germany
T. Kushnir, Department of Chemistry, Tel-Aviv University, Tel-Aviv, Israel
E. Lavie, Department of Biological Chemistry, The Hebrew University, Jerusalem, Israel
P. Lu, Département de Biologie moléculaire, Université de Genève, Genève, Switzerland
R.A. Lutz, Institut für Molekularbiologie und Biophysik, Eidgenössische Technische Hochschule, Zürich, Switzerland
G. Maass, Institut für Klinische Biochemie und Physiologische Chemie, Medizinische Hochschule, Hannover, West Germany
J. Majors, Department of Biochemistry and Molecular Biology, Biological Laboratories, Harvard University, Cambridge, Massachusetts, USA
H.G. Mannherz, Abteilung Biophysik, Max-Planck-Institut für Medizinische Forschung, Heidelberg, West Germany
A. Mansouri, Friedrich Miescher-Institut, Basel, Switzerland
A. Maxam, Department of Biochemistry and Molecular Biology, Biological Laboratories, Harvard University, Cambridge, Massachusetts, USA
M.G. McNamee, Department of Neurology, College of Physicians and Surgeons, Columbia University, New York, New York, USA
H. Meissner, Institut für Genetik, Universität Köln, Köln, West Germany
D. Michaelson, Church Laboratory of Chemical Biology, California Institute of Technology, Pasadena, California, USA
J.H. Miller, Département de Biologie moléculaire, Université de Genève, Genève, Switzerland
T. Moody, Church Laboratory of Chemical Biology, California Institute of Technology, Pasadena, California, USA
B. Müller-Hill, Institut für Genetik, Universität Köln, Köln, West Germany
G. Navon, Department of Chemistry, Tel-Aviv University, Tel-Aviv, Israel
M. Nieto, Instituto de Biologia Celular, Madrid, Spain
I. Pecht, Department of Chemical Immunology, The Weizmann Institute of Science, Rehovot, Israel
H.R. Perkins, National Institute for Medical Research, Medical Research Council, London, England
Th. Pfeuffer, Physiologisch-Chemisches Institut, Universität Würzburg, Würzburg, West Germany
T. Platt, Biological Laboratories, Harvard University, Cambridge, Massachusetts, USA
F.M. Pohl, Fachbereich Biologie, Universität Konstanz, Konstanz, West Germany
J.L. Popot, Neurobiologie Moléculaire, Institut Pasteur, Paris, France
H.-J. Prätorius, Gesellschaft für Molekularbiologische Forschung mbH, Braunschweig-Stöckheim, West Germany
M.A. Raftery, Church Laboratory of Chemical Biology, California Institute of Technology, Pasadena, California, USA
B. Ramirez, Fachbereich Biologie, Universität Konstanz, Konstanz, West Germany

A. Revzin, Institute of Molecular Biology and Department of Chemistry, University of Oregon, Eugene, Oregon, USA

D. Riesner, Institut für Klinische Biochemie und Physiologische Chemie, Medizinische Hochschule, Hannover, West Germany

M.J. Ross, Church Laboratory of Chemical Biology, California Institute of Technology, Pasadena, California, USA

M.R. Rossi Fanelli, Istituto di Chimica, Facoltà di Medicina e Chirurgia, Università di Roma, Roma, Italy

J.R. Sadler, Department of Biophysics and Genetics, University of Colorado Medical Center, Denver, Colorado, USA

M. Schlotmann, Institut für Genetik, Universität Köln, Köln, West Germany

U. Schmeissner, Département de Biologie moléculaire, Université de Genève, Genève, Switzerland

A. Schmitz, Institut für Genetik, Universität Köln, Köln, West Germany

J. Steinhardt, Department of Chemistry, Georgetown University, Washington, D.C., USA

R.M. Stroud, Church Laboratory of Chemical Biology, California Institute of Technology, Pasadena, California, USA

H. Sugiyama, Neurobiologie Moléculaire, Institut Pasteur, Paris, France

H. Sund, Fachbereich Biologie, Universität Konstanz, Konstanz, West Germany

I. Triesch, Institut für Genetik, Universität Köln, Köln, West Germany

R. Vandlen, Church Laboratory of Chemical Biology, California Institute of Technology, Pasadena, California, USA

C. Veeger, Department of Biochemistry, Agricultural University, Wageningen, The Netherlands

K. Wagner, Gesellschaft für Molekularbiologische Forschung mbH, Braunschweig-Stöckheim, West Germany

A.C. Wang, Institute of Molecular Biology and Department of Chemistry, University of Oregon, Eugene, Oregon, USA

J.C. Wang, Department of Chemistry, University of California, Berkeley, California, USA

G. Weber, Department of Biochemistry, University of Illinois, Urbana, Illinois, USA

K. Weber, Max-Planck-Institut für Biophysikalische Chemie, Göttingen, West Germany

M. Weber, Neurobiologie Moléculaire, Institut Pasteur, Paris, France

H.G. Weder, Institut für Molekularbiologie und Biophysik, Eidgenössische Technische Hochschule, Zürich, Switzerland

C.L. Weill, Department of Neurology, College of Physicians and Surgeons, Columbia University, New York, New York, USA

W. Weischet, Biozentrum, Universität Basel, Basel, Switzerland

U.-W. Wiegand, Institut für Molekularbiologie und Biophysik, Eidgenössische Technische Hochschule, Zürich, Switzerland

K.H. Winterhalter, Friedrich Miescher-Institut, Basel, Switzerland

R.L. Wiskocil, Biozentrum, Universität Basel, Basel, Switzerland

R.W. Woody, Department of Chemistry, Arizona State University, Tempe, Arizona, USA

J. Wyman, Istituto Regina Elena per lo Studio e la Cura dei Tumori, Roma, Italy

J. Yonath, Department of Biological Chemistry, The Hebrew University, Jerusalem, Israel

Section I. General

Chairman: *J. P. Changeux*

Rome, June 21, 1974

To Members of the Konstanz Symposium:

Allow me to say how much I appreciate having been asked to participate in the Konstanz Symposium on Protein-Ligand Interactions, and how much I regret not being able to be present. My absence is due to involvement in a program of IUPAB lectures in Eastern Europe which come just at the same time and for which I am responsible. Not being able to take part in person, I was planning to submit a general introductory paper to be read by someone else. Unfortunately, however, I fell ill just at the time I had reserved for preparing it; as a result, I am substituting the rather more specialized article enclosed. It represents a joint effort with two of my colleagues in Rome, Prof. M.Brunori and Mr. A. Colosimo.

As it is, I like to think of some of the things I might have talked about had I been present. I might, for instance, have developed and amplified the concept of linkage, both homotropic and heterotropic, which is so basic to the question of regulation in biological macromolecules, showing how very simply it emerges from the concept of the Binding Potential. Thence I might have gone on to point out that the Binding Potential itself, in which the variables are the chemical potentials μ of all the components except the reference component (e.g. the macromolecule), actually corresponds to only one member of a family, or set, of potentials in each of which the variables consist of an arbitrary combination (or mixture) of μ's and the corresponding n's (n denoting the total amount of a component). These mixed potentials are applicable to experiments in which certain components are maintained at constant μ (this means that the system is open with respect to those components), while others are maintained at constant n (this means that the system is closed with respect to those components). They can all be derived from the total energy of the system (or from one another) by a Legendre transformation, just as the Hamiltonian function can be derived from the Lagrangian one in classical mechanics. They are $2^r - 1$ in number, where r is the total number of components (including the reference component) and the -1 arises from the special property of the energy as a first order homogeneous function of its variables (entropy, volume, and the n's).

Another line of thought, possibly of greater interest at this meeting, would have been a review of the allosteric concept as a mechanism to account for the homotropic and heterotropic linkages exhibited by large working proteins and other macromolecules in which the sites are so far apart that the possibility of direct interactions between them can be ruled out. In nearly all cases of polyfunctional molecules the binding potential may be formulated in terms of the logarithm of a binding polynomial, which is in essence a macroscopic form of the grand canonical partition function of statistical mechanics: it shows how the ligands are partitioned among the various forms which the molecule assumes in their presence and provides an admirable means of getting an overall view of the response of the molecule to its environment. In the case of an allosteric system the binding polynomial assumes a rather special form, whose factorability or non factorability is a criterion of heterotropic or homotropic interaction between sites (as indeed it always is). And then, passing on from this, it might have been interesting to point out how the binding polynomial can be formulated equally well either in terms of the free energy changes and equilibrium constants of individual binding events or alternatively in terms of probabilities and conditional probabilities of site occupation. (The probability approach is the subject of a recent paper by Paul E. Phillipson and myself which will soon appear in the Proceedings of the National Academy of Sciences).

Still one more topic might have been the linkages which can arise in a polyfunctional macromolecule, such as an enzyme, which is not in equilibrium but only in a steady state in relation to its environment. Such a system can be represented formally by a multidimensional cube in which various closed pathways have an element in common. This provides a picture of how, as a result of the steady flow of matter through the system and the consequent steady liberation of energy, one process can drive another.

Of course, even had I been present, I could not have hoped to deal with all these topics, but at least I could have touched on some of them in a more lively and personal way than is possible on paper. All the same I do hope that the somewhat more specialized communication I am submitting may not be wholly without interest.

With best wishes for the success of the meeting and deep regrets at not being there.

Jeffries Wyman

Jeffries Wyman

Concerted Changes in an Allosteric Macromolecule

Alfredo Colosimo, Maurizio Brunori and *Jeffries Wyman*

ABSTRACT

This paper reports an analysis of the behaviour of allosteric macromolecules in the framework of the simple Monod-Wyman-Changeux model. The emphasis is on the analysis of the reciprocal influence of the various parameters which enter into the model, with particular reference to the relationship among a number of sites in the macromolecule and the Hill parameter n. This appears of interest for the behaviour of some of the extremely large respiratory proteins containing up to 100, or more, oxygen binding sites, where a very large value of n is often coupled with a relatively small value of the total interaction free energy.

Cooperative binding of a ligand by a macromolecule is conveniently described by a Hill plot in which $\ln(\bar{x}/1-\bar{x})$ is represented as a function of $\ln x$, where $\bar{x}$ is the fractional saturation of the macromolecule with ligand X and x is the ligand activity. Such a graph must always have an asymptote of unit slope at each end unless the interaction between the sites is infinite. The minimum value of the total interaction free energy realized per site in saturating the macromolecule with the ligand is then given by RT $\Delta\ln x$, where $\Delta\ln x$ is the horizontal distance between the asymptotes. Similarly and more generally, the minimum value of the interaction free energy realized in passing from *any* degree of saturation $\bar{x}_1$ to $\bar{x}_2$ is given by the horizontal distance between the lines of unit slope drawn through the two points $[\ln x_1, \ln(\bar{x}_1/1-\bar{x}_1)]$ and $[\ln x_2, \ln(\bar{x}_2/1-\bar{x}_2)]$. At the same time the point value of the interaction free energy at any degree of saturation is given by

$$\frac{RT}{\bar{x}(1-\bar{x})}\left(1 - \frac{1}{n}\right) = \Delta F_I$$

where n is the slope of the Hill plot at the point in question [1]. The earlier propositions just enunciated are obtained from this by integration; and it will be seen that whenever $n > 1$ anywhere, there must be positive (or stabilizing) interactions between sites, *i.e.* the system is cooperative. The parameter n can of course never exceed the number of sites, and when it is equal to the number of sites it means that the interaction free energy between the sites is infinite, *i.e.* that the sites are so strongly coupled that they all react simultaneously to give an n-th order reaction. It should perhaps be pointed out that n is a statistical quantity and that even when n goes to infinity the point value of the interaction free energy is only $\Delta F_I = RT/\bar{x}(1-\bar{x})$.

Cooperativity in biological macromolecules is often the result of conformational (or allosteric) changes induced by the ligands; in other words, it is the expression of ligand-linked conformational equilibria. It is of interest to see how this shows up in the Hill plot, and what the actual behaviour of some of the very large macromolecules as represented in their Hill plots implies as to the underlying conformational changes. Of particular interest here are some of the extremely large respiratory proteins, containing up to 100 or more sites for oxygen, where a very large value of n is often coupled with a relatively small total interaction free energy, *i.e.* where the Hill plot shows a rather sharp upward bend or kink in the middle range of saturations associated with rather closely spaced asymptotes.

The two most commonly invoked allosteric models are the Monod-Wyman-Changeux (M.W.C.) and the Koshland-Nemethy-Filmer (K.N.F.) models, both of which can be shown to be special cases of a more general parent model [2]. Of these, the M.W.C. (or concerted) model is the simplest and involves the smallest number of parameters (only 2). At the same time it offers the greatest possibility of a large but localized value of n coupled with a relatively small value of $\Delta F_{I(total)}$. It is this model, therefore, which we shall consider in what follows.

In the M.W.C. model [3] it is assumed that the macromolecule exists in only two forms (T and R), in each of which the ligand binding sites are all alike and independent. The binding polynomial (or generating function) for this model is given by

(1) $$P = (1+x)^r + L_0(1+\alpha x)^r$$

where r is the number of sites, the same in each of the two conformations, L_0 is the equilibrium constant for the T→R transition in the absence of ligand (*i.e.*, $L_0 = R/T$ when $x = 0$) and α is the ratio of binding constant K for a site in the R form to that of a site in the T form ($\alpha = K_R/K_T$).

In this model the value of $\bar{x}$, the fractional saturation of the macromolecule with ligand, is given by

$$\bar{x} = \frac{1}{r}\frac{d\ \ln P}{d\ \ln x} = \frac{1}{r}\frac{P'}{P}$$

where

(2) $$P' = \frac{d\ P}{d\ \ln x} = r[(1+x)^{r-1}\ x + L_0(1+\alpha x)^{r-1}\alpha x]$$

At the same time n is given by

(3) $$n = d\ \ln(\bar{x}/1-\bar{x})/d\ \ln x = \frac{P''}{P'} - \left(\frac{rP'-P''}{rP-P'}\right)$$

where

$$P'' = \alpha P'/d\ \ln x = P' + r(r-1)[(1+x)^{r-2}x^2 + L_0(1+\alpha x)^{r-2}\alpha^2 x^2]$$

Equation (3) may also be written as

(4) $$n = \left(\frac{P''}{P'} - r\bar{x}\right)/(1-\bar{x})$$

Further, the total apparent interaction energy is given by

(5) $$\Delta F_{I(total)} = RT\ \ln C$$

where

$$(5.1) \qquad C = \frac{(1+L_0)(1+L_0\alpha^r)}{(1+L_0\alpha)(1+L_0\alpha^{r-1})}$$

In carrying out the analysis it is convenient to treat C and α as the independent variables and then, on the basis of (5.1), to obtain L_0 by the relation

$$(6) \qquad L_0 = (B\pm\sqrt{B^2 - 4A})/2A$$

where

$$(6.1) \qquad A = \alpha^r, \; B = \frac{1-C(\alpha+\alpha^{r-1})+\alpha^r}{C-1}$$

These equations make it possible to calculate the Hill plot (or the corresponding binding curve), of the macromolecule for various sets of values of r, C, α, and L_0. What we are primarily interested in is of course the sharpness of the transition in relation to the spacing of the asymptotes, *i.e.* n in relation of $\Delta F_{I(total)}$ as given by C. As we shall see, for the sufficiently large values of r the transition can be extremely sharp for any value of $\Delta F_{I(total)}$, even a very small one, with the result that we run into something approaching a phase change. Although detailed calculations can be made very easily on the basis of the above equations with the aid of a computer, still, as a first step, it is convenient to get an analytical solution for the special case where we impose the conditions of symmetry of the binding curve. This brings the transition into the middle of the Hill plot, where $\bar{x} = 1/2$; also it reduces the number of parameters to one (α) and greatly simplifies the equations.

The conditions of symmetry of the binding curve (or of the Hill plot, for they are the same) can be shown to be met provided only we set

$$(7) \qquad L_0 = \alpha^{-r/2}$$

For this value of L_0, the value of the median ligand activity, x_m, now equal to $x_{1/2}$ (for which $\bar{x} = 1/2$), is given by

$$x_m = x_{1/2} = \alpha^{-1/2}$$

TABLE I

Dependence of n_m on the number of binding sites r, calculated from Eq. (11) for two values of α.

a) $\alpha = 9, \frac{1+\alpha}{(1+\alpha^{1/2})^2} = 0.625$

r	n_m	C	$\Delta F_{I(total)}$[cals]
4	1.75	7.5	1,195
6	2.26	8.8	1,290
8	3.76	9	1,315
50	11.2	9	1,315
100	24.0	9	1,315

b) $\alpha = 100, \frac{1+\alpha}{(1+\alpha^{1/2})^2} = 0.835$

r	n_m	C	$\Delta F_{I(total)}$[cals]
4	3.01	98	2,700†
6	4.35	100	2,730
8	5.69	100	2,730
50	33.8	100	2,730
100	70.0	100	2,730

†This corresponds very closely with the case of human hemoglobin, for which the binding curve is approximately, if not exactly, symmetrical (see Antonini and Brunori for review [4]). The result is an argument for the applicability of the concerted model to that molecule. The idea that symmetry of function might be associated with symmetry of structure was suggested many years ago [5].

At the same time the equilibrium constant L for the T→R transition, which in the general case is given by

$$(8) \quad L = L_o\left[\frac{1+\alpha x}{1+x}\right]^r$$

becomes

$$(9) \quad L = \left[\frac{\alpha^{-1/2} + \alpha^{1/2}x}{1+x}\right]^r$$

and at the mid point of the binding curve, where $x = \alpha^{-1/2}$, L is unity. At this point the value of n given by Eq. (4) is

$$(10) \quad n \equiv n_m = 2\left[1+(r-1)\frac{(1+\alpha)}{(1+\alpha^{1/2})^2}\right] - r$$

Furthermore

$$(11) \quad C = \frac{2 + \alpha^{r/2} + \alpha^{-r/2}}{2 + \alpha^{(r-2)/2} + \alpha^{-(r-2)/2}}$$

We adopt the convention that $\alpha > 1$, which means that $L_0 < 1$†. It follows from (11) that for sufficiently large values of r, C becomes equal to α. Indeed for $\alpha = 10$ this is already approximately true even for $r = 4$. At the same time the expression $(1+\alpha)/(1+\alpha^{1/2})^2$ in the equation for n_m increases rather slowly from 0.625 to 1 as α goes from 9 to ∞. It is clear therefore that under conditions in which we are interested n_m will be linear in r regardless of the value of C [or ΔF_I*] and will increase without limit as r increases, with the result that the transition becomes a true phase change. This is brought out by the figures shown in Table I (a and b). Another aspect of the situation is revealed in Fig. 1, which shows how the value of n_m/r, *i.e.* the value of n_m, normalized with respect to its maximum value r, increases toward 1 as $\Delta F_{I(total)}$ increases indefinitely. It will be seen that the curves for different values of $r \geq 4$ all lie close together and approach a common bounding curve as r gets bigger and bigger.

It is a simple matter, with the aid of a computer+, to construct full Hill plots for various cases. Such plots (up to r = 36) are shown in Fig. 2. It is also a simple matter to construct Hill plots for more general cases where the constraint of symmetry is relaxed. Such plots are shown in Fig. 3, where the transition (or switch over point, defined as the point

† The opposite assumption would of course give the same results.

* total

\+ Computations were made with a Hewlett Packard Model 9830A computer.

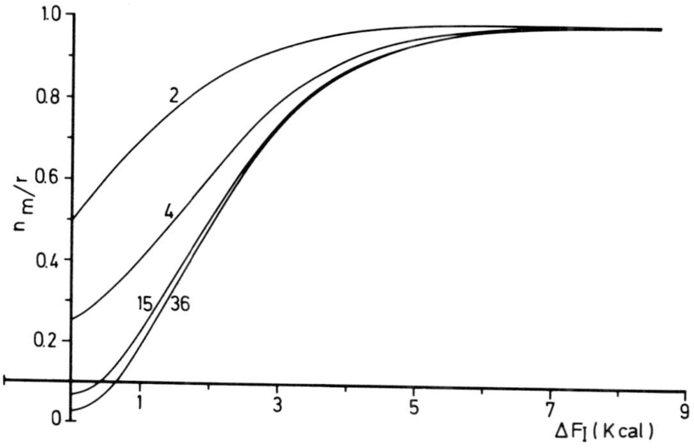

Fig. 1. Values of n normalized to its upper limit r (corresponding to $\Delta F_{I(total)} \to \infty$), *i.e.* n_m/r as a function of $\Delta F_{I(total)}$[cals] for the different values of r shown on the curves. This is for the symmetrical case.

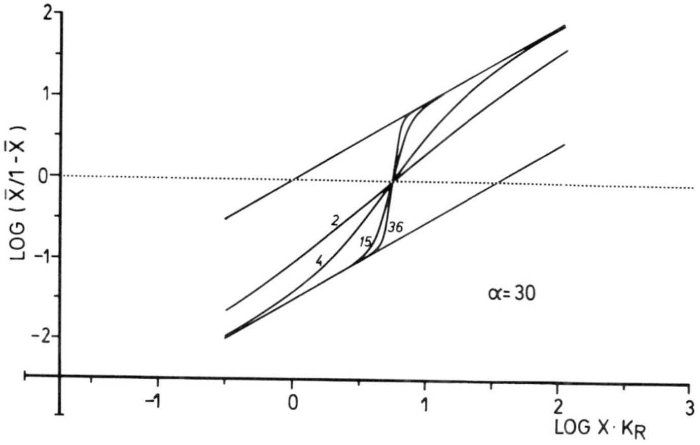

Fig. 2. Hill plots for the symmetrical case calculated for the constant $\alpha = K_R/K_T = 30$ for the different values of r shown on the curves. Except for r = 2, where $\Delta F_{I(total)}$ = 1235 cals, these curves all correspond to $\Delta F_{I(total)}$ = 2000 cals within 1 per cent.

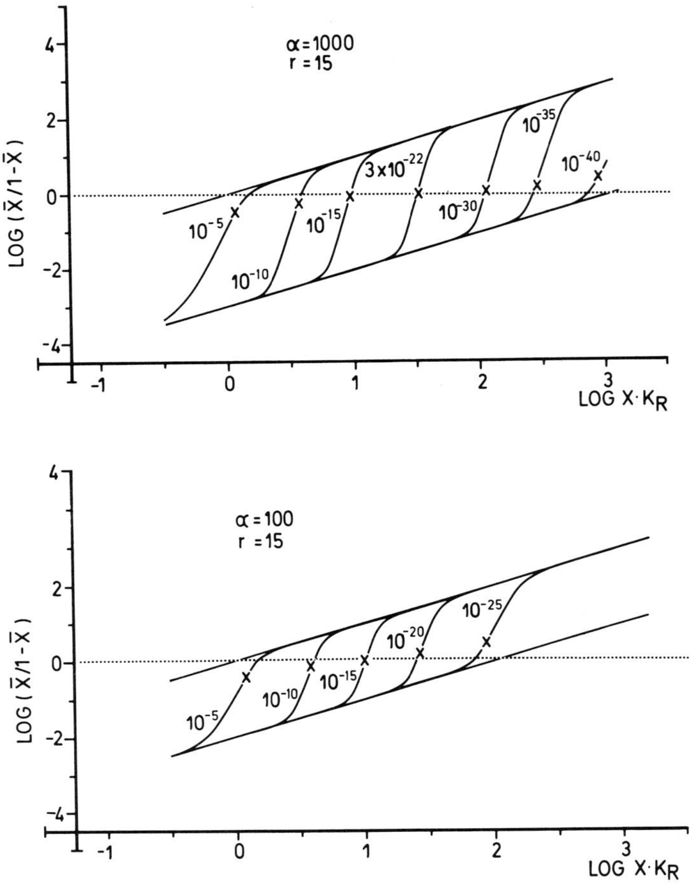

Fig. 3. Hill plots for the more general case where the symmetry condition is relaxed. α = 1000 for top curves and α = 100 for bottom curves. In both cases r = 15 and the values of L_0 are those shown on the curves. Asterisks indicate switch over point. Here we have defined the switch over point as the point where the amounts of the T and R forms are equal. Alternatively we might have defined it as the point where the amounts of *ligand bound* by the T and R forms are equal. The two points are not the same. The former corresponds to L = 1, the latter to $Y_T = Y_R L$.* The latter definition is more apposite to the analysis of kinetic experiments.

---

* Y_T and Y_R are the fractional saturation of the T and R states.

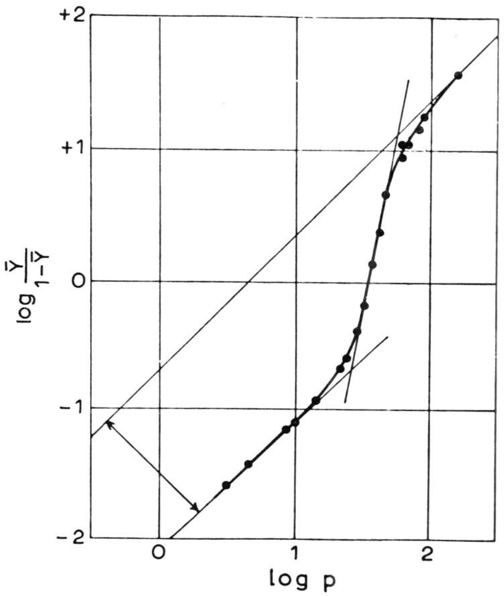

Fig. 4. Hill plot for oxygen binding by the blood of Spirographis spallanzanii at 20° and pH 7 (from Antonini *et al.*, ref. 6).

where R/T = L = 1) is no longer centered at $\bar{x}$ = 1/2. It is interesting to see how, as we depart more and more from the condition of symmetry, the value of n, even at its greatest, decreases slowly but progressively while the region of high values of n moves farther and farther from the point $\bar{x}$ = 1/2, following the switch over point, and eventually comes to be outside the experimentally accessible range of saturation.

The chlorocruorin from the worm Spirographis spallanzanii, which has a molecular weight of ∿ 2,500,000 and contains approximately 72 oxygen binding sites (hemes) per molecule, provides a good example of one of the larger respiratory proteins [6]. Its Hill plot, given in Fig. 4, shows a maximum value of n = 5, which occurs at $\bar{x} \simeq 0.64$, and a value of $\Delta F_{I(total)}$ = 1800 calories, corresponding to C = 25. A glance at Fig. 1 shows that these results are incompatible with the assumption that the whole molecule with its 72 sites behaves like a simple M.W.C. model. If so, the value of $\Delta F_{I(total)}$ = 1800 cals would imply a value of $n_m/r = 0.45$ or $n_m \simeq 72 \cdot 0.45 = 32.4$ as compared with the observed value of 5. We might, as the next simplest hypothesis, assume that the sites interact in independent constellations, each containing r sites. The value of n_m would then be the same as that for a constellation, and we could calculate the value of r, assuming the M.W.C. model to be applicable to each constellation, either with the aid of Fig. 1 or, more directly, by rewriting Equation (10) in the form

$$(12) \qquad r = (n_m+2\lambda-2)/(2\lambda-1)$$

where $\lambda = (1+\alpha)/(1+\alpha^{1/2})^2$

The answer is 10. An alternative hypothesis would seem to be that the interactions involve all the sites, diminishing progressively in some manner with distance. A special case would be that of a number of small constellations of strongly interacting sites with secondary interactions between the different constellations - a kind of hierarchy of interactions. This is an idea which finds an echo in the complicated symmetrical structures revealed by electron microscopy [7].

Very similar considerations apply to an even larger molecule, namely the hemocyanin of Helix pomatia, whose molecular weight is of the order of $9\cdot10^6$ and which contains ∿ 180 oxygen binding sites. Here the value of n_m is approximately 5. In this case it is found that when, in the absence of calcium, the molecular weight drops to (1/10) its full value, all cooperativity is lost although the (1/10) subunits each still contain 18 sites [10]. It would seem that the integrity of the whole molecule is a condition of cooperativity.

It will be seen from this analysis that the M.W.C. model carries, buried within it, the prediction of a ligand linked phase change which is realized more and more completely as the number of sites increases indefinitely. That this is so should not be altogether surprising when it is realized that one of the primary features of the model is the exclusion of all mixed conformations; in applying the model we are in a sense treating the macromolecule as a crystal, and in this connection it is worth recalling that the liganded and unliganded derivatives of hemoglobin crystallize in different forms [8] which are also characterized by different solubilities

[2]. The analysis given here may be compared with an earlier and rather different one [9], which, though physically somewhat unrealistic, is nevertheless suggestive in also predicting a phase change as the number of sites becomes infinite. Actually, although no completely sharp phase change has ever been observed, even in the case of the largest proteins studied, and although, as shown in the last paragraph, the simple M.W.C. model is not as such applicable to them, nevertheless data like those shown in Fig. 4 often reveal a very sudden transition, or switch over, which occurs within a limited critical range of liganding even when the total free energy of interaction is relatively small. At any rate, one thing is clear, and that is that the interactions responsible for the cooperativity in these giant molecules must radiate out to cover a large number of sites, almost certainly substantially larger than the minimum number calculated from Equation (12).

ACKNOWLEDGEMENT

A grant from the National Science Foundation to Jeffries Wyman is gratefully acknowledged.

REFERENCES

1. Wyman, J.: Advan. Protein. Chem. *19*, 223 (1964).

2. Wyman, J.: Curr. Topics Cell Reg. Academic Press, New York, *6*, 209 (1972).

3. Monod, J., Wyman, J., Changeux, J.P.: J. Mol. Biol. *12*, 88 (1965).

4. Antonini, E., Brunori, M.: "Hemoglobin and Myoglobin in their Reactions with Ligands". Frontiers of Biology, Vol. 21, North Holland, Amsterdam, 1971.

5. Allen, D.W., Guthe, K.F., Wyman, J.: J. Biol. Chem. *187*, 293 (1950).

6. Antonini, E., Rossi-Fanelli, A., Caputo, A.: Arch. Biochem. Biophys. *97*, 336 (1962).

7. Guerritore, D., Brunori, M., Antonini, E., Wyman, J., Rossi-Fanelli, A.: J. Mol. Biol. *13*, 234 (1965).

8. Perutz, M.F.: "Barcroft Memorial Conference". Butterworths Sci. Publ., London, 1949, p. 135.

9. Wyman, J.: J. Mol. Biol. *39*, 523 (1969).

10. Van Driel, R.: Biochemistry *12*, 2696 (1973).

Received July 10, 1974

Ligand Interactions in Globular Proteins

Gregorio Weber

FREE ENERGY COUPLING BETWEEN LIGANDS.

The investigations of the last ten years have shown that the functions of proteins as catalysts, specific binding agents or mechanical entities can be greatly modified by the binding of small molecules, so that their properties may best be described as those of complexes of the protein with the small ligands. The functions of proteins as catalysts or as mechanical agents are often too complex to permit at present a complete analysis of the changes in function that result from the binding of different ligands. On the other hand the influence of one ligand upon the binding of another ligand in an isolated protein in solution may be described in simple fashion and its study can provide some of the fundamental information indispensable for the understanding of the more complex cases. I have given elsewhere[1,2] a simple thermodynamic formulation that permits one to extract from binding data the standard free energy coupling, ΔF_{xy}, between two ligands, X and Y, simultaneously bound to the protein. The free energy coupling is positive ($\Delta F_{xy} > 0$) if the binding of one ligand decreases the affinity of the protein for the other ligand, and negative ($\Delta F_{xy} < 0$) if binding of one ligand increases the affinity for the other. These relations between ligands are always mutual, since the conservation of the free energy of binding implies reciprocity of the effects. Few investigations have been directed to determine the free energy coupling between ligands, and from these and from the larger number of, so to speak, unintentional observations recorded in the literature[2] it is possible to draw some general conclusions: 1. Both positive and negative values of ΔF_{xy} have been observed. 2. None of them is larger than 2 kcal./Mole and a figure of 1-1.5 kcal/Mole may be quoted as typical for the absolute value of the free energy coupling between ligands.

LIGAND INTERACTIONS AND CHANGES IN PROTEIN STRUCTURE.

Much attention has been directed towards relating the changes in function of a protein upon ligand binding to the possible changes in structure that are then assumed to take place[3,4]. Energetic and structural changes will bear a direct relation to each other if there is complete reversal to the original properties upon removal of the ligand, a demonstration provided in most, if not in all the cases studied. We can expect each separate ligand, X or Y, to produce changes in the protein structure as compared to the original protein structure in the absence of the ligands. The structural changes relating directly to the ligand interaction are not these, but the probably much smaller changes by which the singly liganded forms PX and YP differ from the doubly-liganded form YPX. The conservation of free energy in the system obliges us to relate directly the value of ΔF_{xy} to these structural changes characteristic of the ternary complex. A small value of ΔF_{xy} bespeaks an equally small change in the energy of interaction linking the protein parts. This is perhaps most easily visualized in the case of positive free energy couplings ($\Delta F_{xy} > 0$) which imply a mutual destabilization or repulsion between the two bound ligands. If this repulsive interaction were to exceed the value of the energy of interaction of the intervening protein parts a compensatory relaxation, or dislocation of the protein structure would then take place reducing ΔF_{xy} to the permissible upper limit. The free energy couplings thus far measured are typically 1-1.5 kcal. The total value of the internal protein interactions is difficult to estimate but assuming individual contributions no greater than 1 kcal/amino acid residue it would still amount to more than 100 kcal for a typical globular protein chain. We can see at once the magnitude of the problem of translating the existence of ligand interactions into structural changes, since the energy of the former would amount to no more than 1% of the total energy of the interactions that maintain the protein structure. To account for such a small change would require knowledge of the average atomic coordinates of the various ligand-protein complexes with a degree of precision which X-ray crystallographic analysis cannot provide today[5], and may never be able to provide. In a recent review Jensen[6] points out that one of the principles used today in the refinement of the X-ray structures of proteins is that of conservation of the 'normal' bond angles and distances.

Evidently if we are to have a structural map sufficiently detailed and accurate to carry out computation of all the energies of interaction among the protein residues, the atomic coordinates must be determined without recourse to auxiliary energetic assumptions if circular arguments are to be avoided. The classical indetermination of Physics arises in the case of large energies and very small particles. Here we seem to be in precisely the opposite case, of very small energies distributed among the many degrees of freedom of a very large particle, and we may have to face the unfeasibility--if not the impossibility--of determining all the distances, and thus all the energies with the required precision.

INSUFFICIENCY OF MECHANICAL MODELS OF LIGAND INTERACTION.

In attempting to explain the decrease in affinity for one ligand when another is bound, various authors have proposed mechanical models which picture the effects as due to the creation of steric barriers of virtually infinite energy. Such models are bound to be poor, and still worse, very misleading representations of the real situation in which a very modest additional free energy is required for the formation of the ternary complexes. The insufficiency of these mechanical models becomes patent when one compares the potential energy of interaction as a function of distance for a pair of atoms, with a similar potential curve describing

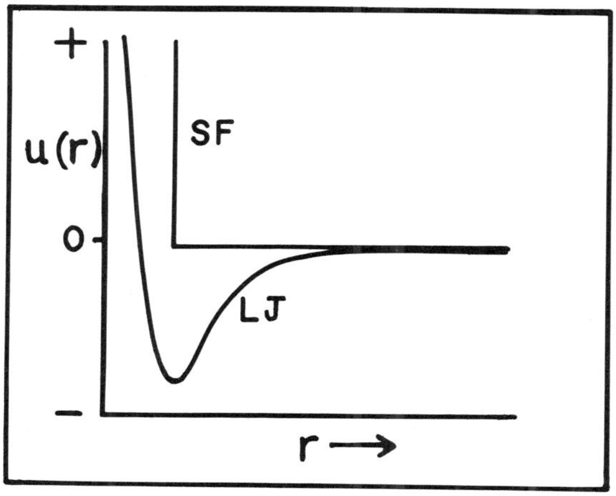

Fig. 1. Comparison of the potential energy of interaction u(r), as a function of the distance r, for a pair of atoms in a space filling model, SF with the Lennard Jones potential (LJ). The depth of the minimum would be 1-3 thermal energy units.

the behaviour of the commonly used space-filling models. As shown in figure 1 the latter cannot represent at all the changes in interaction of the order of a few times the thermal energy, which are the only ones involved in the interligand interactions.

PHYSICAL ORIGIN OF LIGAND INTERACTIONS.

Although the prospects for determining in each particular case the exact structural features responsible for the ligand interactions are almost nil we can nevertheless recognize the main molecular mechanisms that must be responsible for their existence. If two ligands not directly in contact influence each other's binding they must do so through the intervening protein structure. Conceivably such effects arise in two different ways: The first is by electrostatic interactions, which will be stronger across the low-dielectric-constant protein core than when the charged ligands are at the same distance in water solution. These electrostatic effects need not be confined to the case in which both ligands are charged. When only one of them is an ionic ligand it can interact with either charged groups or polarizable groups involved in the binding of the uncharged ligand. Thus 2,3-diphosphoglycerate and each of the four oxygens in hemoglobin are linked by a free energy coupling of +1.3 kcal/Mole[7]. The repeated observation that the most diverse polyanions have effects similar to diphosphoglycerate is one good example of this kind, and in view of the overall prevalence of charged ligands, this mode of influence may well turn out to be the commonest. A second mode of interaction may be envisioned for the case of two uncharged ligands: this will consist of small displacements of the peptide chains, or even only 'strains' in the protein structure, which can be considered for this purpose a portion of a 'hard sphere' fluid, as proposed by Klapper[8].

LIGAND INTERACTIONS IN SINGLE CHAIN PROTEINS AND IN OLIGOMERIC PROTEINS.

Most of the cases of ligand interactions described in the literature refer to proteins made up of several subunits, but in recent years ligand interactions have been conclusively demonstrated in single chain proteins. Kolb & Weber[9,10] have shown the presence of multiple interactions between anilino naphthalene sulfonate (ANS) and 3,5-dihydroxybenzoate when both are bound by bovine serum albumin. Panagou, et al.[11] have observed allosteric behaviour in a single chain protein, ribonucleotide reductase from L.Leishmannii. Moreover, many of the effects of ligand interactions observed in oligomeric proteins appear to be of the type in which the interacting pairs within any single subunit are virtually independent of the others. Kolb[10] has observed this to be the case for the ternary complexes of chicken heart lactate dehydrogenase with NADH and oxalate. These observations are not surprising. From our previous considerations we expect ligand interactions to be characteristic of all compact globular polymers and not to require in any way the existence of an oligomeric aggregate for their presence. However, the prevalence of the observation of these effects in oligomeric proteins can be simply rationalized: The boundary between subunits constitutes a region where only weak interactions are present and binding of a ligand elsewhere in the protein is therefore likely to produce changes in it. In this manner two identical ligands bound to neighbouring subunits may exert influence upon each other. Cooperativity and antagonism in the binding of these two ligands will naturally appear[1] if some intersubunit bonds are broken or formed by the binding of a ligand at either subunit. Independent binding without ligand interaction, but with a change in the free energy of subunit association will follow, if binding at either subunit breaks equal and independent bonds at the subunit boundary (Figure 2). Evidently these subunit boundary changes, corresponding to free energy changes which are small compared to the total free energy of interaction of each peptide chain, or even to the total free energy of boundary interactions, need not introduce by themselves very conspicuous modifications in the structure. It is not therefore immediately obvious in what way the crystal structures observed for the liganded and unliganded forms are related to the mechanism of interligand coupling. Other places at the boundary, besides those involved in common bonds, may be independently modified by binding of each ligand. It is also increasingly recognized[12] that crystal forces arising out of the different packing of the molecules in the liganded and unliganded forms in the crystal must be taken into account.

An important energetic characteristic of oligomeric proteins is that a single ligand at the boundary (e.g., 2,3-diphosphoglycerate in hemoglobin) may couple with the several ligands bound each to one subunit. In this

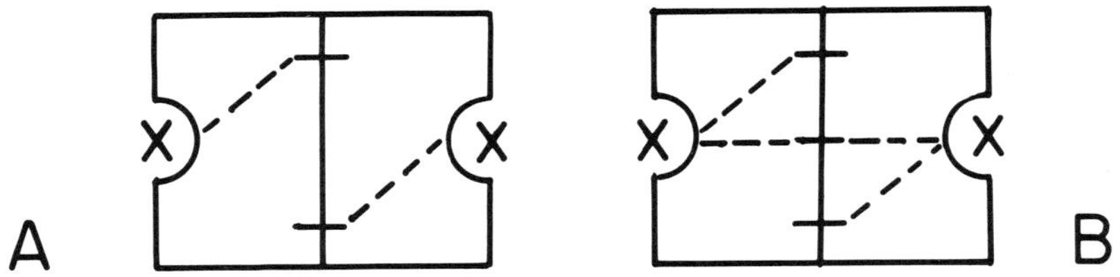

Figure 2. Ligand interaction effects and subunit association: In A binding of X at either side breaks (forms) independent and equal bonds. Ligand binding promotes dissociation (association) of subunits without appearance of ligand interaction. In B common bonds are broken or formed by binding at either side. Ligand interactions are observed together with changes in free energy of association of the subunits.

manner the total free energy coupling for the unique ligand at the boundary may increase to several times the value found for isolated ligands, and may therefore reach 5-7 kcal/Mole. I have discussed elsewhere the possible significance of these facts for the interconversion of chemical and osmotic energies[2,13]. It may well be that this energetic characteristic is responsible, more than any other, for the prevalence of oligomeric proteins.

ENERGY TRANSFER TO AND FROM THE PROTEIN STRUCTURE.

When two ligands are simultaneously bound to the protein with $\Delta F_{xy} < 0$, the ternary complex is stabilized with release of the corresponding free energy. When they are bound with $\Delta F_{xy} > 0$ free energy is stored in the ternary complex in the form of 'repulsive' interaction between the ligands. In either case the free energy coupling equals the change in chemical potential required to attain half saturation with one ligand in the absence, and in the presence of an excess, of the other ligand[1]. A change in chemical potential of 1.5 kcal/Mole results from a change in concentration of one order of magnitude. It follows that there is a direct relation between the strength of the energies of interaction of the protein parts that keep it in the globular form, and the physiological range of concentration of those ligands that regulate the metabolic functions of the protein. From the energetic point of view this relation may be considered to arise from the transfer of free energy to and from the protein structure. As in all other cases of energy transfer, impedance matching between the 'donor' and 'acceptor' systems is strictly required.

The donor system is here the protein, and the acceptor the surrounding solution, if ΔF_{xy} is negative, and conversely the donor is the surrounding solution and the acceptor the protein, if the free energy coupling is positive. Impedance matching corresponds here to chemical potential matching. On the protein side it is the standard chemical potential that changes, while on the ligand side it is the concentration, or variable part of the chemical potential that is involved. It will be noticed that actual 'storage' of free energy in the protein structure takes place only when $\Delta F_{xy} > 0$, that is in the case of repulsive effects between the bound ligands. I believe that this constitutes the only mode of free energy storage in the protein structure, apart from actual formation of covalent bonds.

STRUCTURAL FLUCTUATIONS IN PROTEINS.

While the structural details responsible for ligand-ligand interactions may be virtually inaccessible to our present-day methods, the protein dynamics that they imply and the changes of these dynamics under various conditions may furnish information of importance towards the understanding of biological processes. In assessing the dynamics of the protein structure relevant to our case we shall be governed by a principle which is supported by virtually every line of evidence, namely that nearest neighbour interactions are the most important factor in maintaining the conformation of the protein in its average state, and that there is no delocalized form of energy capable of furnishing additional stabilization. Given these premises we can expect that local fluctuations of the structure will recur with a frequency $Z = A \exp-(E/RT)$, where A is a pre-exponential factor and E the energy required to produce the structural fluctuation. We expect E to be of the same order as ΔF_{xy}, perhaps two or three times as large, while A is determined by the rate of energy exchange among interacting neighbouring structures in the protein. Various lines of thought[2] would suggest for A values of 10^{12}-10^{14} sec^{-1}. If the lower limit is used and E = 2-4 kcal/Mole we have $Z = 10^8$-10^9 sec^{-1}. It is in fact possible to show by direct experimental measurements the reality of the rates and energies of activation that we have assigned to these fluctuations.

DEMONSTRATION OF PROTEIN STRUCTURAL FLUCTUATIONS BY FLUORESCENCE METHODS. The excited states for allowed emission of visible or ultraviolet light last for times of the order of 1-20 nsec. in the majority of cases. The postulated fluctuations of the protein structure should therefore modify the character of the emission and be readily detectable from the properties of the intrinsic protein fluorophores, or of fluorescent probes included in the protein structure. Both types of experiments have yielded positive results. Lakowicz & Weber[14] have studied the quenching of the fluorescence of proteins by oxygen. The fluorescence from the tryptophan residues of the twelve proteins studied was quenched by oxygen at rates equal to 1/2 to 1/5 of the diffusion controlled rate, which applies to tryptophan and simple indole derivatives in water solution. Observations of fluorescence lifetime have shown that the process is not due to static association of oxygen with the tryptophan residues and therefore that it requires the effective diffusion of oxygen within the protein structure to take place in the 2-6 nanoseconds which the excited state lasts. Such rapid diffusion process must necessarily require rapid breathing of the protein through the kind of fluctuations that we have deduced from the small values of the free energy coupling between ligands. More recently we have been able to obtain further evidence for the character of the energy fluctuations proceeding from an original observation of Brand & Gohlke[15]. These authors showed that the solvent relaxation, or rearrangement of the solvent molecules around an excited fluorophore, has a counterpart in the rearrangement of the surroundings when the fluorophore is anilino naphthalene sulfonate (ANS) bound to bovine serum albumin. This observation effectively shows that rearrangement of the protein structure about the excited fluorophore takes place in a time of the order of nanoseconds. To extend this observation we have used the method of differential phase fluorometry[16], which is particularly well adapted to the measurements of subnanosecond time intervals. We employed for the interpretation of the results a two-state model according to which the fluorophore may be effectively surrounded by only two kind of environment, one corresponding to equilibrium in the ground state, the other to equilibrium in the fluorescent, or lowest singlet, state. The experimental data permit the determination of the two rates for conversion of each state into the other. In ANS-Bovine serum albumin complexes[17] we found that conversion from 'ground state' equilibrium surroundings to 'excited state' equilibrium surroundings took place at a rate of 10^9 sec^{-1} while the opposite rate was ten times slower. It may be noticed that the latter rate, which is the one most directly and accurately measured by the experiments, closely corresponds to a fluctuation about equilibrium, of the type that we expect to take place in the intact protein structure. Measurements at different temperatures have indicated for this rate an energy of activation of 3.5 kcal. It is of interest to notice that binding of a second ligand, 3,5 dihydroxybenzoate, which is known to couple with ANS[9], could be followed by its effect upon these relaxation parameters of the ANS fluorescence[17]. Similarly we have observed recently that a complex of bis-ANS and aspartate transcarbamylase[18] undergoes similar relaxation behaviour, and changes it upon the binding of carbamyl phosphate to the enzyme.

LIMITATIONS OF THE DYNAMIC OBSERVATIONS.

I have already pointed out the limitations that the small energies of ligand interaction impose upon the derivation of their causes from the structural X-ray data. Limitations of a different kind but of similar origin apply to the study of the protein dynamics by the fluorescence methods. From very general considerations we expect that the fluctuations in the protein properties will be increasingly difficult to reveal as their energy decreases and they become faster, in accordance with the exponential relation between the rate and the energy of activation. The possibility of direct measurement of low-energy, spontaneous fluctuations, after recognizing them among the many other low-energy changes taking place in the system seems well-nigh impossible. In our experiments this difficulty is circumvented by creating a high energy state within the protein by light absorption and deducing the existence of the protein fluctuations from the changed properties of such easily observable fluorescent state. However, the creation of such state is not without influence upon its surroundings and we are witnessing not so much a spontaneous process characteristic of the 'intact' proteins as an induced change from which the properties of the structure are indirectly deduced. We need not be unduly pessimistic or apologetic about our conclusions, but they must certainly take into account the particular properties of the method of observation and the uncertainties that it necessarily introduces. For it is a characteristic of all observations that they interfere with the object of the observations, and for the time and energy ranges that we are considering the interference can no longer be considered as negligible.

REFERENCES

1. Weber, G.: Ligand Binding and Internal Equilibria in Proteins. Biochemistry 11, 864-878 (1972).
2. Weber, G.: Energetics of Ligand Binding to Proteins, Adv. Prot. Chemistry. vol. 29, (1975), In press.
3. Perutz, M.F.: Stereochemistry of cooperative effects in hemoglobin. Nature (London) 228, 726-739 (1970).
4. Evans, D.R., Warren, S.G., Edwards, B.F.P., McMurray, C.H., Bethge, P.H., Wiley, D.C., Lipscomb, W.N.: Aqueous Central Cavity in Aspartate transcarbamylase from E. Coli. Science 179, 683-685 (1973).
5. Watenpaugh, K., Sieker, L.C., Herriott, J.R. & Jensen, L.H.: The structure of a non-heme iron protein: rubredoxin at 1.5 A resolution. Cold Spring Harb. Symp. Quant. Biol. 36, 359-367 (1971).
6. Jensen, L.H.: Protein model refinement based on X-ray data. Ann. Rev. Biophys. Bioeng. 3, 81-93 (1974).
7. Tyuma, I., Shimizu, K. & Imai, K.: Effect of 2,3-diphosphoglycerate on the cooperativity of oxygen binding by human adult hemoglobin. Biochim. Biophys. Res. Com. 43, 423-428 (1971).
8. Klapper, M.H.: Apolar bonding: a reevaluation. Progress in Bioorganic Chemistry 2, 55-132 (1973).
9. Kolb, D.A. & Weber, G.: Cooperative binding in single chain proteins. Fed. Proc. Abs. 31, 432 (1972).
10. Kolb, D.A., Doctoral Thesis. University of Illinois, (1974).
11. Panagou, D., Dunstone, J.R., Blakley, R. & Orr, M.D.: A monomeric, allosteric enzyme with a single polypeptide chain. Biochemistry 11, 2378-2388 (1972).
12. Lipscomb, W.N.: Enzymic aspects of carboxypeptidase A's in solution and in crystals. Proc. Ntl. Acad. Sci. U.S. 70, 3797-3801 (1973).
13. Weber, G.: Addition of chemical and osmotic energies by ligand-protein interactions. Ann. N.Y. Acad. Sci. 227, 486-497 (1974).

14. Lakowicz, J.R. & Weber, G.: Quenching of protein fluorescence by oxygen. Detection of structural fluctuations in proteins in the nanosecond time scale. Biochemistry 12, 4171-4179 (1973).

15. Brand, L. & Gohlke, J.R.: Nanosecond time resolved fluorescence spectra of a protein-dye complex. J. Biol. Chem. 246, 2317-2319 (1971).

16. Weber, G. & Mitchell, G.W.: Demonstration of anisotropic molecular rotations by differential polarized phase fluorometry. Proc. Intl. Meeting on Excited states of biological molecules. Lisbon, April, (1974), In press.

17. Kolb, D.A. & Weber, G.: To be published. Some of the observations are reported in reference 10.

18. Chien, Y. & Weber, G.: To be published.

Received July 19, 1974

DISCUSSION

Steinhardt: By way of clarification, would you comment on the difference between the small interaction energies you have been discussing and the rather large free energy changes that accompany the binding of certain ligands, such as lipid-like substances, by proteins (ΔF up to about 9 kcal). With long chain fatty acids (Goodman) binding constants up to 10^8 have been measured. Would you say that such binding greatly alters the globular structure?

G. Weber: I was referring to the free energy of ligand-ligand interactions. These are only a fraction (10 - 20 %) of the total standard free energy of binding of either of the interacting ligands.

Helmreich: I just would like to add that the small energy changes involved in ligand binding to proteins of course agree with the very fast rates of ligand binding to proteins which are usually diffusion controlled.

G. Weber: It happens that - for very good reasons- the free energies of binding are directly determined by the rates of dissociation of the ligands, and that the rates of association are relatively constant and often close to diffusion controlled. This, whether the standard free energy of binding is small, like in fumarate-fumarase (Massey) or very large like in avidin-biotin complexes (N.M. Green).

von Hippel: What is molecular interpretation of impedance matching? Is it that something in the protein is capable of changing "one kcal worth"?

G. Weber: Exactly. You put it better than I did.

Antonini: What would be the size of the movements associated with the fast fluctuations?

G. Weber: Probably very small and localized - I imagine, a group rotating, or moving a few Å units, etc. Certainly nothing major.

Navon: In the experiment of fluorescence lifetime, when several ANS molecules are bound to each serum albumin molecule is it possible that there will be energy transfer between the unrelaxed, blue-shifted ANS molecules and the relaxed red-shifted ones? How will your results be affected by this phenomena?

G. Weber: Energy transfer does take place among the various ANS molecules, but since the free energy of binding of the first-bound four molecules is equivalent and the fluorescence spectrum does not change with number bound there is no reason to suppose that there is a class of molecules of ANS which relaxes and one which does not. The temperature dependence of the observed relaxation is another good argument to disregard the effect of energy transfer in the observed results.

Navon: In your lecture you have referred to the large fluorescence shift in the ANS molecule as due to solvent reorganization. What is your opinion on the possibility (E.M. Kosower, private Communication) that it is caused by a planar-non-planar conformational change in the excited state of the ANS molecule?

G. Weber: Similar relaxation effects are observed with 2-diethylaminonaphthalene-5-sulfonate and 1-8-naphthosultam. In the former only a rotation of the amino group could take place. In the latter there is no possibility of a non-planar conformation.

Veeger: Is it possible to discriminate between the two possibilities: a) Fluctuations between the two states as described in your paper. b) Fixed binding in two states with different microenvironment.

G. Weber: Two different microenvironments occupied by the ANS molecules would give the observed results if the "red-shifted environment"-ANS had a longer lifetime of fluorescence than the unshifted one. However we would not expect to observe a temperature dependence. The temperature-dependence experiment is crucial to distinguish between the two possible origins, but the presence of heterogeneity, as you suggest, could somewhat falsify the results. I believe that, if present, it plays here only a very minor part.

Ligand Binding to Enzyme Complexes

Kasper Kirschner, Wolfgang Weischet and *Robert L. Wiskocil*

MULTIENZYME COMPLEXES

The existence of specific aggregates of different polypeptide chains capable of catalyzing a number of distinct metabolic reactions indicates that heterologous protein-protein interactions can be important for expressing enzyme function (1). The oligomeric structure (i.e. multiple identical subunits) may be essential for the allosteric regulation of enzyme activity as, for example, in aspartokinase-homoserinedehydrogenase (2). When the different enzyme activities correspond to sequential metabolic steps, direct cooperation between distinct active sites becomes possible. Composite sites may be created by bringing different active sites into close contact at the subunit-subunit interface, thus providing for molecular "channels" or "compartments" for labile intermediates (3). Finally, the accumulating evidence suggests that heterologous protein-protein interactions may only be a step towards the evolution of more sophisticated enzymes. Gene fusion may provide a way for the conversion of multienzyme complexes to single polypeptide systems (4,5).

We have studied tryptophan synthase from E. coli as a type-case of simple multienzyme complexes (6). This enzyme is a complex of two α- and two β- chains. It is capable of catalyzing the following reactions (6):

1) Indoleglycerol-P + L-ser ⟶ D-glyceraldehyde-3-P + L-trp

2) Indoleglycerol-P ⇌ D-glyceraldehyde-3-P + indole

3) Indole + L-ser ⟶ L-trp

The complex may be dissociated reversibly to α- and β_2- subunits. They can also be obtained as individual proteins from suitable mutants (6). The interaction of α- and β- subunits is characterized by an apparent intrinsic dissociation constant on the order of 10^{-8}M, but depends on the presence of PLP* and serine (15). Reaction 2 is catalyzed also by the α- subunit alone but only 1 - 2 percent as efficiently as an equivalent amount of the $\alpha_2\beta_2$ complex. Similarly, reaction 3 is catalyzed by the β_2- subunit, which contains two equivalents of PLP. Again, the complex is approximately 50 times more efficient in this reaction than the isolated subunit (6). The physical basis for the phenomenon of mutual activation may be of general interest.

The syntheses of indoleglycerol-P and tryptophan from indole are Bi-Uni reactions. Steady-state kinetic studies may shed light on possible differences in the mechanisms as being responsible for the phenomenon of mutual activation. More specifically, the binding of the crucial intermediate indole to α- and β_2 - subunits as well as to the $\alpha_2\beta_2$ - complex might be used as a probe for mapping out the putative composite active site (3,8,9,16). Finally, the synthesis of indolepropanol-P, a nonreactive analogue of indoleglycerol-P (7), provided an additional tool for probing site-site interactions in the complex. The ease with which the complex can be reversibly dissociated into functional components (6) sets the stage for detailed equilibrium and kinetic studies of subunit assembly.

* PLP, pyridoxal 5'-phosphate

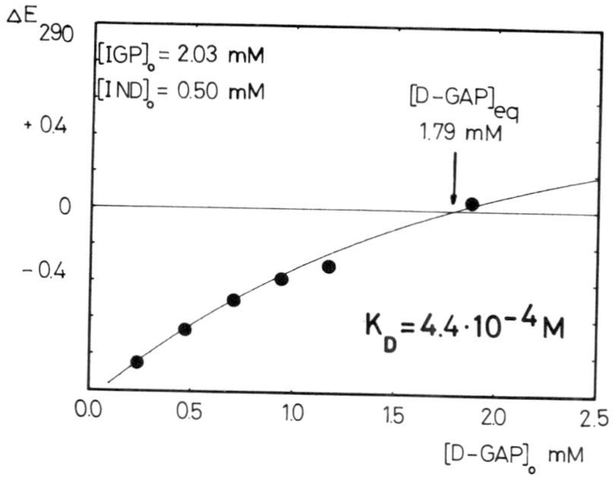

Fig. 1. Determination of the equilibrium constant of indoleglycerol-P cleavage (reaction 2). Various solutions with constant total concentrations of indoleglycerol-P (IGP) and indole (IND) and variable total concentrations of D-glyceraldehyde-3-P (D-GAP) as indicated in the figure were monitored at 290 nm after adding a catalytic amount of $\alpha_2\beta_2$ complex. The ordinate represents the total absorbance change with respect to a blank solution containing only indoleglycerol-P and indole after thermodynamic equilibrium has been reached. Buffer: 0.05 M Tris·HCl, 1 mM EDTA, 0.2 mM DTE, pH 7.8. Temperature: 25°C, d = 1 cm, K = $(IGP)^{-1}$ (IND) (D-GAP).

STEADY-STATE KINETICS (INDOLE ⟶ INDOLEGLYCEROL-P)

The condensation of indole with glyceraldehyde-3-P (reaction 2) can be followed at 290 nm. Using this convenient handle and the $\alpha_2\beta_2$ complex as catalyst, the equilibrium constant of the reaction was determined by "titration" of known fixed concentrations of indole and indoleglycerol-P with varying concentrations of glyceraldehyde-3-P (Fig. 1). That concentration of glyceraldehyde-3-P, at which the addition of enzyme causes no further absorbance change, corresponds to the equilibrium concentration. The magnitude of the equilibrium constant emphasizes the fact that synthesis of indoleglycerol-P is the thermodynamically preferred direction of reaction 2. Since the synthesis of tryptophan (reaction 3) is virtually irreversible (6), a transient accumulation of indole is ruled out if the

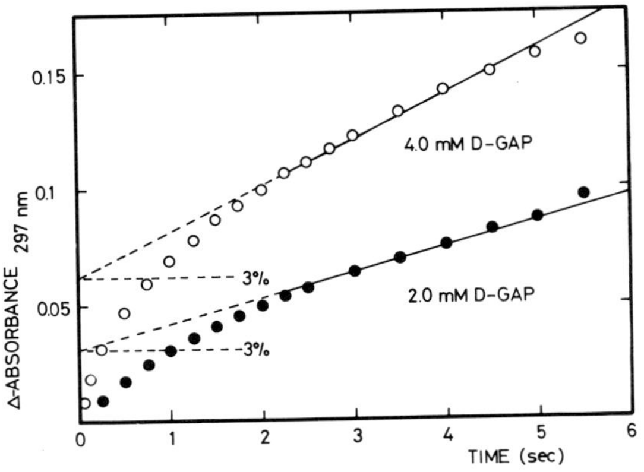

Fig. 2. Biphasic consumption of glyceraldehyde-3-P during synthesis of indoleglycerol-P (reaction 2). $\alpha_2\beta_2$ complex was mixed with solutions of D-GAP and indole and the synthesis of IGP was monitored at 297 nm. o - o: 4 mM D-GAP, 0.53 mM IND, 0.04 mM α- and 0.07 mM β-equivalents. ● - ●: 2 mM D-GAP, 0.53 mM IND, 0.02 mM α- and 0.035 mM β-equivalents. Buffer: 0.1 M K-phosphate, 1 mM EDTA, 0.2 mM DTE, pH 7.6. Temperature: 20°C. The concentration of GAP consumed in the transient phase was calculated from the ordinate intercepts with $\Delta\varepsilon_{297\ nm} = 0.457$ $mM^{-1}cm^{-1}$.

"physiological" reaction 1 is indeed a sequence of reactions 2 and 3.

For chemical reasons, only the carbonyl form of glyceraldehyde-3-P is expected to participate in reaction 2. Fig. 2 shows a stopped-flow experiment in which indole and glyceraldehyde-3-P were mixed with high concentrations of the tryptophan synthase complex.

It is apparent that the observed "burst" corresponds to the consumption of only 3 % of the total concentration of glyceraldehyde-3-P. This amounts to the known fraction of the carbonylform in equilibrium with the hydrate (10). The subsequent zero-order phase is determined by the slow dehydration of the aldehyde hydrate (10).

Table 1. Selected steady-state and rate constants for the synthesis of indoleglycerol-P. Comparison of α-subunit and $\alpha_2\beta_2$ -complex as catalysts.

constant	α-subunit	$\alpha_2\beta_2$ complex
k_1 ($M^{-1}sec^{-1}$)	3.6×10^4	4.2×10^5
K_{IND} (mM)	5.0	2.2
k_{cat} (sec^{-1})	3.3	121
K_{IGP} (mM)	0.040	0.007
K_{IPP} (mM)	0.05	0.01

Detailed studies of the steady-state kinetics of reaction 2 catalyzed by the $\alpha_2\beta_2$ complex in the absence and presence of indolepropanol-P support previous conclusions (8) that the mechanism corresponds to strictly ordered addition with glyceraldehyde-3-P binding first:

Scheme 1 $$E\cdot IPP \rightleftharpoons E \overset{k_1}{\rightleftharpoons} E^{GAP} \rightleftharpoons E^{GAP}_{IND} \overset{k_{cat}}{\rightleftharpoons} E+IGP$$

The results of the analysis are summarized in the righthand column of Table 1. The mechanism implies that only the binary enzyme: glyceraldehyde-3-P complex is capable of binding indole in a productive mode. The second-order rate constant of glyceraldehyde-3-P recombining with the enzyme (k_1) yields another clue. Its value is 2 - 3 orders of magnitude below diffusion control, indicating that it represents an overall process involving at least one additional step. This could be an isomerization creating a productive indole binding subsite.

Similar studies employing the α- subunit cannot yet be interpreted in the same unequivocal manner. The values in the lefthand column of Table 1 have been obtained under the assumption that the same catalytic mechanism applies to the α- subunit.

This procedure appears to be justified by the fact that indolepropanol-P is a competitive inhibitor towards glyceraldehyde-3-P but a noncompetitive inhibitor towards indole in both cases (cf. scheme 1).

STEADY-STATE KINETICS (INDOLE ⟶ TRYPTOPHAN)

Kinetic data obtained for reaction 3 catalyzed by the $\alpha_2\beta_2$ complex under steady-state conditions surprisingly led to non-intersecting double reciprocal plots obeying equation 1a.

eq. 1a $$\frac{1}{v_i} = \frac{1}{V} \left(1 + \frac{K_{SER}}{SER} + \frac{K_{IND}}{IND} \right)$$

This is in contrast to an earlier report (14). The discrepancy is probably due to previously undetected substrate inhibition by indole at concentrations above 0.2 mM. In similar experiments the ratio of indole to serine concentration ($X = \frac{IND}{SER}$) was kept constant while the total concentration was varied (eq. 1b).

eq. 1b $$\frac{1}{v_i} = \frac{1}{V} \left(1 + \frac{1}{IND} \left(X \cdot K_{SER} + K_{IND} \right) \right)$$

The straight lines shown in Fig. 3 prove the validity of equations 1a and 1b. This behaviour is formally identical to Bi-Bi Ping-Pong mechanisms (17). However, since we are dealing with a Bi-Uni reaction, equations 1a and 1b must be the result of a negligible $\frac{K}{SER \cdot IND}$ - term. The kinetic data (Fig. 3) are best explained by the strictly ordered addition mechanisms presented in scheme 2, top line.

In this mechanism, serine adds first and a step (k_3) after serine binding and before indole binding is largely irreversible

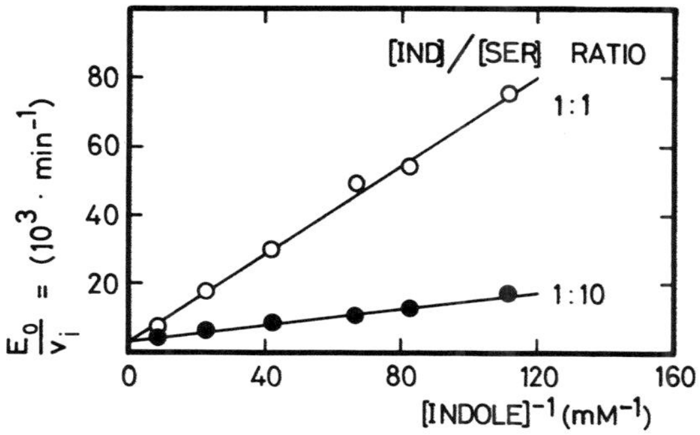

Fig. 3. Initial velocity pattern for the synthesis of tryptophan from indole and serine catalyzed by the $\alpha_2\beta_2$ - complex. Indole and serine were varied simultaneously, keeping constant ratios as indicated in the figure. Buffer as in Fig. 2. Temperature: 25°C. Enzyme concentration: 10^{-7} M β-equivalents, 4 x 10^{-7} M α-equivalents.

Scheme 2

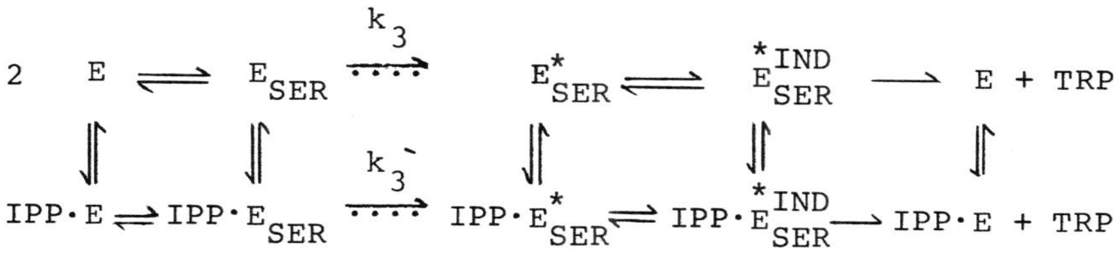

and rate determining. k_3 probably characterizes the formation of the aminoacrylate Schiff's base intermediate (13).

The mechanism of reaction 3 remains qualitatively the same in presence of saturating concentrations of indolepropanol-P, although there are quantitative changes. The dependence of the three steady-state kinetic parameters of eq. 1a on the concentration of indolepropanol-P are shown in Fig. 4. V_{max} decreases by a factor of 3 but the apparent inhibition constant of indolepropanol-P (K_I = 1 μM) is approximately tenfold smaller than the dissociation constant measured directly (7). This means that serine bound to the $\alpha_2\beta_2$ complex enhances the affinity of indolepropanol-P and vice versa. This conclusion was confirmed by direct measurement of binding of indolepropanol-P

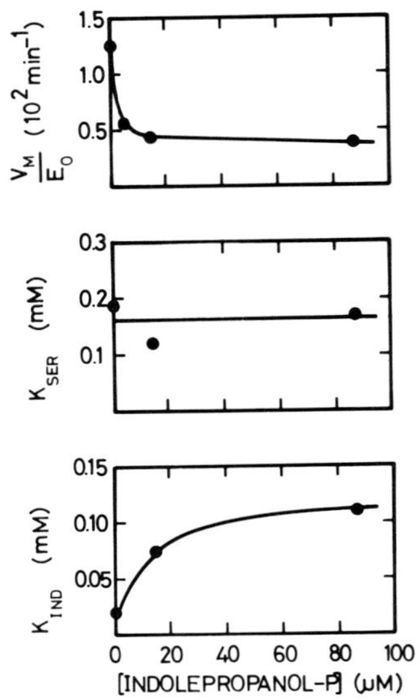

Fig. 4. Non-competitive inhibition of tryptophan synthesis by indolepropanol-P (cf eq. 1a). Conditions as in Fig. 3.

in presence of serine.

These findings are interesting with regard to the postulated composite active site (8,9), since indolepropanol-P is known to bind to the site catalyzing the cleavage of indoleglycerol-P. In a general way they reflect both positive and negative interactions between two different catalytic sites contributed by two different subunits. This is certainly compatible with the idea that the catalytic sites are juxtaposed, but gives no direct information on the fate of indole during the course of reaction 1 (6,9).

It is surprising that neither of the two sites catalyzing condensation reactions with indole are capable of binding indole alone in a productive fashion. This raises the question whether lack of indole binding can be demonstrated directly by equilibrium dialysis.

5a

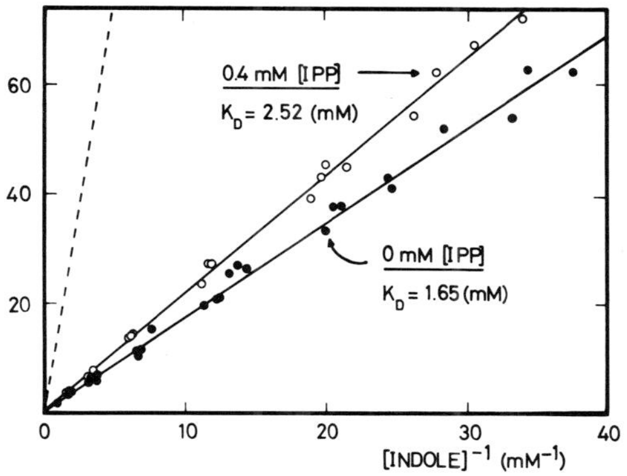

5b

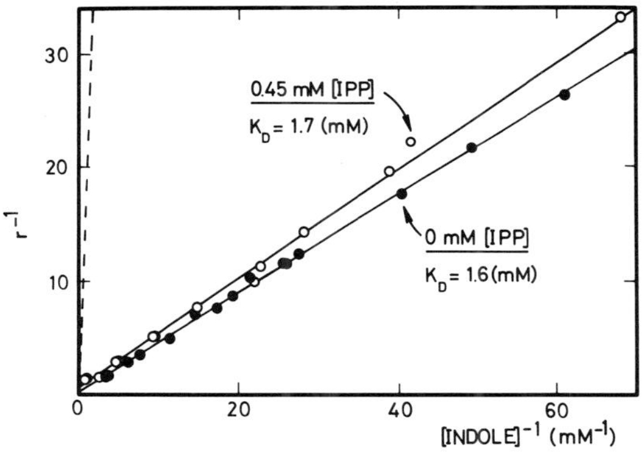

Fig. 5. Binding of indole to α-subunit (a) and the $\alpha_2\beta_2$ - complex (b) in presence and absence of indolepropanol-P. Double reciprocal representation of equilibrium dialysis with 14 C-indole. Dashed curves have been calculated for competition by indolepropanol-P with dissociation constants listed in Table 1 (K_{IPP}). Buffer and temperature: same as in Fig. 1. Enzyme concentrations: a) 0.40 mM and b) 0.1 mM α-subunit equivalents.

The data (most conveniently presented in double reciprocal form) confirm earlier reports (11) that weak binding occurs to the α- subunit (Fig. 5a) and the complex (Fig. 5b). Moreover, the isolated β_2 - subunit binds indole very unspecifically ($K_{IND} = 10^{-2}$M). The data further show that saturation of the enzymes with indolepropanol-P has only a minor and noncompetitive influence on the indole saturation curves. (The dashed curves have been calculated for strict competition between the two ligands).

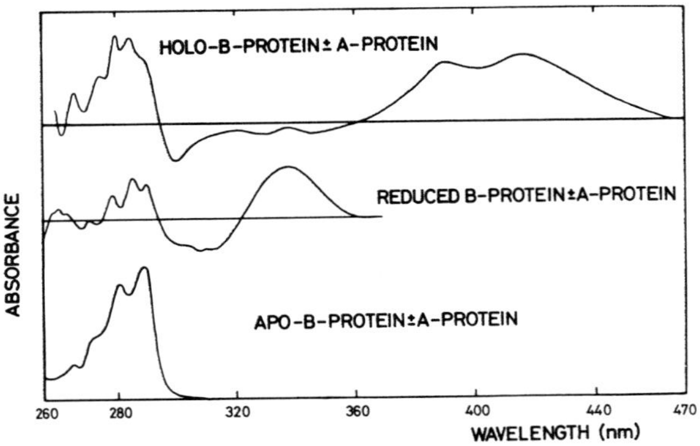

Fig. 6. Difference spectra of α-subunit forming a complex with holo- and apo-β_2 subunit and with holo-β_2 subunit reduced by $NaBH_4$. (0.1 M K-phosphate buffer pH 7.6, 25°C). Concentration of proteins: 0.1 mM β-equivalents, 0.2 mM α-equivalents

The results are in accord with the steady-state kinetics because if indole were capable of productive binding in absence of glyceraldehyde-3-P, strict competition by indolepropanol-P with regard to indole should have been observed.

KINETICS OF ASSEMBLY

One explanation for the mutual activation experienced by the α- and β_2 - subunits upon assembly is based on conformational changes. Assembly might induce or stabilize active conformations of each subunit (7). The application of rapid reaction techniques has led to direct evidence that both the β_2 - (14) and the α- subunits (7) are indeed flexible proteins. When the β_2 - subunit is complexed with α- subunit, the rapid isomerization (indicative of a conformational equilibrium in the β_2 - subunit) disappears (12).

We have used stopped-flow techniques in studying the kinetics of assembly of the multienzyme complex. Fig. 6 presents the difference spectra between the complex and the separated sub-

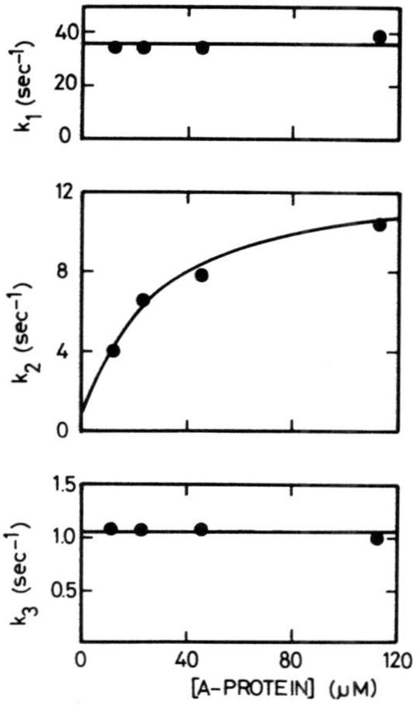

Fig. 7. Kinetics of assembly of $\alpha_2\beta_2$ - complex. Holo-β_2-subunit (0.02 mM β-equivalents) was mixed with excess α-subunit (concentrations indicated in the figure) in a Durrum stopped-flow spectrometer. The rate constants of the three superimposed exponential decay processes were obtained from the progress curve of absorbance at 405 nm vs. time. Buffer and temperature same as in Fig. 2.

units. Apparently the environment of both the pyridoxal-P bound to the β_2 - subunit and the aromatic aminoacid side chains of one or both of the subunits are perturbed in the complex. Similar difference spectra can be obtained using the reduced or the apo- β_2 subunit.

Using light of either 288, 334 or 405 nm for detection, rapid mixing of β_2 - subunit with excess α- subunit leads to complex progress curves. These can be easily interpreted in terms of at least three exponential processes well separated on the time axis. The dependence of the three time constants on the concentration of excess α - subunit are shown in Fig. 7. The time constant of the intermediate process (k_2) increases towards a plateau value. This behaviour is indicative of an isomerization step following a more rapid binding process by which the first complex between α- and β_2-subunits is formed:

eq. 3
$$\alpha + \beta \underset{\text{rapid}}{\overset{k_o}{\rightleftharpoons}} \alpha\,\beta \underset{\text{slow}}{\overset{k_2}{\rightleftharpoons}} (\alpha\,\beta)$$

Eq. 3 is formulated for a β- monomer. This appears to be justified because there is no evidence that the α- binding sites on the β_2 - subunit are non-identical or interact (15).

k_o is probably too rapid for resolution by stopped-flow experiments. The unequivocal interpretation of k_1 and k_3 in terms of a more complicated mechanism must also await further experiments. However, since the latter processes are first order with respect to β_2 - subunit, they must correspond to further isomerization steps. Reaction 1 (indoleglycerol-P→tryptophan) can now be used to answer the interesting question, at which stage during the dynamics of assembly the catalytic activity appears. Once again, the difference spectrum between tryptophan and indoleglycerol-P at 297 nm can be used to follow the turnover of substrate spectrophotometrically. Fig. 8 shows the progress curves obtained for two rapid mixing experiments in which the order of addition differs. In the control, the preformed complex is mixed with the substrates. Zero-order kinetics are established after a short transient period. However, when α- subunit and serine is mixed with β_2 - subunit (10 μM β- monomers) and indoleglycerol-P, a lag-time of 1.2 sec is observed before the steady-state is reached. This proves that the initial complex formed (cf. eq. 3) <u>is enzymically inert</u>.

The step responsible for the lag period can be identified using the assumption that it corresponds to about 10 half-lives of a firstorder process. With $t(\frac{1}{2}) = \frac{\ln 2}{k} = 0.12$ sec, $k = \frac{0.7}{0.12} = 6\ \text{sec}^{-1}$, in good agreement with the value of k_2 at the concentration of α- subunit (20 μM) used in the experiment

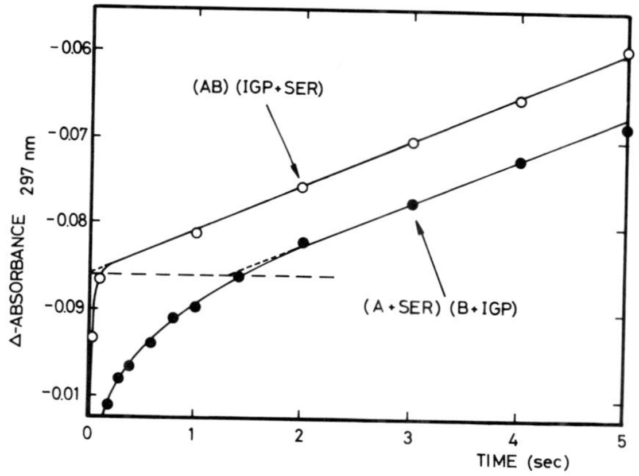

Fig. 8. Initial rates of tryptophan synthesis from indoleglycerol-P and serine (reaction 1). For description of experiment see text. Buffer and temperature as in Fig. 2. 0.25 mM indoleglycerol-P, 5 mM L-serine, 0.011 mM β-equivalents (B), 0.023 mM α-equivalents (A). o - o: Control in which the preformed $\alpha 2\beta 2$ complex is mixed with substrates. ● - ●: α- and β_2-subunits separated before mixing with substrates. -- Baseline for zero reaction between indoleglycerol-P and serine.

(Fig. 7). Thus it appears that a conformational change of one or both subunits is responsible for generating full enzyme activity of the complex in reaction 1.

Similar experiments conducted with reaction 3 (indole ⟶ tryptophan) as test reaction have led to essentially the same results. Unfortunately, analogous experiments involving reaction 2 are technically difficult to do because of the inconvenient hydration equilibrium of glyceraldehyde-3-P (cf. Fig. 2). It remains to be seen whether the activation of the α- subunit by complexformation with β_2- subunit also requires a conformational change.

ACKNOWLEDGEMENTS

We are grateful for support of part of this work by the Schweizerische Nationalfonds zur Förderung der wissenschaftlichen Forschung (Gesuch 3.8600.72) and for expert technical assistance by Martha Foehn.

REFERENCES

1. Reed, L.J.: Multienzyme complexes. Acc. Chem. Res. 7, 40 (1974).

Ginsburg, A., Stadtman, E.R.: Multienzyme systems. Ann. Rev. Biochem. 39, 429 (1970).

2. Veron, M., Saari, J.C., Villar-Palasi, C., Cohen, G.N.: The treonine-sensitive homoserine dehydrogenase and aspartokinase activities of E.coli K_{12}: Intra- and inter-subunit interactions between the catalytic regions of the bifunctional enzyme. Eur. J. Biochem. 38, 325 (1973).

3. Davis, R.H.: Channeling in Neurospora metabolism. "Organizational synthesis" (H.J. Vogel, J.O. Lampen, V. Bryson, eds.) Academic Press, N.Y. (1967), p. 303 ff.

4. Bonner, D.M., de Moss, J.A., Mills, S.E.: The evolution of an enzyme. "Evolving genes and proteins" (V. Bryson, H.J. Vogel, eds.) Academic Press, N.Y. (1965), p. 305 ff.

5. Rossmann, M.G., Moras, D., Olsen, K.W.: Chemical and biological evolution of a nucleotide binding protein. Nature 250, 194 (1974).

6. Yanofsky, C., Crawford, I.P.: Tryptophan Synthetase. "The Enzymes" 3rd ed. (P.D. Boyer, ed.) Academic Press, N.Y. (1972), p. 1 ff.

7. Kirschner, K., Wiskocil, R.L.: Enzyme-enzyme interactions in tryptophan synthetase from E.coli. "Protein-protein interactions". (R. Jaenicke, E. Helmreich, eds.) Springer Verlag, Heidelberg (1972), p. 245 ff.

8. Creighton, T.E.: A steady-state kinetic investigation of the reaction mechanism of the tryptophan synthetase of Escherichia coli. Eur. J. Biochem. 13, 1 (1970).

9. Matchett, W.H.: Indole channeling by Tryptophan Synthase of Neurospora. J. Biol. Chem. 249, 4041 (1974).

10. Trentham, D.R., McMurray, C.H., Pogson, C.I.: The active chemical state of D-glyceraldehyde-3-phosphate in its reactions with D-glyceraldehyde-3-phosphate dehydrogenase, aldolase and triosephosphate isomerase. Biochem. J. 114, 19 (1969).

11. Freedberg, W.B., Hardman, J.K.: Structural and functional roles of the cysteine residues in the α-subunit of the E.coli tryptophan synthetase. J. Biol. Chem. 246, 1449(1971).

12. Faeder, E.J., Hammes, G.G.: Kinetic studies of tryptophan synthetase. Interaction of L-serine, indole and tryptophan with the native enzyme. Biochemistry 10, 1041 (1971).

13. Miles, E.W., McPhie, P.: Evidence for a rate determining proton abstraction in the serine deaminase reaction of the β_2-subunit of tryptophan synthetase. J. Biol. Chem. 249, 2852 (1974).

14. Faeder, E.J., Hammes, G.G.: Kinetic studies of tryptophan synthetase. Interaction of substrates with the B subunit. Biochemistry 9, 4043 (1970).

15. Creighton, T.E., Yanofsky, C.: Association of the α-and β_2-subunits of the tryptophan synthetase of E.coli. J. Biol. Chem. 241, 980 (1966).

16. De Moss, J.A.: Studies on the mechanism of the tryptophan synthetase reaction. Biochim. Biophys. Acta 62, 279 (1962).

17. Plowman, K.W.: Enzyme kinetics. McGraw-Hill, New York (1972).

Received August 19, 1974

DISCUSSION

Buehner: Have you, or has anybody else, obtained crystals suitable for X-ray crystallographic structure analysis?

Kirschner: Schultz and Creighton have published preliminary work with crystals of the α-subunit (Eur.J.Biochem., 10, 195 (1969)). However, progress has been impeded by the strong tendency of the crystals to twin. Recently, Edith Miles has succeeded in crystallizing both the β_2-subunit and the complex. I expect that this will encourage a number of groups to tackle the problem of structure determination.

Maass: The binding of glyceraldehyde-3-P to the enzyme in the $\alpha_2\beta_2$-form is by several orders of magnitude far off diffusion controlled. The binding of the substrate to the α-subunit is still by another order of magnitude slower. What is the reason for this additional retardation?

Kirschner: A simple (but not unique) explanation would be that the postulated rapid isomerization accompanying the binding of glyceraldehyde-3-P is characterized by different equilibrium constants in the α-subunit and the complex. The different extent to which the binding of the first substrate overcomes an unfavourable isomerization equilibrium of the enzyme may be directly related to the difference in k_{cat} (cf. Table 1).

Helmreich: Can you see the Schiffbase formed with L-serine by transamidination? In this case you might be able to decide directly the compulsory order of addition postulated in your scheme.

Kirschner: Transient kinetics observed at 405 nm after rapidly mixing β_2-subunit or the $\alpha_2\beta_2$-complex with serine reveal a number of intermediates before attainment of the steady-state (York, S., Biochemistry 11, 2733 (1972), and our own unpublished observations). One of these steps ($k = 30 - 40\ sec^{-1}$) has been assigned to the deprotonation of the enzyme-bound Schiff's base of serine (ref. 13). One of the two faster processes preceding this step ($k > 500\ sec^{-1}$, $k \sim 200\ sec^{-1}$) might reflect the formation of the Schiff's base from the first, noncovalent complex of serine with the enzyme. It is also true that the presence of indole does not qualitatively change the transient progress curve. However, this does not necessarily mean that indole binding can only occur after Schiffbase formation.

Helmreich: Have you considered using the reduced enzyme containing pyridoxaminephosphate as a reporter group to study the conformational changes accompanying subunit interactions?

Kirschner: The pyridoxaminephosphate group bound covalently to the active site is indeed being used to study the binding of the α-subunit to the β_2-subunit. We do not only observe changes in the absorption spectrum of this reporter group (cf. Fig. 6) but also an enhancement of the quantum yield. It is not clear whether these changes are caused by direct or indirect (i.e. conformational) interactions between the two proteins.

Veeger: Since the rate constants of the $\alpha_2\beta_2$ recombination reaction are independent of the wavelengths, is it justified to conclude that the isomerization reaction (k_2) occurs mainly within the α-subunit?

Kirschner: I would draw another conclusion: Changes observed at 405 and 334 nm certainly imply that β-subunit undergoes the isomerization both as the PLP-complex and in the reduced form. Changes observed at 288 or 297 nm must be due to the perturbation of aromatic side chains, but it is impossible to decide from the data whether the α-subunit is also involved or not.

Veeger: Does the pyridoxamine-modified β-subunit also recombine with the α-subunit?

Kirschner: Yes, as Crawford observed many years ago, the reduced β_2-subunit also can form a complex with the α-subunit. Fig. 6 presents spectral evidence for this interaction. Moreover, reduced β_2-subunit is fully competent in stimulating the activity of the α-subunit in reaction 2.

Veeger: I would like to make the following remark about slow isomerization rates: it is very important to measure the temperature dependence of the isomerization reaction.

Kirschner: The temperature dependence of the rate constants in Fig. 7 has not been determined. I agree that potentially interesting information could be obtained from such studies.

Riesner: Since the fast step k_1 (Fig. 7) is not identical with the binding step, it has to be another step preceeding the conformational change in step k_2. What is the interpretation?

Kirschner: Any attempt of an interpretation of the data in terms of a comprehensive mechanism is obviously premature. However, Faeder et al. (cf. ref. 14) observed that the β_2-subunit alone exists in an equilibrium between two states. This is also supported by the earlier observation of Hatanaka et al. (Arch. Biochem. Biophys. 97, 596 (1962)) that the β_2-subunit can be stimulated from 3 % to about 70 % of the activity in the complex by the addition of 3 M ammonium sulfate. One might speculate that the α-subunit binds to only one of the pre-existing forms of the β_2-subunit and that this state is replenished from the other state with a rate constant of $k_1 = 40\ sec^{-1}$.

Antonini: Similar mechanisms appear to operate in the activation of enzymes associated with recombination of subunits in so far as the processes are dominated by a rather slow (k = 1 to 10 sec^{-1}) monomolecular process. The appearance of cooperativity in hemoglobin on mixing isolated α- and β-chains is also rate limited by a first order process having a similar rate.

von Hippel: Is the time of conformational changes in proteins responding to ligand binding of $\sim$1 to 10 sec^{-1} general? In lac repressor fluorescence changes due to a change of tryptophan environment after inducer binding takes about this time.

Kirschner: I am keenly aware of the analogous situation in the kinetics of hemoglobin assembly and would be delighted if someone would work out the detailed mechanism for it. With regard to the similarity of first-order-rate-constants found for different systems I think it is too early to make any generalizations. Almost nothing is known about the kind and magnitude of conformational changes involved in other systems.

Hinz: You refer to a strictly ordered binding mechanism in the reaction catalyzed by tryptophan synthetase. When assuming equilibrium conditions it is impossible to have a totally compulsory order of addition of the components, since that would imply an infinitely high Gibbs free energy barrier for one of the reactants which could not be overcome by addition of the second ligand either; therefore I would like to know, what ratio of the components one needs, in order to detect deviations from the order of addition of the ligands.

Kirschner: It is true that Bi-Uni ordered addition mechanisms require that one of the binding steps is not at thermodynamic equilibrium during the steady-state. In the case of reaction 2 catalyzed by the $\alpha_2\beta_2$-complex it is the binding of glyceraldehyde-3-P which is not at thermodynamic equilibrium (k_{-1} = 20 sec^{-1} by comparison with k_{cat} = 121 sec^{-1}, cf. Table 1). Your comment therefore does not apply in this case. In the case of tryptophan synthesis from indole and serine, the strict order of addition results from the necessary dehydration of serine before indole can form the new carbon-carbon bond. If the concentration of indole is increased above 0.5 mM, one observes substrate inhibition. Thus forcing the second ligand to bind first has deleterious results which are not interpretable in an unequivocal way.

Properties of Polymeric Proteins Bound to Solid Matrices: Subunit Exchange Chromatography

Eraldo Antonini, M. Rosaria Rossi Fanelli and *Emilia Chiancone*

Many proteins undergo reversible association-dissociation processes into subunits of lower molecular weight; the oligomer-protomer equilibrium is often attained very rapidly and generally depends on pH, ionic strength, and presence of specific protein ligands. The subunit interaction is very specific and the "recognition" by one protein subunit of its partner/s is generally not affected by the presence of other protein components in the system (Klotz et al., 1970, Cann, 1970).

It will be shown by the present work that the specificity of the subunit interaction in oligomeric proteins may be used as a basis for a chromatographic procedure which may be called "subunit exchange chromatography", since the process consists of an exchange of protein subunits between a liquid and a solid phase.

The general principle of the method is as follows: the subunits of a polymeric protein are attached covalently to a solid matrix; then a solution of the polymeric protein under conditions

1975

where there is a finite association-dissociation equilibrium is exposed to the solid matrix containing the protein subunits. A new equilibrium will now be established between the soluble subunits and those linked to the matrix and as a result part of the protein initially in solution will remain attached to the matrix. If the solid matrix is contained in a column and the protein solution is percolated through it, a retardation in the elution of the protein should be expected. The retardation will depend on the relative concentrations of the protein material attached to the matrix and in solution and on the degree of dissociation of the system: it will thus be a complex function of the protein concentration in solution and will reflect the relative tendency to associate of the soluble subunits with themselves and with those covalently bound to the matrix.

The similarities and differences between this procedure and other types of chromatography, notably affinity chromatography (Cuatrecasas and Anfinsen, 1971) are self-evident and will become even more clear after the description of the results.

The potentialities of the method have been explored mainly with hemoglobin, in view, among other things, of the good knowledge of the association-dissociation equilibria of this protein (Kellett, 1971, Chiancone et al., 1974).

A quantitative analysis of the system may be attempted as follows.

Consider the simplest case of a reversible dissociation of a dimer into monomers. Two equilibria will be established in the presence of monomers attached to a solid matrix:

(1) $X + X \underset{}{\overset{L_1}{\rightleftharpoons}} X_2, \qquad (X)^2 L_1 = (X_2)$

$$(2) \quad R - X + X \underset{}{\overset{L_2}{\rightleftharpoons}} R - X_2, \qquad (R - X)\ (X)\ L_2 = (R - X_2)$$

where X and X_2 are the monomer and dimer in solution, R - X is the matrix bound monomer and R - X_2 is the dimer formed by the binding of a monomer in solution to R - X, L_1 and L_2 are the association constants in dl/g.

If (y) is defined as the concentration of the protein initially in solution, which is bound at equilibrium to the matrix, and (R - X_t) as the total concentration of the monomer, initially bound to the matrix:

$$(3) \quad (y) = (R - X_2) \;/\; 2$$

$$(4) \quad (R - X_t) = (R - X) + (R - X_2) \;/\; 2 = (R - X) + (y)$$

from equation 1 and 2 the following relation can be obtained

$$(5) \quad (y) = \frac{(R - X_t) \;/\; 2}{\dfrac{L_1}{L_2} \, \dfrac{(X)}{(X_2)} + 1/2}$$

The ratio $\frac{(X)}{(X_2)}$ corresponds to the ratio $\frac{\alpha}{1 - \alpha}$ for the protein in solution, α being the degree of dissociation, which is a function of the total concentration in solution at equilibrium (c):

$$(6) \quad \frac{(1 - \alpha)}{\alpha^2} = L_1 c$$

Thus equations 5 and 6 allow to relate the quantity (y) to the various variables of the system. Similar treatments may be developed in the case of polymerization equilibria beyond the dimer stage, i.e. for values of n, the degree of polymerization, higher than 2.

Experimentally, the quantity (y) may easily be determined in a chromatographic experiment as follows. A column is filled with the solid matrix containing the covalently bound subunits. To obtain this, the coupling of the protein to the matrix may be performed under conditions of large degrees of dissociation; however, it appears that polymeric proteins are bound to the matrix largely in the form of monomers even when the dissociation is small since the covalent binding to the matrix would itself tend to produce a depolymerization. The amount of monomers bound per unit volume corresponds to $(R - X_t)$. On the column, previously equilibrated with the appropriate buffer, a solution of the protein, in the same buffer, is percolated till the protein concentration in the effluent reaches a value corresponding to that in the eluting solution. If there is interaction between the monomers in solution and those bound to the matrix, the front of the protein will be retarded with respect to the void volume of the column, i.e. the elution volume of an inert component of the solution. The retardation may be indicated by $\Delta V = V_X - V_0$, the difference between the elution volume of the protein and the void volume. V_X and V_0 may be determined by the hypersharp boundary representation of the elution profiles of the interacting and non interacting components.

In an experiment of this kind assuming that the system equilibrates very rapidly in comparison with the rate of flow through the column, (y) in equation 5 corresponds to

$$(7) \qquad (y) = \frac{(V_X - V_0)\ c}{v}$$

where v is the volume of the column and c is the concentration of the protein in the plateau region, which is the same as the c term appearing in equation 6. In turn, the retardation measured by the difference in the position of the hypersharp boundaries, $\Delta V = V_X - V_0$, is given by $\frac{(y)\ v}{c}$.

For a monomer-dimer equilibrium values of (y) as a function of c may be computed, on the basis of equations 5 and 6, for various values of L_1 and L_1/L_2, the quantity $(R - X_t)$ being directly determined by analysis of the protein monomer bound to the matrix.

The treatment of course assumes that all the monomers bound to the matrix have the same tendency to associate with the monomers in solution, which is not likely to be true in the real case due to the possibility of different sites of bonding of the protein to the matrix.

EXPERIMENTAL

Coupling of proteins to the solid matrix: Sepharose 4B was activated with cyanogen bromide and oxyhemoglobin was coupled to it at pH9. After the reaction the resin was washed with 0.3M $KHCO_3$ and with 2.0M NaCl in 0.01M phosphate buffer at pH 7. After the washings, the amount of HbO_2 bound to the resin resulted to be about 4 - 5 mg per ml of packed resin. In the case of other proteins similar conditions were used, and similar amounts of protein were bound to the resin.

The protein containing Sepharose was packed in thermostated columns having bed volumes of 7-9 ml. The flow through the column was maintained at 13-15 ml/hr with the use of a peristaltic pump. The optical density of the effluent was recorded at different wavelengths with a Gilford 2000 spectrophotometer equipped with flow cells.

RESULTS

Hemoglobin

Fig. 1 illustrates the elution profiles of HbO_2 and MbO_2 from a column containing Sepharose bound oxyhemoglobin* and equilibrated with 0.1M phosphate buffer at pH 7.0.

The elution volumes of HbO_2, in various buffers, are shown in Fig. 2 as a function of protein concentration. It is evident

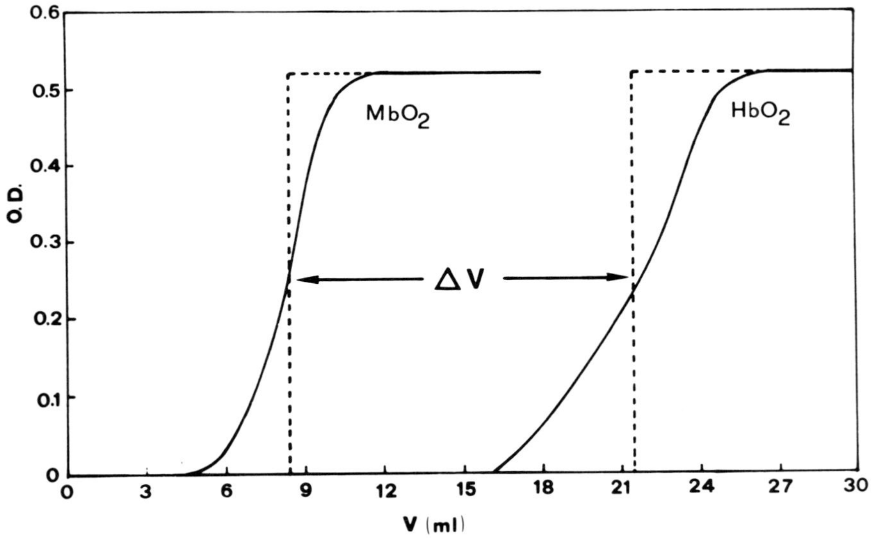

Figure 1 - Elution profiles of MbO_2 and HbO_2 from a column containing Sepharose bound oxyhemoglobin. The protein concentration in the solid phase was 4.7 mg/ml, that in the eluting solution 1 mg/ml. The column, of 8.5 ml bed volume, was equilibrated with 0.1M phosphate buffer at pH 7.0 + 10^{-4}M EDTA. Dashed lines represent the position of the hypersharp front of the boundaries.

---

* There is evidence that oxyhemoglobin is bound to Sepharose in the form of αß dimers.

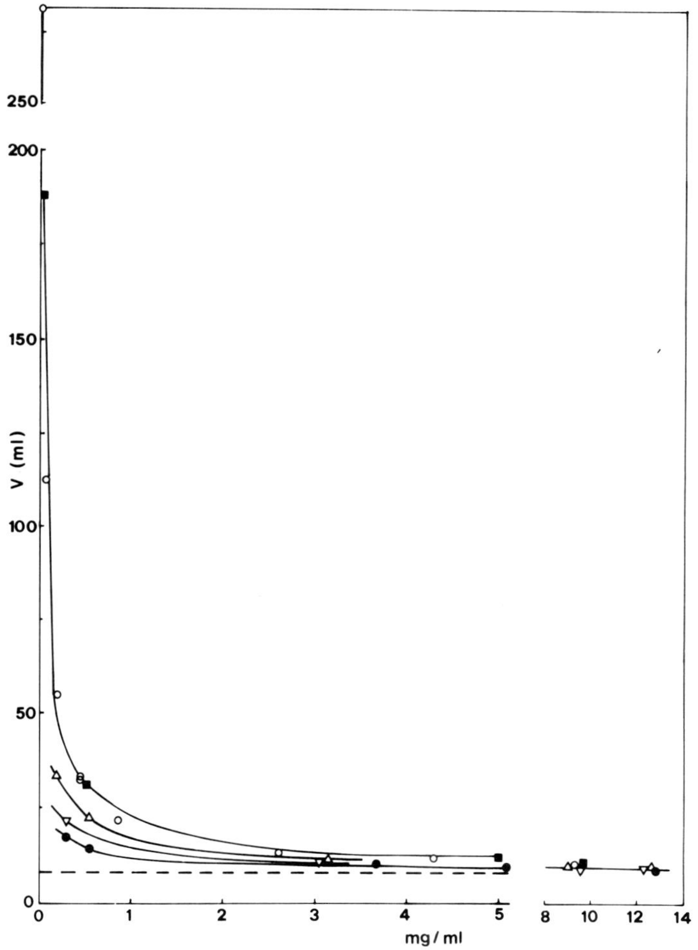

Figure 2 - Elution volume of HbO_2 from a column containing Sepharose bound oxyhemoglobin as a function of protein concentration. The protein concentration in the solid phase was 4.7 mg/ml and the bed volume 8.5 ml. Buffer: (O) 0.1M sodium phosphate, (■) 0.1M Na^+ (0.09M as Cl^- + phosphate), (Δ) 0.5M Na^+ (0.49M as Cl^- + phosphate), (∇) 1.0M Na^+ (0.99M as Cl^- + phosphate), (●) 2M Na^+ (1.99M as Cl^- + phosphate). The buffers were all at pH 7.0 and contained 10^{-4}M EDTA. Dashed line represents the elution volume of an inert protein (MbO_2).

that, as expected, at any protein concentration, they decrease as the ionic strength is increased, paralleling the decrease in the association constant for the dimer-tetramer equilibrium.

Figure 3 shows the same data plotted as y, the total amount of protein, initially in solution, bound to the column, versus log c. The curve is flatter than that computed from eqns 5 and 6 on the basis of L_1 = 1500 dl/g and corresponds to values of L_1 / L_2 in the range 5-20. (Figure 4).

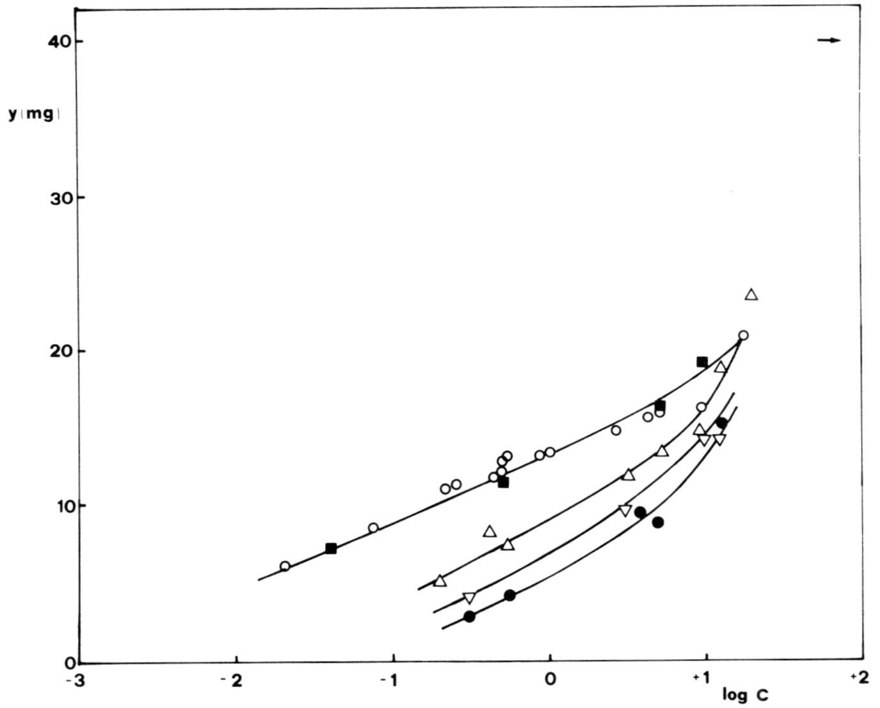

Figure 3 - Amount of HbO_2 (y) bound to the column containing Sepharose bound oxyhemoglobin as a function of protein concentration. Conditions and symbols are as in Fig. 2. Arrow indicates the amount of hemoglobin covalently bound to the column.

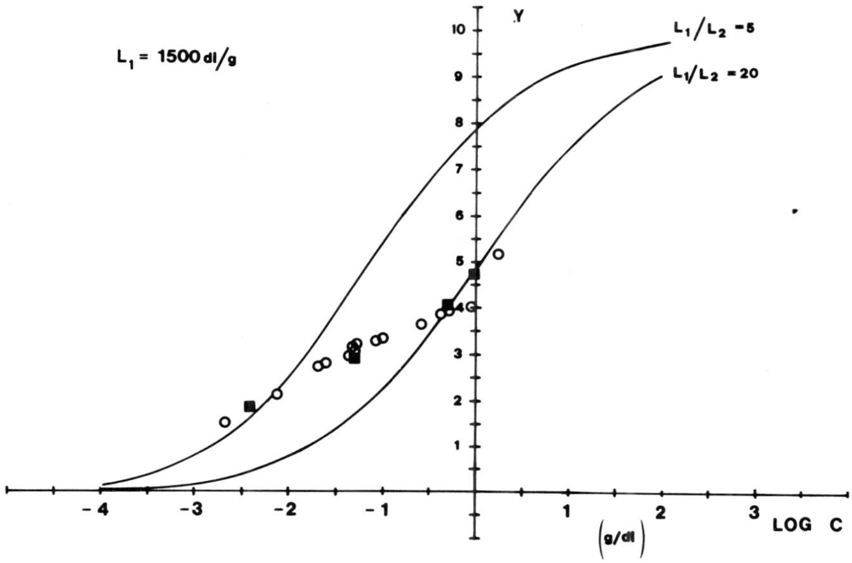

Figure 4 - Amount of HbO_2 (y) bound to a column containing Sepharose bound oxyhemoglobin as a function of protein concentration. Lines are theoretical ones and were computed according to equations 5 and 6. The experimental points are taken from Fig. 3 and refer to oxyhemoglobin in 0.1M sodium phosphate (O) and 0.1M Na^+ (0.09M as Cl^- + phosphate) (■).

Chromatographic experiments which exemplify the potentialities of the method for separation purposes are shown in Figs. 5 and 6.

Other Proteins

Experiments similar to those described for hemoglobin have been performed with several other proteins namely, lysozyme, α-chymotrypsin, insulin.

For all these proteins, under some conditions, a retardation in

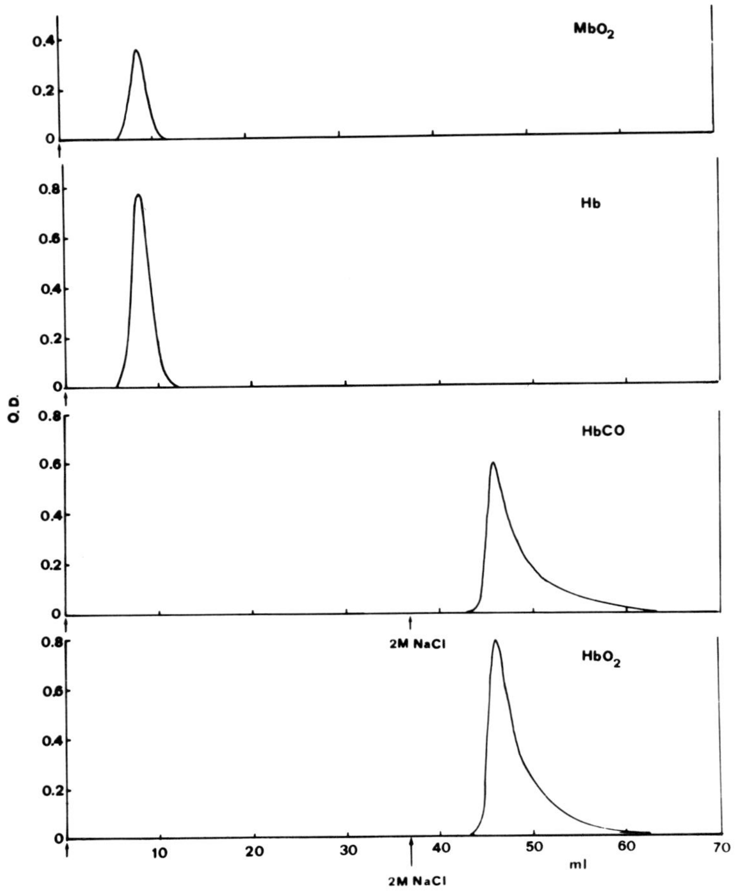

Figure 5 - Elution profiles of oxymyoglobin and several derivatives of hemoglobin from a column containing the same derivative of Sepharose bound hemoglobin. Initial protein concentration in the eluting solution 1 mg/ml. Buffer: 0.1M sodium phosphate at pH 7.0. Under these experimental conditions the association constants of the dimer-tetramer equilibrium are: 1500 dl/g for HbO_2, 6000 dl/g for HbCO; $> 10^5$ dl/g for Hb.

the elution due to specific interaction of the soluble subunits with those attached to the matrix could be shown to occur.

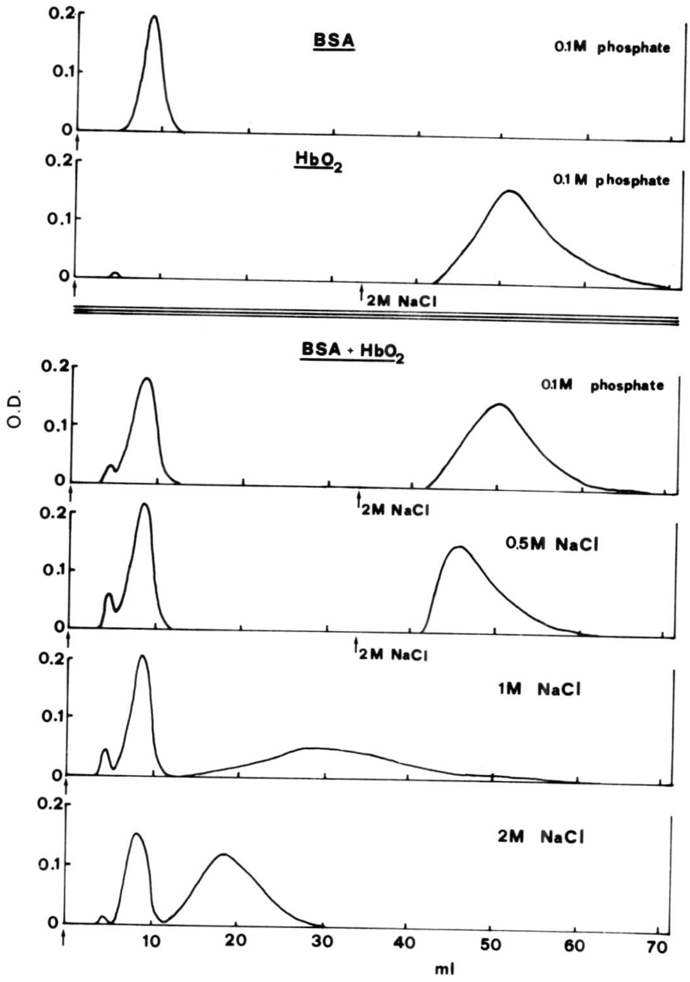

Figure 6 - Elution profiles of hemoglobin, bovine serum albumin and their mixtures from a column containing Sepharose bound oxyhemoglobin as a function of different eluting buffers.

DISCUSSION

The chromatographic procedure described here may be applied, in general, to all proteins (and even non protein substances) which undergo a reversible association dissociation process. It may be useful for a number of preparative and analytical purpo ses. As a preparative tool, separation of the specific protein from other components of the solution may be achieved by adsorption of the protein under conditions where the dissociation constant is low and by elution under conditions which enhance dissociation.

It is noteworthy that the method may be applied in different and eventually more general cases than affinity chromatography. It is also useful to note that relatively greater resolution is obtained by this procedure when the protein concentration in so lution is low, a condition which is generally encountered in the isolation of proteins from raw material.

The potentialities of subunit exchange chromatography for analy tical purposes appear also promising and are presently being exploited. The procedure can be used to study association-disso ciation processes in polymeric proteins under conditions which are not easily accessible with other methods. It also could be of value in evaluating the relative tendency to associate of monomers of homologous proteins, i.e. the relative tendency to form hybrids.

The behaviour of the systems experimentally studied is more complex than expected on the basis of the simple treatment outlined in the introduction. The complexities may arise, on one hand, from the heterogeneity in the properties of the covalently bound subunits (in turn reflecting various sites of attachment) and, on the other hand, from aspecific interactions between the soluble proteins and the solid matrix beyond the specific ones reflecting subunit association.

REFERENCES

Klotz, I.M., Langerman, N.R. and Darnall, D.W.: Quaternary structure of proteins. Ann. Rev. Biochem. 39, 25-62 (1970).

Cann, J.R.: Interacting Macromolecules, Academic Press, New York (1970).

Cuatrecasas, P. and Anfinsen, C.B.: Affinity chromatography. Ann. Rev. Biochem. 40, 259-278 (1971).

Kellett, G.L.: Dissociation of hemoglobin into subunits. Ligand-linked dissociation at neutral pH. J. Mol. Biol. 59, 401-424 (1971).

Chiancone, E., Anderson, N.M., Antonini, E., Bonaventura, J., Bonaventura, C., Brunori, M. and Spagnuolo, C.: Effect of heme and non-heme ligands on subunit dissociation of normal and carboxypeptidase digested hemoglobin: gel filtration and flash photolysis studies. J. Biol. Chem. in press (1974).

Received July 24, 1974.

DISCUSSION

Sund: Do you know if in the reaction between Sepharose and the α,ß-subunit primarily the α-polypeptide chain or the ß-polypeptide chain reacts or is the reaction statistical?

Antonini: We do not know yet exactly but I would expect that it would be essentially random.

Steinhardt: On the basis of some early work would one not expect rather different behavior on your columns between methemoglobin and cyanmethemoglobin?

Antonini: Certainly yes.

Pohl: Is there a way to obtain kinetic information of subunit interaction by varying the flow rate and do you have such experimental results?

Antonini: The procedure can be used to obtain kinetic information and we are collecting data at present.

Helmreich: You correctly pointed out one of the major problems in studying subunit interactions with matrix-bound enzymes; i.e. the heterogeneity of the preparation caused by the lack of chemical selectivity of CNBr used for the activation. I wonder whether one should not consider to activate selectively an amino acid side chain of the protein rather than the matix, this would give a more homogeneous reaction product provided that the reaction of the amino acid residue does not seriously perturb structure and function of the protein. It would, though, allow to bind selectively α- or ß-subunits to the matrix and to decide the question raised by Dr. Sund.

Antonini: We have already started experiments along the lines you suggest.

Wagner: I wonder whether myoglobin is a good reference for your quantitative measurements with hemoglobin. Because of possible differences in the gel filtration properties of these two proteins, would it not be better to take a column of the same dimensions but without protein fixed on the Sepharose?

Antonini: In a column of Sepharose the elution volumes of myoglobin and hemoglobin are the same and the gel appears to be completely permeable to both proteins.

G. Weber: What is the lowest macromolecular dissociation constant that can be studied by your method?

Antonini: Very low dissociation constants can be measured provided that the system reaches equilibrium during the experiments.

G. Weber: Why does one fail to see dissociation of deoxyhemoglobin?

Antonini: With deoxyhemoglobin there is no retardation because the deoxytetramer does not dissociate appreciably into dimers within the time of the experiment.

Kirschner: Do I understand you correctly that your columns contain 5 - 10 mg αß-dimers per ml?

Antonini: Yes.

Kirschner: Then what are the oxygen-binding properties of this material?

Antonini: It binds oxygen but without cooperativity. Yet it exhibits a Bohr effect.

Jaenicke: If a subunit is fixed to the solid matrix one would expect different side chains to be involved in binding giving rise to different affinities of other subunits to the fixed subunit. How can you distinguish between this kind of heterogeneity and the true microheterogeneity you are planning to analyze with your method?

Antonini: The problem can be solved performing experiments

with fractions obtained from the microheterogeneous material.

Di Iorio: Do you have any data on methemoglobin and on pH dependence of your effect?

Antonini: Yes.

Sund: Do you assume that the Sepharose-bound α,ß-subunit is located at the surface and is freely available to the α,ß-subunit in solution?

Antonini: Yes.

Sund: How did you determine the concentration of the Sepharose-bound hemoglobin?

Antonini: By spectrophotometric analysis of the packed gel.

Contributions of Aromatic Amino Acid Residues to the Optical Activity of Protein-Ligand Complexes

Robert W. Woody

The study of protein-ligand interactions requires the application of all available physical and chemical probes of structure. Circular dichroism is a spectroscopic probe of great sensitivity and wide applicability. In the last decade, CD has become almost as commonly used in characterizing proteins as absorption or fluorescence spectroscopy. However, the very sensitivity of CD to conformation and stereochemistry which has made it so useful for detecting conformational changes has made it difficult to accurately interpret them.

Although most of the early work on the CD of proteins focussed on attempts to analyze the secondary structure, there has been increasing interest[1,2] in the CD bands displayed by coenzymes, prosthetic groups, inhibitors and other chromophoric groups bound to proteins. Many such protein ligands are optically inactive when free in solution, or exhibit only weak CD bands. However, the protein-bound form frequently exhibits strong CD bands, referred to as induced CD bands or extrinsic Cotton effects. Since protein-ligand interactions are the sine qua non for these induced CD bands, a thorough elucidation of their mechanism should provide important information about the stereochemistry of the bound ligand and the nature of the binding site.

In this paper, I will discuss the interpretation of the induced CD of hemes in heme proteins and of FMN in a flavodoxin.

THEORETICAL BACKGROUND

When we measure circular dichroism, we are measuring the difference in absorbance between left- and right-circularly polarized light:

(1) $CD \equiv \Delta\varepsilon = \varepsilon_{\ell} - \varepsilon_{r}$; Molar ellipticity $\equiv [\theta] = 3300\Delta\varepsilon$

The contribution of a given electronic transition to the CD is determined by the rotational strength, R:

(2) $R = \mathrm{Im}\{\underset{\sim}{\mu}_{oa} \cdot \underset{\sim}{m}_{ao}\}$.

Here $\underset{\sim}{\mu}_{oa}$ and $\underset{\sim}{m}_{ao}$ are, respectively, the electric and magnetic dipole transition moments for the transition o → a. Im refers to the imaginary part of the scalar product of $\underset{\sim}{\mu}_{oa}$ and $\underset{\sim}{m}_{ao}$. A crude physical picture which illuminates this definition is that the electric dipole transition moment is a measure of a linear displacement of charge upon excitation, while the magnetic dipole transition moment is a measure of a circular displacement. The superposition of a linear and a circular displacement of charge corresponds to a helical displacement, which will interact differently with left- and right-circularly polarized light.

Experimentally, we determine the rotational strength by determining the area under the CD curve, just as we determine the oscillator strength of a transition from the integrated extinction coefficient:

(3) $$R = 0.2477 \int \Delta\varepsilon \frac{d\lambda}{\lambda} \cong \frac{0.2477}{\lambda_{max}} \int \Delta\varepsilon d\lambda$$

where R is in Debye-Bohr magnetons (1 DBM = 0.9273×10^{-38} cgs units).

Theoretically, we can calculate the rotational strength of a transition if we know enough about the geometry of the chromophore and its surroundings, and the electronic structure of the chromophore. The formalism for carrying out such calculations in the framework of perturbation theory was developed by Tinoco[3], whose theory unified the earlier theories of Kirkwood[4] (coupled-oscillator theory) and of Condon, Altar and Eyring[5] (one-electron theory). The complete formula will not be written out here but we can indicate the various contributions to the rotational strength of a transition o → a occurring in the ith chromophore as:

(4) $$R_{ioa} = \begin{pmatrix}\text{Intrinsic}\\ \text{rotational}\\ \text{strength}\end{pmatrix} + \begin{pmatrix}\text{coupled-oscillator}\\ \text{interactions}\end{pmatrix} + \begin{pmatrix}\text{one-electron}\\ \text{interactions}\end{pmatrix}$$

Let us see how this scheme applies to the treatment of CD bands induced in a ligand chromophore by binding to a protein. The intrinsic rotational strength is the scalar product of the zeroth-order electric and magnetic dipole transition moments for the isolated chromophore. If, as is the case with most chromophores of interest, there is a plane of symmetry in the chromophore, this term vanishes. However, a few chromophores such as skewed dienes, di- and triphenylmethyl dyes, etc., in which the chromophore is inherently dissymmetric[6], can have large contributions from this term.

The one-electron terms, which were emphasized by Condon et al.[5], result from the mixing of zero-order states of differing local symmetry under the influence of the static electric field of the surroundings. This mixing of states permits magnetically-allowed and electrically-forbidden transitions such as $n\pi^*$ and dd transitions to acquire electric dipole strength and become optically active. Conversely, electrically-allowed but magnetically-forbidden transitions ($\pi\pi^*$) acquire magnetic dipole character through such mixing.

The third type of contribution is the coupled-oscillator term. Because this is the most important term in our subsequent discussion, an explicit expression[3] will be given for it:

$$(5) \quad R_{i,c.o.} = -\frac{2\pi}{c} \sum_{j\neq i} \sum_{b\neq a} \frac{V_{ioa;job}\, \nu_a \nu_b\, (\underset{\sim}{R}_j - \underset{\sim}{R}_i) \cdot (\underset{\sim}{\mu}_{job} \times \underset{\sim}{\mu}_{ioa})}{h(\nu_b^2 - \nu_a^2)}$$

$$- 2 \sum_{j\neq i} \sum_{b\neq a} \frac{\mathrm{Im}\, V_{ioa;job}\, (\underset{\sim}{\mu}_{ioa} \cdot \underset{\sim}{m}_{jbo}\nu_a + \underset{\sim}{\mu}_{job} \cdot \underset{\sim}{m}_{ioa}\nu_b)}{h(\nu_b^2 - \nu_a^2)}$$

As the name implies, this term results from a coupling of oscillators or transition moments in different chromophores. The first term is the contribution emphasized in Kirkwood's theory[4]. Here $V_{ioa;job}$ is the energy of coupling between the two oscillators; ν_a and ν_b are the frequencies of the transitions o → a and o → b; $\underset{\sim}{\mu}_{ioa}$ and $\underset{\sim}{\mu}_{job}$ are the electric dipole transition moments for the transitions in groups i and j; $\underset{\sim}{R}_i$ and $\underset{\sim}{R}_j$ are the positions of the centers of the ith and jth groups; and c and h have their standard meanings of the velocity of light in vacuo and Planck's constant, respectively. Physically, this term corresponds to the fact that a linear displacement of charge ($\underset{\sim}{\mu}_{job}$) in group j leads to a circular displacement of charge about the center of group i and hence leads to a magnetic dipole transition moment. The second term in Eqn. (5) has been termed by Schellman[7] the μm mechanism. It results from the coupling of the electric dipole transition moment in one group with the magnetic dipole transition moment in another group.

CD OF HEME PROTEINS

The circular dichroism of heme proteins presents a rich collection of induced CD bands. Heme proteins are attractive model systems for studying such effects for several reasons. Myoglobin, hemoglobin and the cytochromes have been studied extensively by a variety of techniques. Most importantly, the three-dimensional structures of these proteins have been elucidated by X-ray diffraction. Further, the π-electronic structure of porphyrins has been extensively investigated. Although porphyrins are complex structures, their high symmetry makes theoretical work more tractable. The studies of Gouterman and co-workers[8] have provided a rather satisfactory picture of the lower $\pi\pi^*$ excited states. The wide variety of heme proteins available means that insights gained in the study of hemoglobin and myoglobin may be useful in the analysis of a large number of systems. Finally, if we can learn how to interpret the Cotton effects in heme proteins, we may gain important insights into the interpretation of other interesting systems such as enzyme-coenzyme complexes, dye-protein complexes, visual pigments, etc.

The CD and absorption spectra of human hemoglobin in various states of ligation are shown in Figure 1. This figure is taken from the work of

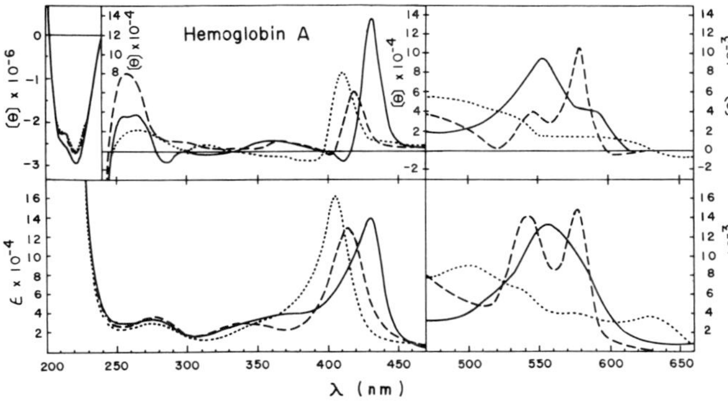

Fig. 1. CD and absorption spectra of human hemoglobin. Deoxyhemoglobin, _____; oxyhemoglobin, - - -; methemoglobin, ... Reproduced from the paper of Sugita et al.[9] by permission of the authors and of the American Society of Biological Chemists, Inc.

Sugita and co-workers[9]. Other mammalian hemoglobins and myoglobins are qualitatively similar and differ only quantitatively.[9]

The absorption spectrum of oxyhemoglobin (HbO_2, broken curve) is typical of closed-shell metalloporphyrins, with two closely spaced bands of moderate intensity in the 500-600 nm region, which are called the α and β bands or Q bands; an extremely intense band at about 400 nm called the γ, Soret, or B band, and a rather broad feature near 350 nm which is the δ or N band. There is another band in closed-shell metalloporphyrins near 260 nm (the ϵ or L band) but this is obscured by the tyrosine and tryptophan absorption bands near 280 nm. The absorption spectra of the other two hemoglobin derivatives shown here are qualitatively similar except in the visible region. For deoxyhemoglobin, the α and β bands overlap strongly and appear as a single band. The methemoglobin visible spectrum is more complex, with evidence of 4 bands. The two longer-wavelength transitions are assigned to charge transfer transitions, while the two short-wavelength bands are the α and β bands.[10]

Turning now to the CD spectra, we see that each of the absorption bands we have mentioned has associated with it a CD band. These CD bands are generally positive. The α and β bands in deoxyHb are well resolved and the 260 nm ϵ band is clearly defined in CD. These features illustrate the fact that CD spectra generally show greater resolution than do absorption spectra.

Several years ago Dr. Ming-Chu Hsu and I set out to determine the origin of the CD associated with the porphyrin $\pi\pi^*$ transitions. We considered only the $\pi\pi^*$ transitions initially because these are rather well understood, in contrast to the charge transfer transitions and possible metal dd transitions, which are still poorly characterized. All of the $\pi\pi^*$ transitions have a substantial electric dipole transition moment in the porphyrin plane. Thus the problem is to account for the magnetic dipole transition moment, which is zero in an isolated porphyrin. Three possibilities come to mind.

(1) The porphyrin may be bound in a distorted conformation and thus be an inherently dissymmetric chromophore. The work of Hoard[11] and of

Fleischer[12] on simple porphyrins has shown that the porphyrin plane is not as rigid as one might think and that it can undergo ruffling and doming with several of the atoms out of the mean plane by as much as several tenths of an Å. We suggest that small distortions of this magnitude are probably not significant.

(2) The $\pi\pi^*$ transitions of the porphyrin ring can mix with metal dd transitions of the same symmetry through the one-electron mechanism. The charge state and nature of the ligand have profound effects on the energy and magnetic dipole transition moments of metal d → d transitions and should therefore strongly influence the $\pi\pi^*$ rotational strengths if the one-electron mechanism were at work. However, from Figure 1 it can be seen that the CD spectra of oxy-, deoxy- and metHb have comparable intensities. This argument is greatly strengthened by the subsequent observation[9,13] that porphyringlobin has a Soret CD band comparable in intensity to that of hemoglobin.

(3) The porphyrin $\pi\pi^*$ transitions can couple with the electric and magnetic dipole transition moments of various protein groups through the coupled-oscillator mechanism. In fact, it had previously been suggested[14] that coupling of heme transitions with helical arrays of peptide groups might be responsible for the CD bands of heme proteins. Another suggested mechanism[15] involved the proximal histidine, the imidazole group of which is directly bonded to the heme iron.

Through the courtesy of Dr. John Kendrew[16], we obtained the detailed atomic coordinates of sperm whale metmyoglobin at 2Å resolution and carried out a systematic calculation of the coupled-oscillator interaction of the heme $\pi\pi^*$ transitions with various types of protein backbone and side-chain groups. The results of these calculations have been published[17,18] and these papers and the thesis[19] of Dr. Hsu should be consulted for detail.

Briefly, the major result of this study was that the sign and approximate magnitude of all four detectable heme $\pi\pi^*$ transitions can be accounted for by coupled oscillator interactions with the $\pi\pi^*$ transitions of aromatic side-chains (including histidines). The peptide groups make only minor contributions to the heme rotational strengths.

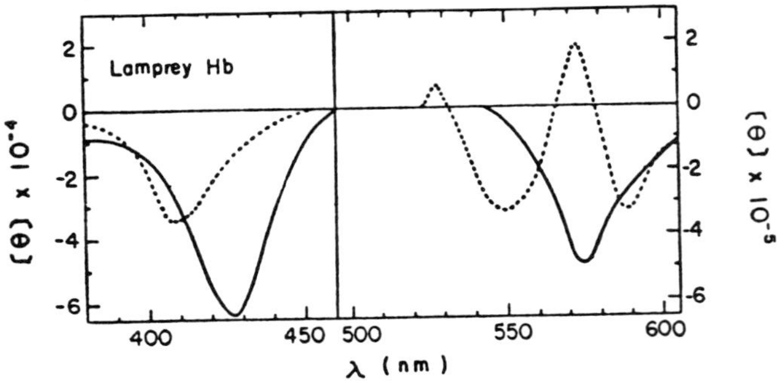

Fig. 2. Circular dichroism of lamprey hemoglobin. Deoxyhemoglobin, ____; oxyhemoglobin, - - -. Reproduced from the paper of Sugita et al.[9] by permission of the authors and of the American Society of Biological Chemists, Inc.

Calculations were also performed on hemoglobin[18,19] using the coordinates of horse oxyhemoglobin kindly provided by Dr. Max Perutz[20]. Again, coupling of heme transitions with aromatic $\pi\pi^*$ transitions provided rotational strengths agreeing in sign and approximate magnitude for all four observable porphyrin $\pi\pi^*$ transitions. One additional point of interest is that coupling of aromatic transitions in one subunit with the heme in another subunit is not negligible.

Given the success of these calculations on mammalian myoglobin and hemoglobin, we might ask how general this mechanism of heme-aromatic interactions as a source of heme Cotton effects really is. Fortunately, in the last few years, X-ray structural studies on two invertebrate hemoglobins have been completed. These are a hemoglobin of *Chironomus thummi thummi* studied by Huber's group[21] and lamprey hemoglobin studied by Hendrickson *et al.*[22] These hemoglobins have strikingly different CD spectra from those of mammalian hemoglobins. The CD of lamprey hemoglobin, reported by Sugita *et al.*[9] is shown in Figure 2. In particular, we note that the Soret band is negative, as it is also for *Chironomus* hemoglobin.[15]

We might note that these observations provide direct empirical support for one conclusion from our myoglobin calculations--that interaction of the heme with the polypeptide chain is not important in determining the CD. Both Chironomus[21] and lamprey[22] hemoglobins have essentially the same tertiary structure as do mammalian Hb and Mb. Therefore one cannot attribute the changes in sign to differences in chain folding. However, there are major differences in the distribution and location of aromatic residues in the heme locality.

Fleischhauer and Wollmer[23] have recently reported calculations on the Soret rotational strength of Chironomus Hb. These calculations of the heme-aromatic side-chain coupled oscillator interactions yield a negative Soret rotational strength, in agreement with experiment. Fleischhauer and Wollmer did not report calculations for the other heme transitions, but their results provide a further verification of the importance of this mechanism for the induced CD of heme proteins.

In order to test the mechanism further, calculations have now been performed for lamprey hemoglobin. Dr. Wayne Hendrickson[24] kindly provided the X-ray coordinates of the cyanomet form. The contributions of the various aromatic side-chains to the Soret CD are given in Table I. The R_x and R_y refer to the rotational strengths of the two nearly degenerate components of the Soret band. These components have transition moments which are perpendicular to one another. Their orientation in the plane is unknown but was assumed to be along the N-N line for this calculation. This orientation affects the magnitude of the individual components but not their total.

From the data of Sugita _et al._ we estimate a value of -0.20 DBM for the Soret rotational strength of oxyhemoglobin and -0.36 DBM for deoxyhemoglobin. No data are available for the cyanomet form but it is probable that the oxy form provides the best basis for comparison, and thus the calculated value of -0.19 agrees extremely well.

From the values in the table, it can be seen that although nearby aromatic groups make the largest contributions, there are significant contributions from more distant ones, up to nearly 20 Å away. The interaction energy $V_{\underline{ioa;job}}$ decreases roughly as R^{-3}, but there is a factor of R_{ij} in the numerator of the coupled-oscillator expression (Eqn. 5) which leads to a slower R^{-2} distance dependence. In addition, the large magnitude of the transition moments of both the porphyrin (especially the Soret) and the side-chain $\pi\pi^*$ transitions makes it possible for these transitions to

Table I Contributions to the Soret rotational strength of lamprey hemoglobin (in DBM)

Residue	Distance (Å)	R_x	R_y	$R_x + R_y$
Phe 41	7.2	-0.0167	0.4227	0.4060
Phe 42	9.4	0.0483	0.1198	0.1681
Phe 51	9.9	0.1586	-0.2588	-0.1002
Phe 52	6.2	0.4812	-0.6047	-0.1235
Phe 55	10.7	0.0801	-0.0250	0.0551
Phe 108	8.4	0.1134	-0.4667	-0.3533
Phe 115	7.7	-0.5211	0.3049	-0.2162
Phe 133	17.9	-0.0158	0.0969	0.0811
Tyr 27	17.6	0.0815	-0.0222	0.0593
Tyr 30	15.4	-0.1135	0.0877	-0.0258
Tyr 114	12.6	-0.1969	-0.0015	-0.1984
Tyr 148	11.0	0.2643	-0.1557	0.1086
His 73	5.6	0.3488	-0.4924	-0.1436
His 104	2.8	-0.6125	0.6388	0.0263
Trp 23	15.4	0.1144	0.0135	0.1279
Trp 72	15.9	-0.1800	0.1187	-0.0613
Total		0.0341	-0.2240	-0.1899

couple significantly across large distances. This means that the Soret CD band is sensitive to the distance and orientation of most of the aromatic side-chains in a molecule of this size. Conversely, a change in the heme CD cannot be readily ascribed to the alteration of any single group.

The proximal histidine (His 104) makes large contributions to the individual components, but its net contribution is small. As a result, its orientation can affect the shape of the Soret CD band but not its overall rotational strength. This is the case also with mammalian hemoglobin and myoglobin.

The calculations predict the following rotational strengths for the other heme $\pi\pi^*$ transitions: -0.010 DBM for the visible bands; -0.07 DBM for the δ or N band; and +0.06 DBM for the ε or L band. From the data of Sugita et al.[9,25] we obtain values of -0.015 DBM for the visible bands and +0.16 for the 260 nm band. No data are available for the δ band. We see that the coupling of aromatic transitions with porphyrin bands can account for the sign and approximate magnitude of the induced CD bands of lamprey hemoglobin, as well as the other systems previously studied.

CD OF FLAVODOXIN

The flavoproteins are a widespread class of proteins with a prominent chromophore as a prosthetic group. The CD of a number of flavoproteins has been reported by Tollin and co-workers.[26,27] The flavodoxins are particularly favorable systems for investigating the induced CD of flavoproteins. The structure of 2 flavodoxins has been determined by X-ray diffraction: one from Desulfovibrio vulgaris by Watenpaugh et al.[28] and one from Clostridium MP by Ludwig and co-workers.[29] The Desulfovibrio flavodoxin[28] has a tyrosine side chain parallel to and in essentially van der Waals contact with the plane of the flavin. On the other side of the flavin ring is an indole. In the clostridial flavodoxin[29], an indole replaces the tyrosine and a methionine the indole of the Desulfovibrio species.

The CD spectrum of D. vulgaris and C. pasterianum flavodoxins are shown in Figure 3, taken from the work of D'Anna and Tollin.[27] There are at least 3 distinct CD bands with extrema at ca. 450, 375 and 305 nm. The two long-wavelength bands correlate with the familiar flavin absorption bands. The 305 nm band is not clearly defined in absorption but has been observed in linear dichroism[30] and predicted theoretically.[31,32]

Calculations of the coupled-oscillator interaction between the flavin and the two adjacent aromatic residues of Desulfovibrio flavodoxin have been performed. Coordinates for these groups were provided through the courtesy of Dr. K. Watenpaugh. The transition moment directions of the flavin are only known approximately.[32] Wavefunctions were obtained using the method of Fox et al.[31] for the flavin, indole and phenol. The results of these calculations are shown in Table II.

The rotational strengths for the three bands are not in agreement with experiment. Two of the bands are predicted to have the wrong sign. There are two possible explanations for these discrepancies. It may be that

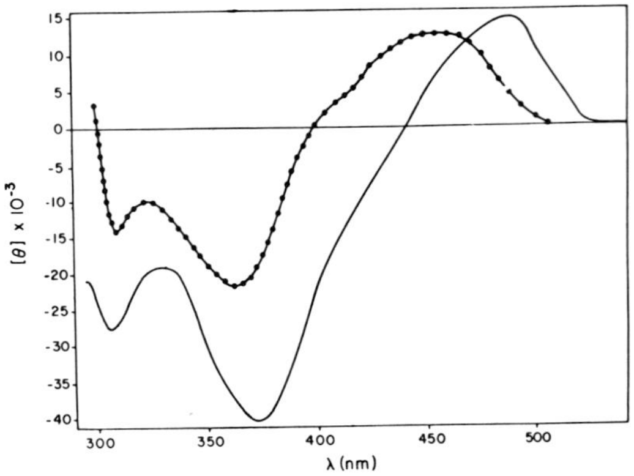

Fig. 3. Circular dichroism of flavodoxins from D. vulgaris ———; C. pasteurianum ••••. Reproduced from the paper of D'Anna and Tollin[27] by permission of the authors and of the American Chemical Society.

Table II Aromatic contributions to flavin rotational strengths (in DBM)

	450 nm	375 nm	305 nm
Trp 60	0.0273	-0.0749	-0.0271
Tyr 98	-0.0264	-0.0434	0.0253
Tyr 100	-0.0128	-0.0067	0.0306
Phe 101	-0.0513	-0.0150	0.0314
Total Calcd.	-0.0633	-0.1400	0.0602
Obsd.[a]	+0.12	-0.57	-0.20

a. Estimated from data of d'Anna and Tollin.[27]

other groups not yet considered, including nearby peptide groups, make a significant contribution through a coupled-oscillator interaction. On the other hand, there may be one or more $n\pi^*$ transitions underlying these bands which could make large one-electron contributions. An $n\pi^*$ transition has been predicted[32] at 380 nm and this would be expected to couple strongly with the 450 nm band. If such a coupling were of the correct sign, this could reverse the sign of the 450 nm band and make the 375 nm band more negative. Until further calculations have been performed, we cannot be certain whether the coupled-oscillator mechanism which seems adequate for the induced CD of hemes will also account for flavin CD. It is possible that one-electron contributions may play a significant role in flavoprotein CD.

CONCLUSIONS

As a result of these theoretical studies, it is clear that the coupled-oscillator interaction of the prosthetic group with nearby aromatic chromophores is largely responsible for determining the sign and magnitude of the induced CD bands in heme proteins. In the case of the Soret band, the range of this interaction is sufficiently large that many aromatic groups are involved, making structural interpretation of observed changes in CD difficult. In flavoproteins, we cannot yet be sure whether the dominant mechanism is of the coupled-oscillator or of the one-electron type. Even though a detailed structural interpretation may not be possible, conformational changes which alter the relative positions of the prosthetic group and the aromatic side chains can be readily detected by CD. Furthermore, we can anticipate that these mechanism will also be of importance in other ligand-protein systems.

ACKNOWLEDGEMENTS

I wish to acknowledge Dr. Ming-Chu Hsu's major contributions to the earlier work on myoglobin and hemoglobin. I am very grateful to Drs. John Kendrew, Herman Watson, Max Perutz and Keith Watenpaugh who provided the X-ray coordinates that have made these studies possible. I also thank Drs. Richard Frankel and Gordon Tollin for helpful discussions of heme and flavoprotein CD spectra, respectively. Permission to reproduce the figures was graciously granted by the authors and by the American Society of Biological Chemists (Figures 1 and 2) and the American Chemical Society (Figure 3). This work was supported in part by U.S. Public Health Service Grants, GM 13910 and GM 17850.

REFERENCES

1. Ulmer, D. D., Vallee, B. L.: Extrinsic Cotton effects and the mechanism of enzyme action. Advan. Enzymol. 27, 37-104 (1965).

2. Perrin, J. H., Hart, P. A.: Small molecule-macromolecule interactions as studied by optical rotatory dispersion-circular dichroism. J. Pharm. Sci. 59, 431-448 (1970).

3. Tinoco, I.: Theoretical aspects of optical activity: Polymers. Advan. Chem. Phys. 4, 113-160 (1962).

4. Kirkwood, J. G.: On the theory of optical rotatory power. J. Chem. Phys. 5, 479-491 (1937).

5. Condon, E. U., Altar, W., and Eyring, H.: One-electron rotatory power. J. Chem. Phys. 5, 753-775 (1937).

6. Moscowitz, A.: Some applications of the Kronig-Kramers theorem to optical activity. Tetrahedron 13, 48-56 (1961).

7. Schellman, J. A.: Symmetry rules for optical rotation. Accts. Chem. Res. 1, 144-151 (1968).

8. Weiss, C., Kobayashi, H., Gouterman, M.: Spectra of porphyrins. III. Self-consistent molecular orbital calculations of porphyrin and related ring systems. J. Mol. Spectry. 16, 415-450 (1965).

9. Sugita, Y., Nagai, M., Yoneyama, Y.: Circular dichroism of hemoglobin in relation to the structure surrounding the heme. J. Biol. Chem. 246, 383-388 (1971).

10. Eaton, W. A., Hochstrasser, R. M.: Single-crystal spectra of ferrimyoglobin complexes in polarized light. J. Chem. Phys. 49, 985-995 (1968).

11. Hoard, J. L., Hamor, M. J., Hamor, T. A.: Configuration of the porphine skeleton in unconstrained porphyrin molecules. J. Amer. Chem. Soc. 85, 2334-2335 (1963).

12. Fleischer, E. B.: The structure of nickel etioporphyrin-I. ibid. 85, 146-148 (1963).

13. Ruckpaul, K., Rein, H., Jung, F.: Zirkulardichroismus von Porphyrin-Globinen. Naturwiss. 57, 131-132 (1970).

14. Stryer, L.: A conformation-dependent Cotton effect in the Soret band of hemin: poly-L-lysine. Biochim. Biophys. Acta 54, 395-397 (1961).

15. Formanek, H., Engel, J.: Optical rotatory dispersion of a respiratory hemeprotein of Chironomus thummi. Biochim. Biophys. Acta 160, 151-158 (1968).

16. Kendrew, J. C., Watson, H. C.: Private Communication (1968).

17. Hsu, M.-C., Woody, R. W.: Origin of the rotational strength of heme transitions in myoglobin. J. Amer. Chem. Soc. 91, 3679-3681 (1969).

18. Hsu, M.-C., Woody, R. W.: The origin of the heme Cotton effects in myoglobin and hemoglobin. ibid. 93, 3515-3525 (1971).

19. Hsu, M.-C.: Optical activity of heme proteins. Ph.D. Thesis, Univ. of Illinois, Urbana, Ill. (1970).

20. Perutz, M. F.: Private Communication (1968).

21. Huber, R., Epp, O., Steigemann, W., Formanek, H.: The atomic structure of erythrocruorin in the light of the chemical sequence and its comparison with myoglobin. Eur. J. Biochem. 19, 42-50 (1971).

22. Hendrickson, W. A., Love, W. E., Karle, J.: Crystal structure analysis of sea lamprey hemoglobin at 2Å resolution. J. Mol. Biol. 74, 331-361 (1973).

23. Fleischhauer, J., Wollmer, A.: The influence of the aromatic amino acid side-chains on the sign of the Soret Cotton effect in Chironomus hemoglobin (CTTIII). Zeit. Naturforsch. 27b, 530-532 (1972).

24. Hendrickson, W.: Private Communication (1972).

25. Sugita, Y., Dohi, Y., Yoneyama, Y.: Circular dichroism of human and lamprey hemoglobins. Biochem. Biophys. Res. Commun. 31, 447-452 (1968).

26. Edmondson, D. E., Tollin, G.: Circular dichroism studies of the flavin chromophore and of the relation between redox properties and flavin environment in oxidases and dehydrogenases. Biochemistry 10, 113-124 (1971).

27. D'Anna, Jr., J. A., Tollin, G.: Studies of flavin-protein interaction

in flavoproteins using protein fluorescence and circular dichroism. Biochemistry 11, 1073-1080 (1972).

28. Watenpaugh, K. D., Sieker, L. C., Jensen, L. H.: The binding of riboflavin-5'-phosphate in a flavoprotein: Flavodoxin at 2.0Å resolution. Proc. Natl. Acad. Sci., U.S. 70, 3857-3860 (1973).

29. Anderson, R. D., Apgar, P. A., Burnett, R. M., Darling, G. D., Lequesne, M. E., Mayhew, S. G., Ludwig, M. L.: Structure of the radical form of clostridial flavodoxin: A new molecular model. Proc. Natl. Acad. Sci., U.S. 69, 3189-3191 (1972).

30. Siodmiak, J., Frackowiak, D.: Polarization of fluorescence of riboflavin in anisotropic medium. Photochem. Photobiol. 16, 173-182 (1972).

31. Fox, J. L., Laberge, S. P., Nishimoto, K., Forster, L. S.: Electronic structures of lumazines and isoalloxazines. Biochim. Biophys. Acta 136, 544-550 (1967).

32. Sun, M., Moore, T. A., Song, P.-S.: Molecular luminescence of flavins. I. The excited states of flavins. J. Amer. Chem. Soc. 94, 1730-1740 (1972).

Received August 7, 1974

DISCUSSION

Heyn: Some of the aromatic amino acids are very close to the porphyrin ring, which is quite a large group. How do you define the distances in such cases and what effect does it have on the calculated rotational strengths?

Woody: The distance of the perturbing groups from the porphyrin is measured from center to center. This distance would be very critical for the calculation if a dipole-dipole approximation were used for the coupling energy. However, we use a monopole-monopole approximation because the dipole-dipole approximation is poor when the distances are comparable to the dimensions of the groups involved. The only place the center-to-center distance enters explicitly is in the numerator of the coupled-oscillator expression, and the results do not appear to be very sensitive to this factor.

Brahms: How is the contribution of far-UV transitions of the polypeptide backbone reflected in the contribution to the rotational strength of the Soret band?

Woody: We have not taken peptide transitions other than the $n\pi^*$ and $\pi\pi^*$ into account. As I mentioned, these latter do not contribute significantly. This is probably due in part to the fact that there are so many peptide groups and they are distributed more or less isotropically around the heme. For the aromatic side-chains, however, we have included the fully-allowed $\pi\pi^*$ transitions which lie at or below 200 nm, and these are much more important than the relatively weak near-UV transitions.

Navon: The system you are dealing with consists of a sum of many contributions with different signs, so that the accumulated error may be large. Do you know of any simpler system which for instance contains only one pair of chromophores at large distances, where such theoretical calculations were compared with experiments?

Woody: I do not know of any such models involving distances of 10 Å or more. There have been successful tests of this approach at smaller distances. For example, S.F. Mason has succeeded in accounting for the CD of calycanthine, an alkaloid containing two aniline chromophores. It is certainly true that the presence of many opposing contributions is a cause for concern. However, the fact that results in semiquantitative agreement with experiment have been obtained for four different transitions in four different proteins provides the best vindication of the calculations.

von Hippel: Can you use a hemoglobin mutant where an aromatic is replaced by an aliphatic amino acid to isolate effect of one aromatic by looking at a difference spectrum for mutant and wild-type?

Woody: This is a potentially useful approach. However, there are two difficulties. First, relatively few mutant hemoglobins have been described in which an aromatic group near the heme has been replaced. Such an alteration is rather drastic. Second, a change in one amino acid is generally not confined to that site, but leads to more or less significant readjustments throughout the neighborhood of the site.

The one reported case where a mutation involving a nearby aromatic side-chain has been studied by CD (or actually ORD) is hemoglobin M_{Boston}. In this case, the distal histidine in one chain has been replaced by a tyrosine. The rotational strength of the Soret band appears to be about twice that of hemoglobin A. We actually did some calculations on this. Assuming that the tyrosine ring simply replaces the histidine ring, with no other changes, the CD is predicted to decrease slightly, in contrast with experiment. However, one only has to rotate the plane of the phenolic ring by about

45° about the C^{β}-C^{γ} bond to bring about a doubling of the Soret rotational strength. Recently, the structural changes in hemoglobin M_{Boston} have been studied by Fourier difference methods and there are significant alterations in structure near the heme. It would be interesting to re-examine this case in light of the new X-ray data.

Di Iorio: In relation to Dr. von Hippel's question - we have recorded CD spectra of hemoglobin$_{Zürich}$ ($\beta_{His}63 \rightarrow$ Arg) and no deviation was observed from the CD properties of hemoglobin A in the far-UV.

Woody: However, this reflects only the gross polypeptide chain conformation. What is relevant for these calculations is the rotational strength of the Soret and other porphyrin bands. It would be very interesting to look at these transitions. Our calculations would predict that if other perturbations can be neglected, the Soret rotational strength should be larger for hemoglobin$_{Zürich}$ than for hemoglobin A, since the distal histidine makes a negative contribution.

Chiancone: The CD spectrum in the Soret region of earthworm erythrocruorin is similar to that of human hemoglobin in the liganded form, but is inverted upon deoxygenation. Incidentally, the visible spectrum of the deoxy form is split. Do you have any suggestions as to the mechanisms involved?

Woody: I can only suggest that this must reflect a substantial change in conformation on going from the liganded to unliganded state. The splitting of the visible bands in absorption might reflect a coupling between hemes, but such a coupling should also lead to a split Soret band. In addition, coupling between hemes can lead to substantial changes in the shape of the Soret CD band, but it is not likely to alter the net rotational strength, especially to the extent of reversing its sign.

Wagner: Some people synthesized flavin derivatives with aromatic amino acid covalently fixed to them. Could these compounds be useful as models to study the influence of aromatic amino acids upon the circular dichroic properties of the flavin component?

Woody: Yes, I believe that they could be useful. However, unless the connecting link is rather restrictive, there may be a problem with defining the geometry. Unless the aromatic group and the flavin are confined to a nearly unique geometry, analysis of the CD of such models would probably not be very rewarding. A combination of CD and NMR measurements might be useful in such systems.

Veeger: I wonder whether the discrepancy between the calculated and observed data is due to the uncertain MO calculation. A difficulty in the calculations could be that the flavin absorption bands are split upon binding to the protein. For instance, we have always been surprised by the fact that the flavoprotein lipoamide dehydrogenase does not show a 450 nm CD band.

Woody: The necessity of relying upon MO calculations for transition moment directions, particularly in a complex heterocyclic molecule such as a flavin, is certainly a possible source of error. However, I have compared three different parameter choices and they all give transition moment directions which agree within about 10 degrees for the three long-wavelength bands. Furthermore the directions of the second and third transitions relative to the first are in general agreement with fluorescence polarization and linear dichroism measurements.

With regard to your second point, I can only suggest that the failure of lipoamide dehydrogenase to give a 450 nm CD band must be due to a coincidental cancellation of various contributions.

Section II. Enzymes

Chairmen: *S. Bourgeois* and *G. Weber*

The Architecture of the Coenzyme Binding Domain in Dehydrogenases as Revealed by X-Ray Structure Analysis

Manfred Buehner

For a long time dehydrogenases have been regarded as members of a family of enzymes. This is not only due to the rather trivial fact that they are using the same coenzyme, but also because of striking similarities in their molecular properties. Thus, lactate dehydrogenase (LDH), D-glyceraldehyde-3-phosphate dehydrogenase (GAPDH) and alcohol dehydrogenase (ADH) from yeast all have the same molecular weight and the same gross quaternary structure. They are tetramers of subunits of identical size and similar (isoenzymes) or identical amino acid sequence. ADH from liver (LADH) and soluble malate dehydrogenase (s-MDH) are dimers with about the same subunit size as the aforementioned tetramers. As in any big family there are of course some members whose similarity is less pronounced. Glutamate dehydrogenase for example is much bigger and homoserine dehydrogenase from E. coli is connected with another enzyme, aspartokinase, within one single polypeptide chain.

Similarities prevail, however, as is borne out by comparing the redox potentials of the bound coenzyme or its dissociation constants. Based on this rather superficial evidence, underlying primary principles of structural similarity have been expected. However, proof of this similarity by conventional chemical and physical methods of structure analysis appeared to be very difficult and it was only X-ray crystallography which turned out to be powerful enough to solve this problem.

Five years ago, Carl Brändén and Michael Rossmann presented low resolution balsawood models of LADH and LDH, respectively, at the Dehydrogenase Symposium here in Konstanz. Since then, X-ray analysis of dehydrogenases has made enormous progress. LDH [1,2], s-MDH [3], LADH [4,5] and GAPDH [6,7] and several complexes of these enzymes [8,9] are now analyzed at high resolution and numerous other dehydrogenases and complexes are at various stages of crystallographic investigation.

Expected structural similarities have been confirmed to a surprising extent, their range could even be extended beyond the class of dehydrogenases. The NAD binding site of the four enzymes mentioned above exhibits near identity of most structural features among these dehydrogenases and many of these features are also paralleled in flavodoxins [10,11] and, surprisingly, in kinases [12,13,14].

I will here restrict myself to outline the basic features of the NAD binding sites of LDH, LADH and GAPDH, which are the only dehydrogenases where both amino acid sequence and three-dimensional structure are presently known. Apart from a difference in the length of the polypeptide chain (which seems to be responsible for the quaternary structure difference) the tertiary structure of s-MDH is almost identical with that of LDH [15] and therefore, for most purposes within this paper, s-MDH can be represented by LDH.

COENZYME CONFORMATION

Physico-chemical analysis of the conformation of bound coen-

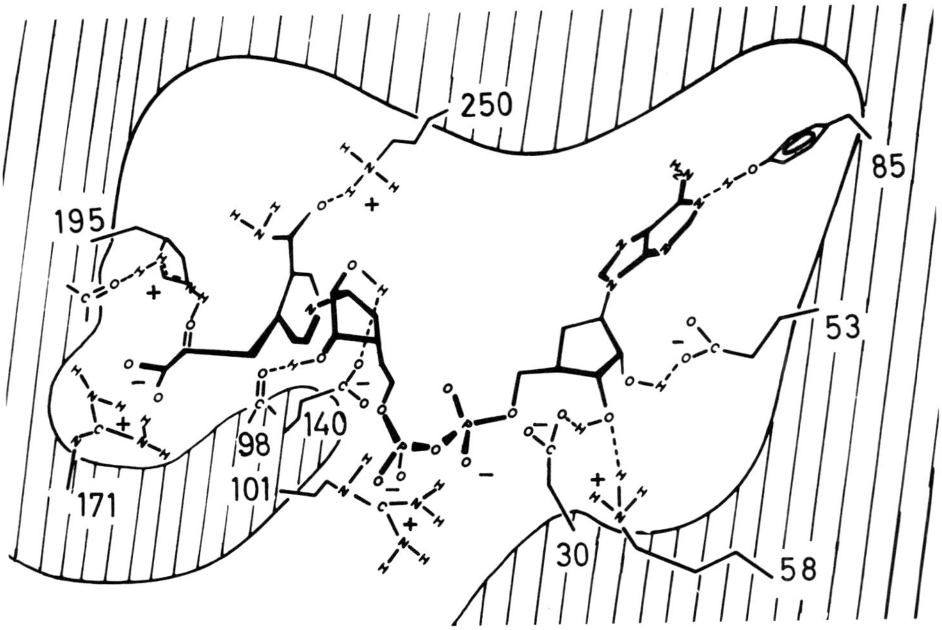

Fig. 1 Diagrammatic representation of the coenzyme binding site in the LDH:NAD-pyruvate complex.

zymes has often been restricted to distinguish between 'open' and 'closed' conformations. A more sophisticated discussion is now possible in the light of high resolution structure analysis of a number of complexes of dehydrogenases with nucleotides. Fig. 1 represents the active center of LDH in the abortive complex LDH:NAD-pyruvate [16]. The coenzyme is expanded, in an 'open' conformation (for details cf. [17]). After a similar analysis of the holo-enzyme of s-MDH, small differences in the conformation of s-MDH-bound NAD^+ and LDH-bound NAD were reported [18]. However, these differences might represent different interpretations of similar electron densities rather than real differences in the underlying structures. This will be better understood when the sequence of s-MDH is known, side chains can be assigned and estimates of the forces between enzyme and coenzyme can be made.

A view into the active center of GAPDH is shown in fig. 2

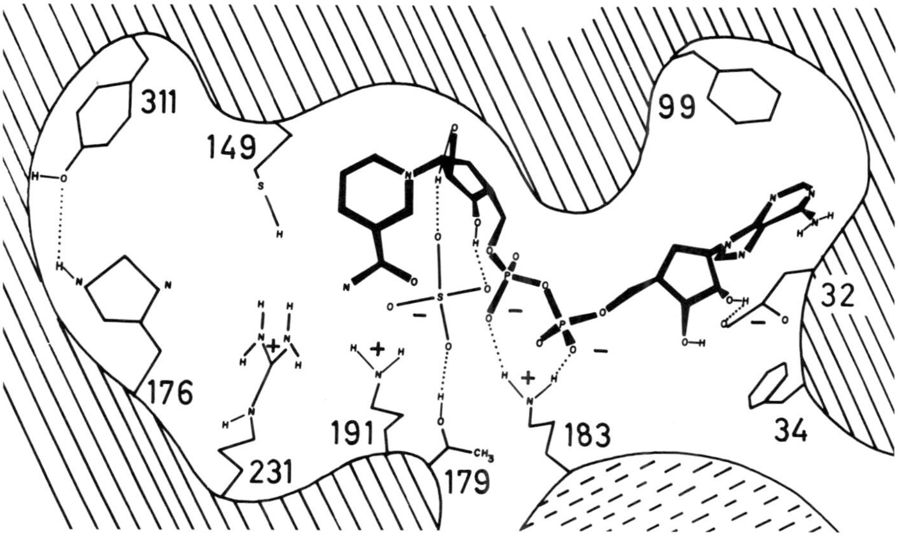

Fig. 2 Diagrammatic representation of the coenzyme binding site in the holo-enzyme of GAPDH.

[19]. The similarity of the coenzyme conformation as compared to fig. 1 is evident. All conformational angles of the coenzymes are the same within the limits of experimental error, except for a change by 180° of the torsion angle of the ribose-nicotinamide glycosidic bond. The nicotinamide relative

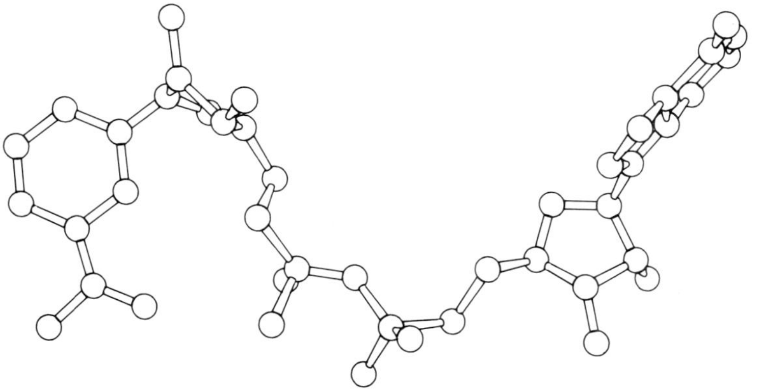

Fig. 3 Drawing of GAPDH-bound NAD^+ (Program ORTEP).

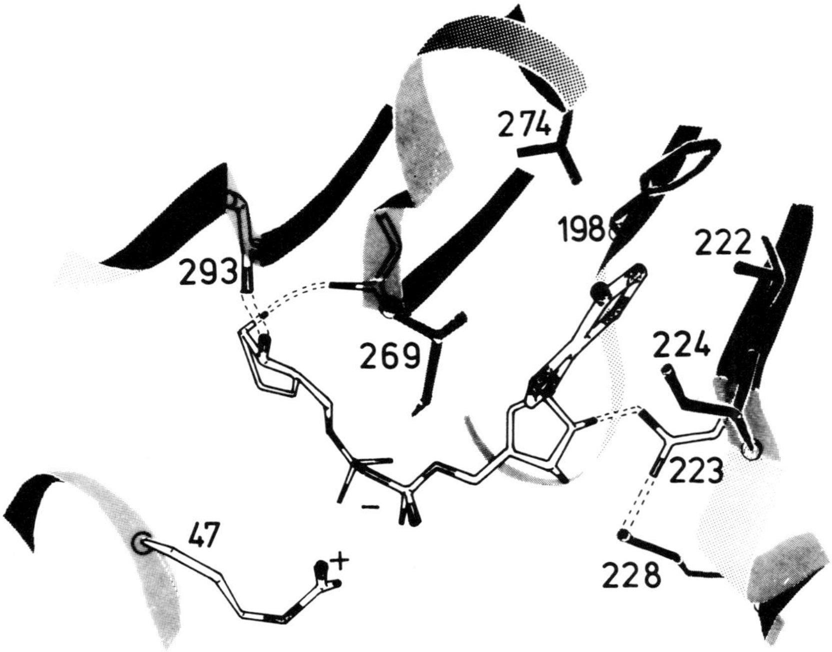

Fig. 4 Diagrammatic representation of the coenzyme binding site in the LADH-ADPRib complex.

to its ribose is thus in an 'anti' conformation in LDH and 'syn' in GAPDH. However, since the substituent on the pyridine ring is in the 3-position, the difference in conformational energy is small. This different orientation of the nicotinamide ring explains why LDH is A-specific and GAPDH is B-specific with respect to the hydride transfer onto C-4 of the nicotinamide. Fig. 3 shows a computer drawing of NAD^+ bound to GAPDH. The 'syn' orientation of the nicotinamide and the 'anti' conformation of the adenosine are easily recognized.

For LADH the picture is less complete at present as the study of the holo-enzyme has not yet reached sufficiently high resolution. However, a high resolution difference map of the

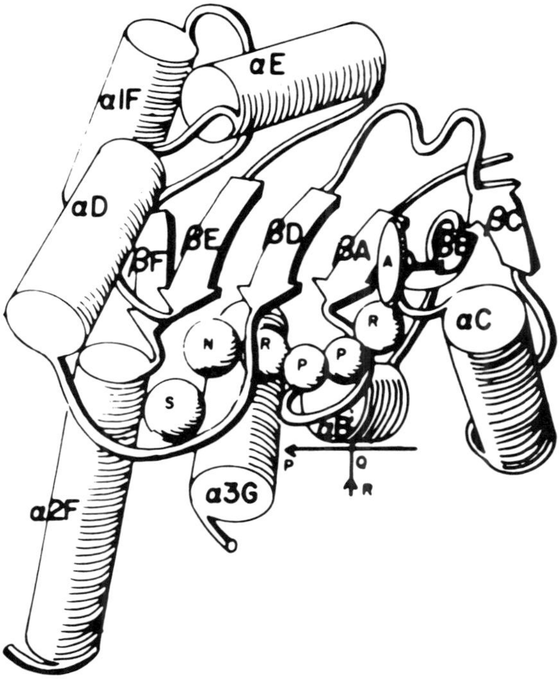

Fig. 5 Schematic drawing of the coenzyme binding domain of LDH. The labelled balls indicate the parts of the coenzyme molecule, S is the substrate site.

binary complex of LADH with ADPRib [9] yields quite some information which can be extrapolated to the whole coenzyme. Fig. 4 is a schematic representation of ADPRib bound to LADH. It is clearly seen that the conformation of the coenzyme fragment is quite similar to that of its counterparts in figs. 1, 2 and 3. The similarity in conformation of NAD bound to different dehydrogenases ought to be reflected in similarities of the respective binding sites in these enzymes.

PROTEIN CONFORMATION

The dominant feature in the structure of the LDH subunit is a parallel six-stranded pleated sheet which is the core of the

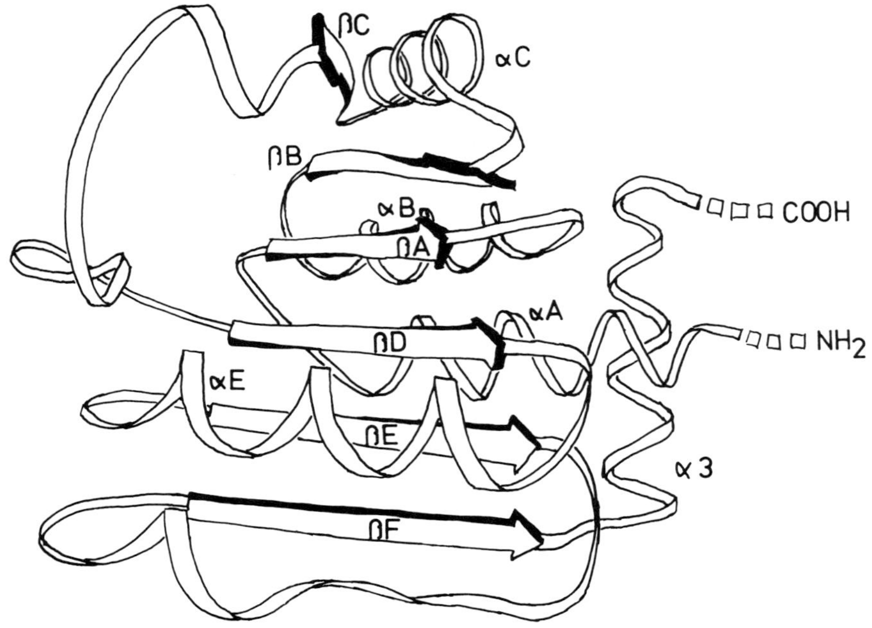

Fig. 6 Schematic drawing of the coenzyme binding domain of LADH.

coenzyme binding domain. The coenzyme is bound at the edge formed by the C-terminal ends of the strands (fig. 5, for the nomenclature of structural elements cf. [8]). Such a six-stranded parallel pleated sheet with the same folding pattern has been found in all the other dehydrogenases analyzed, s-MDH, LADH (fig. 6) and GAPDH (fig. 7), and is invariably the locus of coenzyme binding. If the amino acid sequence follows the order of the alphabet, then the topological sequence of the strands is C,B,A,D,E,F. The strands are mostly connected by α-helices whose orientation is invariant in some but not all positions.

The essentials and non-essentials of the NAD binding domain are summarized in table 1. A rough schematic representation of the spatial arrangement of these structural elements is given in fig. 8, emphasizing the symmetry of the dinucleotide

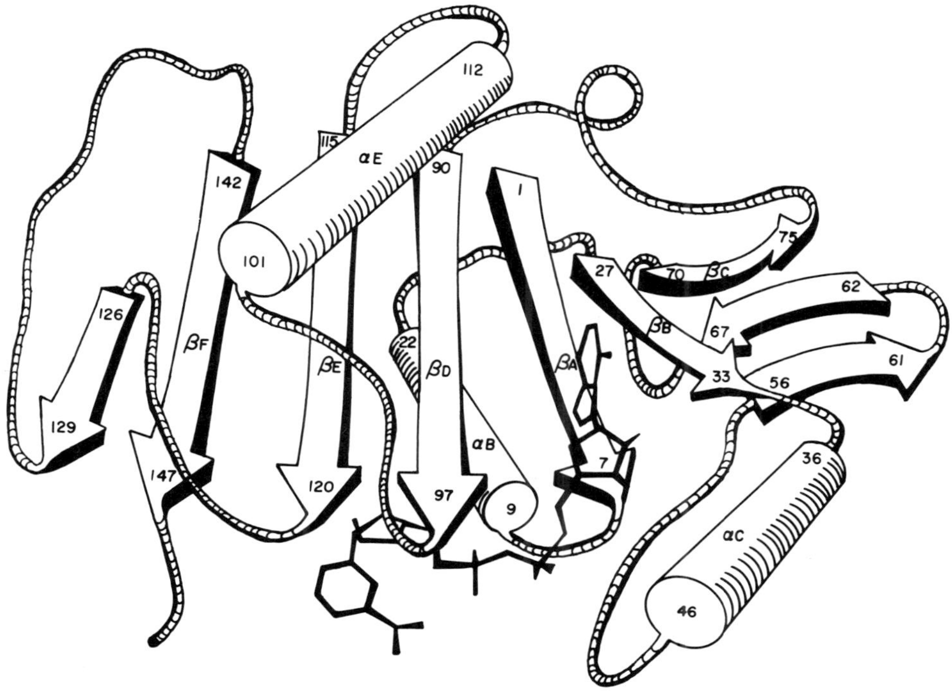

Fig. 7 Schematic drawing of the coenzyme binding domain of GAPDH.

binding site which is composed of two mononucleotide binding sites. Such a mononucleotide site consists of two parallel strands invariably connected by an α-helix, and a third strand which is connected to the second one by some other secondary-structured segment of polypeptide chain. To accommodate a dinucleotide coenzyme this structure is duplicated by adding a second copy of it at the free side of the first strand, thus creating an approximate 2-fold axis of symmetry between strands ßA and ßD, generating a six-stranded parallel pleated sheet.

Quantitative comparisons of coordinates of equivalent amino acid atoms have been made for the dehydrogenases mentioned [15,20]. It is remarkable that in spite of different loops, insertions etc. equivalent α-carbon atoms in the strands and

Table 1 Structural elements of the coenzyme binding domain.

LDH	LADH	GAPDH		Invariant Orientation
βA	βA	βA		βA
αB	αB	αB		αB
βB	βB	βB		βB
αC	αC	αC,β,β		
βC	βC,αCD	βC		βC
βD	βD	βD		βD
αD,αE	αE	αE		αE
βE	βE	βE		βE
α1F	βS,3^{10}S	β		
βF	βF	βF		βF
α2F	α3	α2F	(=αX)	
α3G	αA	$\alpha_{C\text{-term.}}$	(=αY)	αY

in some of the helices deviate by hardly more than the error in measuring coordinates. This accurate match is only possible because the left-handed twist of the pleated sheet of about 100° is essentially constant for all these dehydrogenases. This must be seen in the light of the fact that only 4 of the 126 to 147 amino acids involved are identical in the

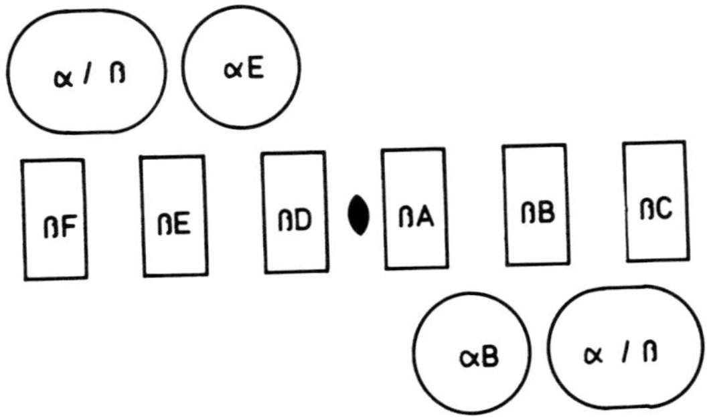

Fig. 8 Invariant components of the dinucleotide binding structure.

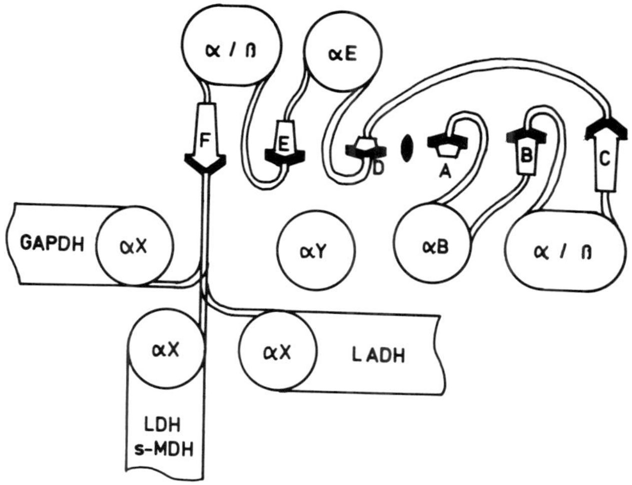

Fig. 9 Connection of the dinucleotide binding structure to the rest of the subunit in LDH, s-MDH, LADH and GAPDH.

three enzymes sequenced. Of course, many of the equivalent positions are occupied by amino acids of similar chemical character [20,21].

In different enzymes the total six-stranded dinucleotide binding domain is joined to the rest of the respective subunit in different ways. There is always one 'connector' helix, αX, which invariably follows ßF in the sequence but assumes different orientations in space (fig. 9). A second 'connector' helix, αY, is reproducibly defined in space, but is made out of different parts of the polypeptide chains. In LADH (αY=αA) it is directly preceding ßA in the amino acid sequence, in LDH (αY=α3G) it follows αX (=α2F) after 'insertion' of 65 amino acids, and in GAPDH (αX=α2F, αY is the C-terminal helix) the whole second domain of the subunit (150 residues) is 'inserted' between αX and αY.

The orientation of αX has significant bearing on the relative spatial arrangement of all structural domains in each enzyme and, indirectly or directly, on the quaternary structure. Due to the special orientation of αX (=α3) in LADH, the space to the left of ßF (in fig. 9) is cleared. Thus, two dinucleotide binding domains can be joined by hydrogen-bonds between their ßF-strands generating the particular dimeric structure of LADH which is based on a 12-stranded pleated sheet.

In all dehydrogenases analyzed thus far the ßC-edge of the six-stranded sheet is pointing out towards the solvent whereas the ßF-edge is involved in contacts to other domains of the same subunit (LDH, s-MDH, GAPDH) or in inter-subunit contacts (LADH).

Slightly differently designed parallel pleated sheets have been found in other nucleotide binding proteins. Flavodoxin (using FMN as a coenzyme) has been shown to contain a five-stranded parallel sheet [10,11] which can be described as a dinucleotide binding structure from which ßC has been deleted. Phosphoglycerate kinase uses a six-stranded parallel sheet to bind its coenzyme ATP [12,13]. Previously reported ambiguities in connecting these strands have now been resolved (C. Blake, personal communication to C. Brändén, and H. Watson, personal communication to M. Rossmann) and found to conform to the same pattern as the dehydrogenases described above. However, the fold of the analogous five-stranded parallel sheet in adenylate kinase appears to be different [14]. This enzyme seems to deviate from other kinases in many properties and might represent a sub-class of its own. The single-molecule double enzyme homoserine dehydrogenase/aspartokinase might in the future provide interesting insights into the relationship between dehydrogenases and kinases.

COENZYME LOCATION

The conformation of both the coenzyme and the protein domain

Table 2 Invariant residues in the coenzyme binding domain.

Amino acid		Gly	Gly	Asp	Gly
Position		ßA	αB	ßB	ßD
Sequence Nr.	in LDH	28	33	53	99
	in GAPDH	7	12	32	97
	in LADH	199	204	223	270
Binds to		A-Rib	A-PO_4	A-Rib 2'-OH	N-Rib

it binds to are so similar among dehydrogenases that it is not in the least surprising to find that the coenzyme always binds in the same way to the same structural elements. Bound coenzyme molecules match closely, upon superposition of equivalent protein parts, without having been included in the equivalences.

The adenine part of NAD binds in a hydrophobic crevice lined by side chains from ßA, ßB, ßD and some connecting segments of polypeptide chain. This subsite appears to be rather unspecific, accommodating nearly any aromatic molecule. On the other hand, the binding of the adenosine ribose seems to be crucial. It is positioned at the C-end of ßA on top of an invariant glycine (LDH 28). An invariant aspartate (LDH 53) on ßB forms a hydrogen-bond to the 2'-OH group. A second invariant glycine (LDH 33) on αB is close to the AMP phosphate. A third invariant glycine (LDH 99) assumes the same relative position at the NMN ribose as does the first one at adenosine. Table 2 shows the relation to the three known dehydrogenase sequences.

The phosphate groups are electrostatically bound by various positively charged side groups. The NMN ribose exhibits variable hydrogen-bonding patterns involving residues from the C-ends of ßD and ßE (LDH, s-MDH, LADH) or the C-3 phosphate group of the substrate (GAPDH). In LDH and LADH, 3'-OH hydro-

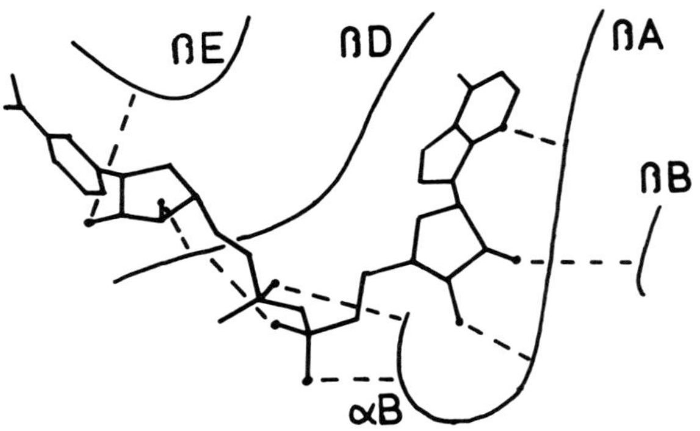

Fig. 10 Diagrammatic representation of some binding interactions between NAD+ and s-MDH.

gen-bonds to the main chain carbonyl of the residue preceding the invariant glycine on ßD (i.e. LDH 98 and LADH 269, respectively) and, similarly, 2'-OH binds to equivalent main chain peptide groups on ßE (LDH 140 and LADH 293). There is no equivalent on the NMN side to the invariant aspartate of the AMP side of the coenzyme.

The nicotinamide ring does not seem to provide many interactions with the enzymes at all. Its position is determined by the ADPRib part of the coenzyme, and its orientation ('syn' or 'anti') by 'passive' steric competition with a few crucial side chains of different space-filling rather than by 'active' interactions. The relative orientation of protein and coenzyme can be seen clearly in figs. 5 and 7 for LDH and GAPDH, respectively. Figs. 4 and 10 indicate that LADH and s-MDH follow the same binding pattern [18].

The approximate 2-fold axis of symmetry in the NAD binding protein domain is thus not reflected in all aspects of the binding and not in the conformation of the coenzyme itself.

COMMON ORIGIN ?

Was there a primordial general-specificity dehydrogenase from which our present dehydrogenases have been evolving in a diverging process?

If we think along the conventional ideas of evolution which are based on the laws of later biological differentiation, the answer to this question has to be NO. The different positions of the coenzyme binding domains in the respective polypeptide chains and even more the dissimilarity of the other domains in these dehydrogenases indicate clearly that linear gene reproduction with occasional minor mutations cannot be responsible for the variety of dehydrogenase structures.

However, looking at chemical and very early biological evolution we may have to change our thinking a little. For these early times we may have to accept a relatively labile genome, a high frequency of cutting and joining genes, and thus a reshuffling of even rather small RNA/DNA segments. We may have to consider interdependences of divergent and convergent evolutionary processes.

We will have to have a closer look into the other domains of these dehydrogenases. There are similarities there as well, but less pronounced than in the coenzyme binding domains. We will also have to include the other domains of kinases in this analysis. We do have to accept the existence of building blocks much smaller than today's proteins, respectively genes, which could have been shuffled and reconnected in different ways and may be used in a wide variety of present proteins for a range of functions.

Today's dehydrogenases certainly have not linearly descended from a common ancestor, but they do contain at least one common building block, in itself a symmetrical twin, which is as big as half the subunit. By its highly conserved three-dimensional structure this building block proves evolutionary relationships between molecules which have been diverging for three billion years [21].

Acknowledgements. I am deeply indebted to my friends and colleagues Drs. C.-I. Brändén, B. Nordström, G.C. Ford, D. Moras, K.W. Olsen and M.G. Rossmann for their generous permission to use unpublished results and for contributions of unpublished figures. The design of various figures by B. Furugren and extended financial support by the Deutsche Forschungsgemeinschaft are gratefully acknowledged.

REFERENCES

1. Adams, M.J., Ford, G.C., Koekoek, R., Lentz, P.J., Jr., McPherson, A., Jr., Rossmann, M.G., Smiley, I.E., Schevitz, R.W. and Wonacott, A.J., Nature 227, 1098-1103 (1970).

2. Rossmann, M.G., Adams, M.J., Buehner, M., Ford, G.C., Hackert, M.L., Lentz, P.J., Jr., McPherson, A., Jr., Schevitz, R.W. and Smiley, I.E., Cold Spring Harbor Symp. Quant. Biol. 36, 179-191 (1971).

3. Hill, E., Tsernoglou, D., Webb, L. and Banaszak, L.J., J. Mol. Biol. 72, 577-589 (1972).

4. Brändén, C.-I., Eklund, H., Nordström, B., Boiwe, T., Söderlund, G., Zeppezauer, E., Ohlsson, I. and Åkeson, Å., Proc. Natl. Acad. Sci. USA 70, 2439-2442 (1973).

5. Eklund, H., Nordström, B., Zeppezauer, E., Söderlund, G., Ohlsson, I., Boiwe, T. and Brändén, C.-I., FEBS-Letters 44, 200-203 (1974).

6. Buehner, M., Ford, G.C., Moras, D., Olsen, K.W. and Rossmann, M.G., Proc. Natl. Acad. Sci. USA 70, 3052-3054 (1973).

7. Buehner, M., Ford, G.C., Moras, D., Olsen, K.W. and Rossmann, M.G., J. Mol. Biol. 82, 563-585 (1974).

8. Adams, M.J., Buehner, M., Chandrasekhar, K., Ford, G.C., Hackert, M.L., Liljas, A., Lentz, P.J., Jr., Rao, S.T., Rossmann, M.G., Smiley, I.E. and White, J.L. in "Protein-Protein Interactions", eds. R. Jaenicke and E. Helmreich (Springer-Verlag, Heidelberg), p. 139-156 (1972).

9. Abdallah, M.A., Biellmann, J.-F., Nordström, B. and Brändén, C.-I., Eur. J. Biochem., in press (1974).

10. Watenpaugh, K.D., Sieker, L.C., Jensen, L.H., LeGall, T.

and Dubourdieu, M., Proc. Natl. Acad. Sci. USA 69, 3185-3188 (1972).

11. Anderson, R.D., Apgar, P.A., Burnett, R.M., Darling, G.D., LeQuesne, M.E., Mayhew, S.G. and Ludwig, M.L., Proc. Natl. Acad. Sci. USA 69, 3189-3191 (1972).

12. Bryant, T.N., Watson, H.C. and Wendell, P.L., Nature 247, 14-17 (1974).

13. Blake, C.C.F. and Evans, P.R., J. Mol. Biol. 84, 585-601 (1974).

14. Schulz, G.E., Elzinga, M., Marx, F. and Schirmer, R.H., Nature 250, 120-123 (1974).

15. Rao, S.T. and Rossmann, M.G., J. Mol. Biol. 76, 241-256 (1973).

16. Adams, M.J., Buehner, M., Chandrasekhar, K., Ford, G.C., Hackert, M.L., Liljas, A., Rossmann, M.G., Smiley, I.E., Allison, W.S., Everse, J., Kaplan, N.O. and Taylor, S.S., Proc. Natl. Acad. Sci. USA 70, 1968-1972 (1973).

17. Chandrasekhar, K., McPherson, A., Jr., Adams, M.J. and Rossmann, M.G., J. Mol. Biol. 76, 503-518 (1973).

18. Webb, L.E., Hill, E.J. and Banaszak, L.J., Biochemistry 12, 5101-5109 (1973).

19. Buehner, M., Ford, G.C., Moras, D., Olsen, K.W. and Rossmann, M.G., J. Mol. Biol., in press (1974).

20. Ohlsson, I., Nordström, B. and Brändén, C.-I., J. Mol. Biol., in press (1974).

21. Rossmann, M.G., Moras, D. and Olsen, K.W., Nature 250, 194-199 (1974).

Received July 17, 1974

DISCUSSION

Kirschner: Have there been any successful attempts to show that peptide fragments comprising the coenzyme binding domains of various dehydrogenases mentioned in your lecture have an autonomous tendency to fold? One would call this pursuit molecular archeology.

Buehner: There have been many attempts to split subunits at strategic positions. However, I am not aware of any successful experiment in which an isolated intact domain was obtained, leave alone its structure determined.

Sund: With respect to the conformation of the coenzyme molecule: does X-ray structure analysis yield the same results for the reduced and for the oxidized coenzyme?

Buehner: It does within the limits of the method. With a resolution of 2 Å (and better is hardly possible for protein crystals) one wouldn't be able to detect subtle differences like for example a slight non-planarity of the dihydronicotinamide ring in NADH. However, the gross conformation for reduced and oxidized coenzyme is the same as far as it has been analyzed.

Beyreuther: What do you think is the sequence of folding for the peptide chain of the dehydrogenase-kinase group? If the sequence is the same for all enzymes belonging to this structural group the nucleation should start at different parts of the primary structure of the individual enzymes.

Buehner: This would indeed be expected. The highly conserved topology of this domain will certainly be reflected in the folding process. To achieve the particular architecture of a domain the folding process must be determined by whatever forces govern the tertiary structure. The chemical character of amino acids is what counts, not their sequence number. I would think it is legal to think about a "domain" as an independently folding substructure with its own nucleation site. In the particular case of the dinucleotide binding domain it would not seem unreasonable to assume polypeptide-linear assembly of the pleated sheet, starting with ßA and ending with ßF.

Riesner: Is there an explanation why in kinases the binding domain for ATP, which is a much shorter molecule than NAD, does also contain the full number of six strands?

Buehner: Not really. At present, the investigation of kinases hat not yet yielded as precise a knowledge of position, orientation and interactions of the coenzyme with its binding domain as is available for dehydrogenases. There might be differences, but it is too early to speculate about that.

Veeger: I don't object comparison of structures of dehydrogenases and flavodoxin , but still there are some differences which have to be explained. In flavodoxin the FMN has no adenine-ribose-phosphate part and is extremely tightly bound (a phenomenon general in flavoproteins). Affinity for the protein increases even a few orders of magnitude upon the uptake of two reducing equivalents. If thus has to be explained why this occurs in comparison with the constant shuttling of the NAD coenzyme. Even in case the FAD in certain flavoproteins might turn out to be in a similar open conformation the question of the tight binding has to be answered.

Buehner: I can't see much of a difficulty to explain the differences. I was talking about the fold of the polypeptide chain: the skeleton of the binding site is very similar. However, the flesh, the side chains which occupy the strategic positions to actually bind the coenzyme, can be very different. The frame is the same in both cases, but different clamps are used to fix different workpieces to it with different tightness. The frame is stable and has proven its ability to hold all sorts of workpieces and the clamps are exchanged and adapted to the individual problem.

Sund: In general the fluorescence intensity of the reduced coenzyme increases upon binding to dehydrogenases except glyceraldehyde-3-phosphate dehydrogenase. If the coenzyme binding domain is similar for all dehydrogenases investigated so far and also - in contrast to earlier suggestions - the conformation of the enzyme-bound coenzyme then the question arises what kind of interaction between the dihydronicotinamide part and glyceraldehyde-3-phosphate dehydrogenase is responsible for the decrease of the fluorescence intensity upon binding. Do you have any information about this interaction?

Buehner: I could venture some speculations, but since I am by no means an expert on optics I wouldn't believe myself too strongly in my own ideas. It is highly improbable that any residue in the coenzyme binding domain would be responsible for the quenching of 340 nm excitation. However, there are two interesting-looking tyrosines and a tryptophan in the catalytic domain which might be optically coupled to the dihydronicotinamide ring and help dissipating its energy in a radiationless way. On the other hand, the same chemical influence of the neighboring residues which induces fluorescence in the oxidized coenzyme might as well express itself in quenching the reduced coenzyme's fluorescence. Somehow, everything seems to be the other way around in the coenzyme binding of GAPDH.

Woody: This question is related to that of Dr. Sund. If there are no significant contacts for the nicotinamide ring and the conformations of oxidized and reduced coenzymes are nearly identical, what accounts for the large difference in binding constants?

Buehner: The positive charge of the nicotinamide ring ought to play a major role, but there might be other effects as well. Since the "handle" of the nicotinamide ring, the ADPRib part of the coenzyme, is so rigidly fixed one might think that small differences in the planarity of the ring can play a bigger role than meets the eye. However, the biggest contribution is probably due to the electric charge. In the case of GAPDH there is mounting evidence that the Racker band is caused by a covalent addition of Cys 149 to C4 of the nicotinamide ring. Such an effect would of course add a new dimension to the problem.

Wagner: Can you grossly picture out what the main contributions to the binding affinity of the coenzyme are?

Buehner: Rather schematically one could say:

- Hydrophobic interactions with the adenine ring.
- Hydrogen bonds to both riboses.
- Electrostatic interactions with the pyrophosphate group.
- Little interaction of the nicotinamide ring, maybe electrostatic repulsion in the oxidized form.

The Interaction of Glutamate Dehydrogenase with a Coenzyme Analog

R. Koberstein, H. Dieter and *H. Sund*

INTRODUCTION

The analysis of the binding between dehydrogenases and their coenzymes is one of the main topics in the elucidation of the mechanism of action of this class of enzymes. X-ray analysis has shown that the coenzyme binding domain of the dehydrogenases investigated so far exhibits striking structural similarities [1]. In this class of enzymes special interest has been directed to mammalian glutamate dehydrogenase (GluDH) [2] because of its very complex coenzyme and nucleotide binding behavior [3,4], which is also expressed in its nonlinear enzyme kinetic properties [5]. Since X-ray data are not available for this enzyme because only microcrystals have been obtained so far, our knowledge of the interaction of GluDH with coenzymes and other nucleotides results mainly from kinetic and equilibrium binding studies.

The smallest enzymatically active subunit of mammalian GluDH contains six identical polypeptide chains (molecular weight 56 000 [6]) which are arranged in two layers of three [7,8]. The enzyme can use NAD and NADP as coenzyme, whereas the enzymes from non-animal sources in general are specific for either, NAD or NADP. The enzymatic activity of mammalian GluDH is affected by a variety of small molecules [9] among which ADP and GTP are of special interest because of their possible regulatory role in vivo.

The quantitative analysis has shown that each subunit of beef liver GluDH has a total of 18 nucleotide binding sites [3]. A first set of six sites, the <u>active sites</u>, is occupied by NADH and NADPH with nearly the same affinity and is characterized by a nonlinear Scatchard plot [4]. A second set of six sites, the nonactive sites, binds the reduced coenzyme too, but less tightly than the active sites and NADPH with a lower affinity than NADH. ADP and α-NADH compete with the reduced coenzymes for this site [3,10]. In the presence of ADP the negative cooperativity of the active NAD(P)H binding sites is abolished and in the ternary complex with GluDH, ADP and NADH decrease each other's affinity to the enzyme compared to the corresponding binary complexes [3]. A third set of six binding sites, the GTP sites, is responsible for the GTP binding [3, 11]. In contrast to ADP and NAD(P)H, GTP and NAD(P)H enhance each other's binding in the ternary complex exhibiting positive cooperativity.

Since the six polypeptide chains within the subunit are identical, the nonlinear binding curves of NAD(P)H to the active sites may be caused (a) by functional nonidentity of the six polypeptide chains, i.e. by different intrinsic affinities for the coenzyme as a result of geometrical arrangement, or (b) by negative cooperative interaction between the binding sites. The basis of such an interaction could be the two-layer arrangement of the polypeptide chains within the subunit. Using this model pairwise cooperativity between neighboring polypeptide chains along the trimer-trimer association axis leading to a 3 + 3 interaction of the active coenzyme binding sites may be responsible for the observed binding phenomena.

For the further analysis of the nucleotide binding sites of beef liver GluDH we studied the 1,N^6-ethenoadenine analog [12] of NAD (ε-NAD) and NADP (ε-NADP). This analog allows fluorescence studies not only in the reduced but also in the oxidized state of the coenzyme [13]. As this analog is active as coenzyme in the reaction catalyzed by GluDH and changes its emission properties upon binding to the enzyme, fluorescence studies should give useful information about the coenzyme binding and the influence of effectors or substrates on enzyme-coenzyme complexes. In addition, the analog may serve to differentiate between the active and the nonactive coenzyme binding sites.

RESULTS AND DISCUSSION

Fluorescence Measurements

The fluorescent coenzyme analogs change their emission properties in the presence of GluDH. Fig. 1A shows the fluorescence emission spectra of ε-NAD^+ in the presence of the enzyme and the competitive inhibitor glutarate together with the respective controls. Besides a marked fluorescence enhancement, a blue shift of about 10 nm of the ε-adenine emission in the ternary complex is observed. From the titration of a mixture of ε-NAD^+ and glutarate with increasing enzyme concentrations, the fluorescence enhancement factor Q for the ternary complex is evaluated by a computer fit as 21. In the binary system enzyme and ε-NAD^+ no fluorescence change is observed indicating that in a binary complex the emission remains unaltered or that such a complex is not formed in a detectable amount. This result is supported by preliminary stopped flow experiments which only for the ternary complex show a rapid fluorescence increase, the amplitude of which is independent of the order of mixing.

ε-NADH contains two fluorescent chromophores, ε-adenine and dihydronicotinamide, with overlapping excitation and emission bands. Fig. 1B shows the fluorescence enhancement of ε-NADH in the presence of GluDH under conditions with predominating dihydronicotinamide emission. A fluorescence enhancement factor of 8.5 is calculated from the titration of ε-NADH with the enzyme. Excitation at 320 nm favors the ε-adenine emission at 400 nm yielding Q = 10.5.

The fluorescence spectrum of ε-$NADP^+$ in the ternary complex

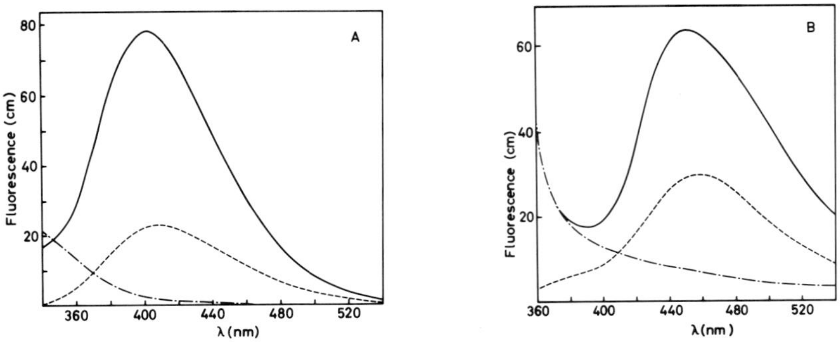

Fig 1A. Fluorescence emission spectrum of ε-NAD+ in the ternary complex with GluDH and glutarate. Measurements in 0.067 M phosphate buffer at pH 7.6 and 20°C, excitation at 320 nm. GluDH 17.8 μM (calculated with respect to the polypeptide chain), ε-NAD+ 37.5 μM, glutarate 0.1 M. (———) the ternary complex, (- - -) ε-NAD+ and glutarate, (—·—·—) GluDH and glutarate or GluDH alone.

Fig. 1B. Fluorescence emission spectrum of ε-NADH in the binary complex with GluDH. Measurements in 0.067 M phosphate buffer at pH 7.6 and 20°C, excitation at 340 nm. GluDH 22 μM, ε-NADH 9.4 μM. (———) the binary complex, (- - -) ε-NADH, (—·—·—) GluDH.

with GluDH and glutarate is very similar to that of ε-NAD+ under the same conditions, however, with ε-NADP+ a higher fluorescence enhancement with Q = 34.3 is obtained. Comparison of the fluorescence properties of free ε-NAD+ and ε-NADP+ shows that the quantum yield of ε-NADP+ is 34% lower than that of ε-NAD+. Therefore, the larger Q-value of ε-NADP+ in the ternary complex is mainly a consequence of the stronger internal quenching in the free ε-NADP+ compared to the free ε-NAD+. From X-ray analysis of other dehydrogenases [1] the coenzyme is known to be bound in the open conformation. Therefore, the observed enhancement factors of ε-NAD+ and ε-NADP+ may be compared to values of the unquenched ε-adenine chromophore. Using the quantum yield of ε-AMP as reference, enhancement factors of 12 and 18 would be expected for the open form of ε-NAD+ and ε-NADP+, respectively. Therefore, the observed much higher Q-values of the ternary complexes most likely reflect additional interactions of the ε-adenine chromophore with the enenzyme environment.

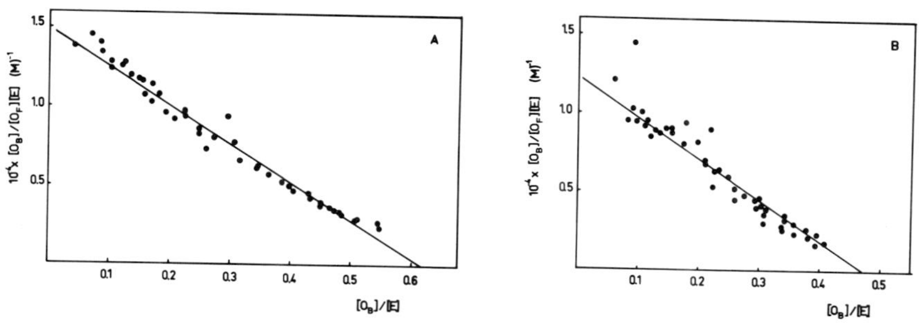

Fig. 2. Scatchard plot of the fluorescence titration of the binary GluDH-glutarate complex with ε-NAD^+ (A) and ε-$NADP^+$(B). Measurements in 0.067 M phosphate buffer at pH 7.6 and 20°C, GluDH 35.7 μM, glutarate 0.1 M. Excitation at 320 nm, emission at 400 nm, Q-values 21 (A) and 34.3 (B), respectively. $[O_B]$ and $[O_F]$ are the concentrations of the enzyme-bound and the free ε-NAD(P)+, [E] is the enzyme concentration with respect to the polypeptide chain.

The fluorescence enhancement factors were used to determine the binding of ε-$NAD(P)^+$ in the ternary complexes by fluorescence titration. Fig. 2A shows the Scatchard plot of a titration of GluDH with ε-NAD^+ in the presence of glutarate. The experimental points can be fitted by a straight line which extrapolates to about three binding sites per subunit. The apparent dissociation constant is found to be 41 μM. The Scatchard plot for the titration with ε-$NADP^+$ (Fig. 2B) closely resembles that obtained with ε-NAD^+. Again a linear plot is obtained extrapolating to three binding sites per subunit and yielding an apparent dissociation constant of 39 μM. For comparison, the natural coenzyme NAD^+ shows nonlinear binding behavior under the same conditions characterized by dissociation constants of 6 and 40 μM [14].

Since ε-$NAD(P)^+$ does not change its emission properties in the presence of GluDH, the fluorescence enhancement reflects exclusively the formation of the ternary enzyme-ε-$NAD(P)^+$-glutarate complex. On the other hand, glutarate concentrations above 40 mM have been shown to saturate the enzyme with glutarate [14]. Therefore, the apparent dissociation constants represent the following equilibrium:

$$\text{GluDH-glutarate} + \varepsilon\text{-NAD(P)}^+ \rightleftharpoons \text{GluDH}\begin{matrix} \diagup \varepsilon\text{-NAD(P)}^+ \\ \diagdown \text{glutarate} \end{matrix}$$

Under the assumption of a constant Q-factor for all binding sites, the Scatchard plots yield a maximum of three sites per subunit indicating that only half of the coenzyme binding sites present are titratable with the coenzyme analog. Similar results suggesting half-of-the-sites reactivity of GluDH have been obtained from stopped flow enzyme kinetics, showing a burst amplitude corresponding to three molecules of NAD^+ reacted per subunit. An explanation can be given on the basis of strong negative cooperativity between the binding sites. This could reduce the affinity for the coenzyme analog to an extent that these binding sites are not detectable although they in principle enhance the coenzyme fluorescence. However, other mechanisms must be considered which could account for the observed fluorescence titration: (a) The Q-factor may depend on the degree of saturation. A decreasing Q-factor with increasing saturation would yield less binding sites than really present. (b) There may exist additional binding sites which do not enhance the coenzyme fluorescence and, therefore, are not detected as in the case of the nonactive NADH binding sites of GluDH [4]. At present a decision is not possible, however, the determination of the binding equilibrium by an independent method like equilibrium dialysis will help to solve this problem.

Enzymatic Activity

Preliminary experiments had shown the ε-analogs to be fully active as coenzymes in the standard assay for GluDH [13]. However, since GluDH exhibits strongly nonlinear Lineweaver-Burk plots with the natural coenzymes, the ε-analogs were studied in some more detail. In Fig. 3 the Lineweaver-Burk plots of $\varepsilon\text{-NAD}^+$ and $\varepsilon\text{-NADP}^+$ are compared to those of NAD^+ and $NADP^+$.

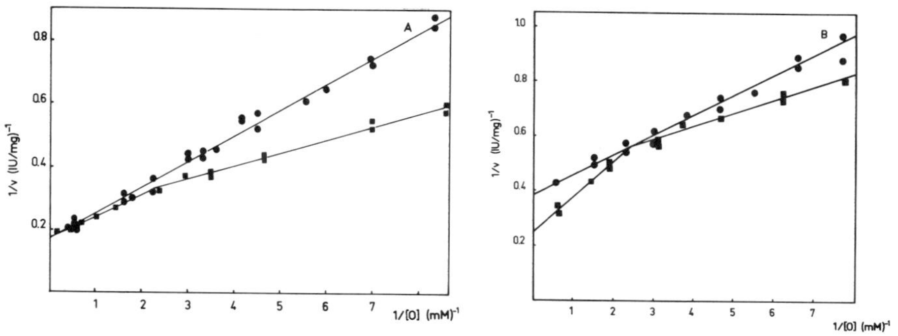

Fig. 3. Lineweaver-Burk plot for the reaction of GluDH with NAD^+ and ε-NAD^+ (A) and with $NADP^+$ and ε-$NADP^+$ (B). Initial rate measurements in 0.067 M phosphate buffer at pH 7.6 and 20.0 ± 0.1°C, glutamate 8.5 mM. (-■-■-) NAD^+ and $NADP^+$ (O), (-●-●-) ε-NAD^+ and ε-$NADP^+$ (O).

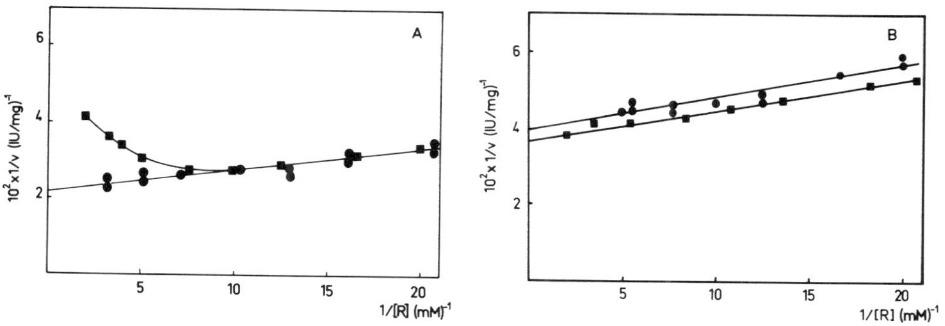

Fig. 4. Lineweaver-Burk plot for the reaction of GluDH with NADH and ε-NADH (A) and with NADPH and ε-NADPH (B). Initial rate measurements in 0.067 M phosphate buffer at pH 7.6 and 20.0 ± 0.1°C, 2-oxoglutarate 2 mM, ammonium chloride 50 mM. (-■-■-) NADH and NADPH (R), (-●-●-) ε-NADH and ε-NADPH (R).

Obviously, the discontinuity of the natural coenzymes, which is well established to occur as a relatively sharp bend [5], is not present with ε-NAD^+ or ε-$NADP^+$, indicating that the coenzyme self-activation has disappeared at least in the investigated concentration range. Since stopped flow experiments did not show any discontinuity of the NAD^+-dependent Lineweaver-Burk plot for the transient blue-shifted GluDH-NADH-oxoglutarate complex, the bends observed in the steady-state phase were ascribed to the rate limiting step, the dissociation of NADH from the product complex [15]. For the acceleration of this latter process by high NAD^+ concentrations a "push-pull" mechanism has been proposed explaining the observed NAD^+ self-activation [15]. Such a mechanism seems to be prevented when ε-NAD^+ or ε-$NADP^+$ is used as coenzyme, possibly due to the accessibility of only three of the six coenzyme binding sites to ε-$NAD(P)^+$ (Fig. 2).

The reductive amination reaction leads to the Lineweaver-Burk plots for NADH and NADPH shown in Fig. 4. The most striking difference is the self-inhibition at high NADH concentrations which is not observed with NADPH. This effect is ascribed to the different affinity of the nonactive sites for NADH and NADPH, respectively, the latter being bound at least ten times less tightly than the former [4]. The Lineweaver-Burk plot of ε-NADPH closely resembles that obtained with NADPH (Fig. 4B) indicating that the rate limiting step of the reductive amination is insensitive to the chemical modification of the adenine part of the coenzyme. This is also supported by the identity of the Lineweaver-Burk plots of NADH and ε-NADH at concentrations below 100 μM (Fig. 4B).

Above concentrations of 100 μM, NADH and ε-NADH behave differently. The absence of self-inhibition of ε-NADH led us to investigate the interaction of ε-NADH with the nonactive sites. Circular dichroism studies in the presence of GTP had shown a specific negative ellipticity for the saturation of these sites by NADH [16]. The corresponding signal is not observed with ε-NADH. In contrast to ADP, ε-ADP fails to show an interaction with the nonactive sites. Finally, preliminary equilibrium dialysis experiments yield only one binding site per polypeptide chain for ε-NADH. The conclusion from these results, that ε-NADH does not bind to the nonactive sites, also explains the absence of ε-NADH self-inhibition in the enzymatic reaction.

Conclusion

From the results it is concluded that the active and the nonactive coenzyme binding sites exhibit different specificity with respect to the adenine moiety of the coenzyme. In this respect, the active sites are relatively unspecific since the ε-analogs are strongly bound and well converted in the enzymatic reaction. The nonactive sites, however, appear to be highly specific requiring the intact adenine moiety of both ADP and NADH. Therefore, using the ε-analogs of NAD(P) differentiation between effects caused by the active and those produced by the nonactive binding sites is possible. The strong fluorescence enhancement of ε-$NAD(P)^+$ upon binding in the ternary complexes allows fluorescence studies of the coenzyme even in the oxidized state. In addition, it provides a useful signal for stopped flow kinetic studies of intermediates during the enzymatic reaction. Such experiments are in progress.

Acknowledgements. We thank Mr. L. Cobianchi for his excellent experimental help. This work was supported by grants from the "Deutsche Forschungsgemeinschaft" (Sonderforschungsbereich 138: "Biologische Grenzflächen und Spezifität") and the "Fonds der Chemischen Industrie".

REFERENCES

1. Buehner, M.: The architecture of the coenzyme binding domain in dehydrogenases as revealed by X-ray structure analysis. This Symposium, pp.

2. Sund, H., Markau, K., and Koberstein, R.: Glutamate dehydrogenase. Subunits in Biological Systems (Vol. 6 of Biological Macromolecules, ed. by Fasman, G.D., and Timasheff, S. N.), Part B. Marcel Dekker, New York, in press.

3. Koberstein, R., and Sund, H.: Studies of glutamate dehydrogenase: The influence of ADP, GTP, and L-glutamate on the binding of the reduced coenzyme to beef-liver glutamate dehydrogenase. Eur. J. Biochem. 36, 545-552 (1973).

4. Krause, J., Bühner, M., and Sund, H.: Studies of glutamate dehydrogenase: The binding of NADH and NADPH to beef-liver glutamate dehydrogenase. Eur. J. Biochem. 41, 593-602 (1974).

5. Dalziel, K., and Engel, P.C.: Antagonistic homotropic interactions as a possible explanation of coenzyme activation of glutamate dehydrogenase. FEBS Letters 1, 349-352 (1968).

6. Moon, K., Piszkiewicz, D., and Smith, E.L.: Glutamate dehydrogenase: Amino-acid sequence of the bovine enzyme and comparison with that from chicken liver. Proc. Natl. Acad. Sci. USA 69, 1380-1383 (1972).

7. Eisenberg, H.: Structure and association of glutamate dehydrogenase solutions. Pyridine Nucleotide-Dependent Dehydrogenases (ed. by Sund, H.). Springer-Verlag, Berlin-Heidelberg-New York, 1970, 294-304.

8. Hucho, F., and Janda, M.: Investigation of the quaternary structure of beef liver glutamate dehydrogenase with bifunctional reagents. Biochem. Biophys. Res. Commun. 57, 1080-1088 (1974).

9. Goldin, B.R., and Frieden, C.: L-glutamate dehydrogenases. Current Topics in Cellular Regulation (ed. by Horecker, B.L., and Stadtman, E.R.), Vol. 4. Academic Press, New York-London, 1971, 77-117.

10. Koberstein, R., Krause, J., and Sund, H.: Studies of glutamate dehydrogenase: The interaction of glutamate dehydrogenase with α-NADH. Eur. J. Biochem. 40, 543-548 (1973).

11. Colman, R., and Frieden, C.: Cooperative interaction between the GTP binding sites of glutamate dehydrogenase. Biochem. Biophys. Res. Commun. 22, 100-105 (1966).

12. Barrio, J.R., Secrist, J.A., and Leonard, N.J.: A fluorescent analog of nicotinamide adenine dinucleotide. Proc. Natl. Acad. Sci. USA 69, 2039-2042 (1972).

13. Dieter, H., Koberstein, R., and Sund, H.: Nicotinamide $1,N^6$-ethenoadenine dinucleotide, a coenzyme for glutamate dehydrogenase. FEBS Letters, in press.

14. Dalziel, K., and Egan, R.R.: The binding of oxidized coenzymes by glutamate dehydrogenase and the effects of glutarate and purine nucleotides. Biochem. J. 126, 975-984 (1972).

15. Markau, K.: Unpublished results.

16. Koberstein, R., and Sund, H.: Circular dichroism studies on the complex between beef liver glutamate dehydrogenase and NADH. FEBS Letters 19, 149-151 (1971).

17. Sund, H., Koberstein, R., Krause, J., and Markau, K.: Physicochemical studies on the interaction of beef liver glutamate dehydrogenase with nucleotides. Proc. 1st Eur. Biophys. Congr. Vol. VI. 97-103 (1971).

Received July 20, 1974

DISCUSSION

Kempfle: How do you explain the "non enhancement" of fluorescence in the binary ε-NAD-GluDH complex although this coenzyme analogue has to be bound as seen by the enzymatic activity?

Koberstein: There are two explanations: Either the fluorescence of the binary complex is not enhanced suggesting different environments of the ε-adenine chromophore in the binary and ternary complex, or the binary complex is so weak that the enhancement is not detectable under the conditions of the experiment. An estimate with a dissociation constant of 1 mM which seems reasonable from data with NAD^+ favours the latter possibility.

Kempfle: How high are the concentrations of glutarate in your experiment to build the ternary complex?

Koberstein: We used 100 mM glutarate in our experiments.

Veeger: The saturation of only half of the active sites by ε-NAD^+ is puzzling to me. According to your kinetic figure and assuming saturation of half of the active sites the analogue is presumably twice as active as NAD^+. If thus can be expected that working at saturated ε-NAD^+ concentration the addition of NAD^+ will increase the rate, because of saturation of the remaining three active sites. According to Dalziel six molecules of NAD^+ are bound. Of course the possibility exists that of the six molecules NAD^+ bound only three are bound to the active site and three to the non-active site, thus explaining the negative cooperative kinetics. Thus ε-NAD^+ would lack in this case this property. This latter possibility seems however rather unlikely.

Koberstein: According to Dalziel, non-linear binding of NAD^+ to GluDH in the presence of glutarate is observed suggesting negative cooperativity. In our case fluorescence titration of the enzyme-glutarate complex with ε-NAD^+ would not enable us to detect additional weaker binding sites. Therefore we cannot at present rule them out. The answer to this question can hopefully be given by using an independent method like equilibrium dialysis. Evidence for a half-of-the-sites reactivity of GluDH does exist from stopped-flow experiments of Dr. Iwatsubo's group and Dr. Markau in our laboratory, however, the detailed description of the overall mechanism including negative cooperative kinetics has not yet been achieved.

Kirschner: There appears to be a contradiction between the statement that the catalytic mechanism is random addition rapid equilibrium (i.e. a binary complex between enzyme and ε-NAD^+ is formed) and the lack of evidence from fluorescence titration for this complex.

Koberstein: There is no evidence from fluorescence titration, however, from the catalytic mechanism one would expect formation of a binary enzyme-ε-NAD^+ complex, which might be too loose to be detected under the experimental conditions.

Kirschner: You should be able to obtain a value for the dissociation constant from steady-state rate data.

Koberstein: Yes, this might be worth looking into in more detail.

Steinhardt: I am not sure I have followed the argument as to the occurrence of binary complexes. It is always difficult to prove a negative, but more so here than elsewhere. The binary complex random mechanism can prevail with the actual concentration of the complex at any extremely low level you care to specify. After all, what are the alternatives? Surely, no one will postulate a mechanism that depends on direct formation of a ternary complex. Do you believe the binary complex exists?

Koberstein: Yes.

Sund: In any case, I am sure, that a binary ε-NAD^+-glutamate dehydrogenase exists but till now with the methods used we couldn't prove it. As with the natural coenzyme, NAD^+, the affinity of the enzyme for ε-NAD^+ is probably weak and, therefore, the quantitative analysis not very easy.

Antonini: We have obtained evidence from static and kinetic measurements for the formation of a low affinity binary complex between glutamate dehydrogenase and NAD^+.

Veeger: Concerning the high activity of the ε-NAD^+ and referring to the X-ray work, I wonder whether it can be expected that the introduction of a bulk ring in the coenzyme does not have any influence. Maybe Dr. Buehner could answer this question.

Buehner: The additional ring can easily be accommodated as long as the ε-adenosine is in the anti-conformation, since it sticks out from the hydrophobic crevice towards the solvent. There might in some cases be additional "hydrophobic contact" areas, but no dramatic steric effects of the additional ring should be expected.

The Interaction of Adenosine Triphosphate and its Analogs with Myosin and the Modification by Actin

R. S. Goody and *H. G. Mannherz*

Introduction

There has been much progress in recent years towards an understanding of the chemomechanical energy transformation process in muscle. Since the elucidation of the sliding filament mechanism by H. E. Huxley and J. Hanson and A. F. Huxley and Niederkerke (1,2) (1954) and the identification of the myosin cross-bridge as the molecular force generator (3), interest has mainly concentrated on the way in which force is produced by the cross-bridge, and the kinetic mechanism of ATP-hydrolysis by myosin. The generally accepted working hypothesis visualises cross-bridges attaching and detaching from the actin filament in a cyclic manner producing mechanical force whilst attached to the actin filament, in a kind of rowing motion (4,5). Concomitant with the movement of one cross-bridge one or even two molecules of ATP are hydrolysed per myosin molecule, the exact mode of chemomechanical energy transformation being still unknown. It is clear that the interaction between ATP and myosin plays a central role in the mechanism of muscle contraction and from a knowledge of the kinetic mechanism of ATP-binding and hydrolysis by myosin and actomyosin, more detailed information about this process can be gained. These will be presented in the following contribution.

Abbreviations: ATP=adenosine triphosphate; ATP($\alpha\beta CH_2$)=α,β-methylene-ATP; ATP($\beta\gamma$NH)=β,γ-imido-ATP; ATP(γS)=adenosine 5'-(3-thiotriphosphate).

The Kinetics of ATP-hydrolysis by Myosin

After the discovery of the early rapid phase of phosphate production in the hydrolysis of ATP by myosin-containing systems (6), a period of 16 years passed before detailed experiments and an interpretation explaining this phenomenon were published by Lymn and Taylor (7) in 1970. Since then, further work using rapid reaction techniques with ATP and ATP-analogs has led to the formulation of the following kinetic scheme for ATP-hydrolysis:

$$M + ATP \underset{}{\overset{K_1}{\rightleftarrows}} M{\cdot}ATP \overset{K_2}{\rightleftarrows} M^{*}{\cdot}ATP \overset{K_3}{\rightleftarrows} M^{**}{\cdot}ADP{\cdot}Pi \overset{K_4}{\rightleftarrows} M^{*}{\cdot}ADP{\cdot}Pi \overset{K_5}{\rightleftarrows} M^{*}{\cdot}ADP + Pi \overset{K_6}{\rightleftarrows} M{\cdot}ADP + Pi \overset{K_7}{\rightleftarrows} M + ADP + Pi \qquad (1)$$

where the stars serve to distinguish complexes which have the same bound nucleotide but are kinetically or physically distinguishable. In Table 1, a list of rate and equilibrium constants for the individual steps is given, together with the appropriate references. As can be seen, there are important gaps at certain stages in the mechanism. In particular, it is

Table 1

Step	k_{+n}	k_{-n}	K_n
1	-	-	$4.5\times10^{3} M^{-1}$ (8)
2	$400 s^{-1}$ (8)	$<0.02 s^{-1}$ (9)	$>2\times10^{4}$ (8)
3	$>160 s^{-1}$ (7)	>18 (9)	9 (9)
4	$0.05 s^{-1}$ (7)	-	-
5	-	-	-
6	$1.4 s^{-1}$ (10)	$400 s^{-1}$ (8)	3.5×10^{-3} (8)
7	-	-	$2.7\times10^{-4} M$ (8)

important to obtain values for the reverse rate constants for substrate binding and for the rate-controlling step.

In our laboratory, we have attempted to do this by studying the reverse reaction i.e. the synthesis of ATP from ADP and Pi in the presence of S1. Since it is clear that the overall rate of ATP synthesis, giving ATP free in solution, must be extremely slow, since the forward rate is slow and equilibrium constant for ATP-hydrolysis very large, a study was made of the synthesis of ATP formed bound to the enzyme. The evidence presented in reference 9, that the equilibrium constant for ATP-cleavage on the enzyme is small, indicated that this approach might be fruitful.

An expression for the rate and extent of ATP formation from ADP and Pi at the S1-active site can be derived (20) using the scheme 2 as given above as a basic kinetic scheme, and given the following conditions:

i) (S1) > $(ADP)_o$ [$(ADP)_o$ = original (ADP) concentration]

ii) (S1) >> K_i for ADP ($K_i \sim 10^{-6}$M)

iii) Equilibria described by K_3 and K_5 are rapidly established compared to K_4.

iv) k_{-2} is small compared to k_{+4} (9)

Conditions can be chosen to comply with i) and ii), and the work described above justifies assumptions iii) and iv). The relationship derived is:

$$(2) \qquad (M^{*}\cdot ATP) = \frac{k_{-4}(ADP)_o}{\lambda(1+K_3)\left(1+\frac{K_5}{(Pi)}\right)} \cdot (1-e^{-\lambda t})$$

where

$$(3) \qquad \lambda = \frac{k_{-4}}{1+\frac{K_5}{(Pi)}} + \frac{k_{+4}K_3}{1+K_3}$$

and from this, the equation relating ATP formed on the enzyme and (Pi) at t = ∞ is:

$$(4) \qquad 1 + K_3(1+K_4) + K_3K_4K_5 \cdot \frac{1}{(Pi)} = (ADP)_o \cdot \frac{1}{(M^{*} \cdot ATP)}$$

from which it can be seen that a plot of $\frac{1}{(M^{*} \cdot ATP)}$ against $\frac{1}{(Pi)}$ should give a straight line with intercepts equal to $\frac{1 + K_3(1+K_4)}{(ADP)_o}$ on the ordinate and $-\frac{\left[1 + \frac{1}{K_4}(1+\frac{1}{K_3})\right]}{K_5}$ on the abscissa, and with a slope of $\frac{K_3K_4K_5}{(ADP)_o}$.

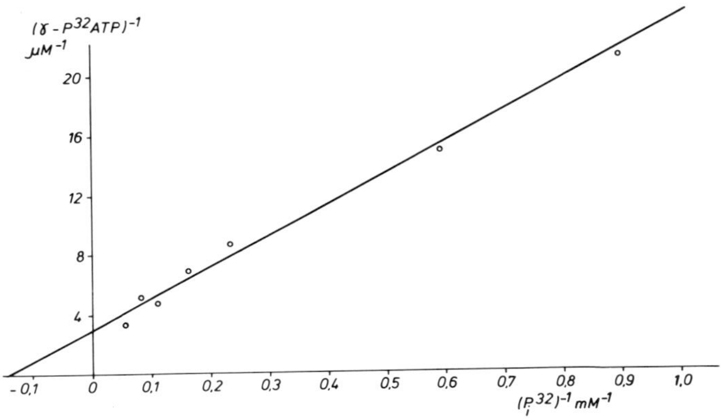

Fig. 1. Synthesis of [γ-^{32}P] ATP from ADP and Pi at the S1-active site. The reaction mixture contained 10^{-4}M S1,5.10^{-5}MADP, 20mM Tris. HCl, 50mM KCl, 2.5mM $MgCl_2$ adjusted to pH 8.0 at 21°. [γ-^{32}P] ATP was measured after quenching with 10% perchloric acid by a two-dimensional chromatographic procedure described in reference 20.

In Fig. 1, the results of an experiment on the synthesis of ATP at the S1-active site at constant ADP (5×10^{-5}M) and S1 (1×10^{-4}M) concentrations and with a variable Pi concentration in the presence of Mg^{++} at pH 8.0 and 21° are shown. From the intercepts, values for K_4, the equilibrium constant for the rate-controlling step of the ATPase of ca. 16, and for K_5, the dissociation constant of Pi from $M^{*} \cdot ADP$, of ca. 7×10^{-3}M are obtained. From the slope of the line, $K_3K_4K_5$ is seen to be ca.

1. K_4 is seen to be surprisingly small (surprising since K_3 is also small, and the equilibrium constant for ATP-hydrolysis under these ionic conditions is ca. 10^7 (11)). This means that some other step in the mechanism must have a very large equilibrium constant, since $K_1K_2K_3K_4K_5K_6K_7 = 10^7$. Using the value of $K_3K_4K_5$ obtained above, and $K_6K_7 = K_i$ for ADP = 10^{-6}M (10), this gives a value for K_1K_2 of $10^{13}M^{-1}$. It should be pointed out that this value for K_1K_2, the ATP binding constant to S1, is independent of most other kinetic data and also of the proposed mechanism. Only the dissociation constant for ADP from S1 needs to be known, together with the equilibrium constant for ATP-hydrolysis under the chosen conditions. A value for the binding constant can be obtained without using the equations given above, which are dependent on the kinetic model used,and also on the values of the individual rate constants. Regarding the equilibrium situation, and knowing the concentrations of S1, ADP, Pi, Mg^{++} and ATP formed on the enzyme, it is possible to calculate the concentrations of free S1 and ADP (given K_1 for ADP) and of free ATP (given the equilibrium constant for ATP-hydrolysis). The concentrations of free S1, free ATP, and S1ATP are then related by the binding constant for ATP to S1. In the experiment used for Fig. 1, an excess of S1 over ADP was used, which makes calculations of the free S1 concentration simpler. Without knowledge of the kinetic mechanism of myosin ATPase, it is possible to derive the following equation relating the concentration of free ATP to the concentration of ATP bound to the enzyme:

$$(5) \qquad \frac{1}{(M\cdot ATP)} = \frac{Ks}{(M)_o(ATP)} + \frac{1}{(M)_o} + \frac{Kc}{(M)_o}$$

using as a basic scheme

$$(6) \qquad M + ATP \underset{}{\overset{Ks}{\rightleftarrows}} \overbrace{M\cdot ATP \rightleftarrows \rightleftarrows \rightleftarrows M\cdot ADP\cdot Pi}^{Kc} \rightleftarrows M + ADP + Pi$$

where Ks is the ATP dissociation constant, Kc the constant

describing the distribution of nucleotide between ATP and ADP on the enzyme, and $(M)_o$ the concentration of enzyme not bound to ADP before ATP synthesis has occurred. From a plot of $\frac{1}{(MATP)}$ against $\frac{1}{(ATP)}$, it is therefore possible to determine Ks and Kc from the slope = $\frac{Ks}{(M)_o}$ and from the intercepts (on the ordinate = $\frac{1+Kc}{(M)_o}$, on the abscissa = $\frac{1+Kc}{Ks}$). Ks is calculated to be ca. 10^{-13}M from this treatment of the results, and Kc = 160. Relating Kc to the more detailed mechanism given in scheme 1, it is seen that Kc = K_3 $(1+K_4)$, which from the values of K_3 and K_4 obtained from Fig. 1 = 153. The two methods of treating the results give values which are clearly in agreement, despite the fact that the second method has the disadvantage that the exact concentration of S1 should be known for the calculations, and that effects such as binding of Pi to free S1 will lead to errors. However, the method might be of use for similar studies on enzymes, possibly other ATPases, where the kinetic mechanism is less well established than for myosin. It would be of interest in this way to see if any other ATPases have a similarly large ΔG^o for binding.

Since ATP appears to have such a high affinity for myosin, it is of interest to compare the properties of ATP-analogs which are used in muscle research. For ATP(βγNH), the situation is simple, since this analog is not cleaved. As shown in Table 2, the rate constant for association is significantly smaller than for ATP. kdiss for this analog is measurable, being $0.02s^{-1}$ at 21°. For ATP(γS), the rate constant for association is similar to that for ATP. Experiments to determine kdiss for this analog have only been done at 4°, where the turnover rate is very slow. As shown in Fig. 2, after mixing [^{3}H] ATP(γS) with excess myosin, the rate of decay to [^{3}H] ADP and thiophosphate is not affected by subsequent addition of a large excess of ATP, showing that kdiss for this analog at 4° is very much slower than kcat.

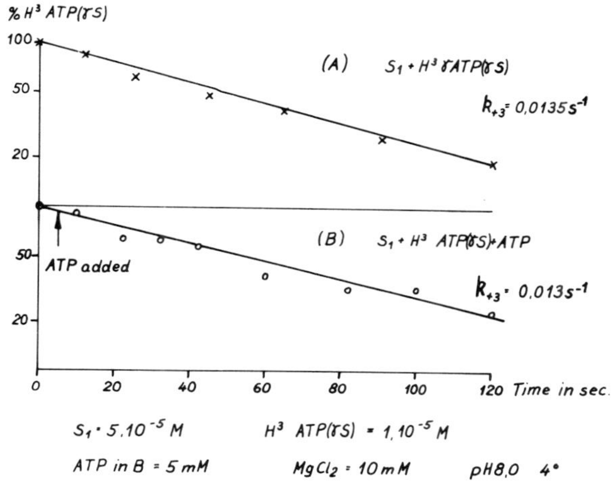

Fig. 2. (A) Kinetics of ATP(γS) cleavage by S1 at 4°.
(B) As in A, with addition of excess cold ATP after 5 seconds.

A similar experiment with ATP($\alpha\beta CH_2$) (Fig. 3) showed that in this case, kdiss is measurably large and has a value of $0.022 s^{-1}$ at 4°.

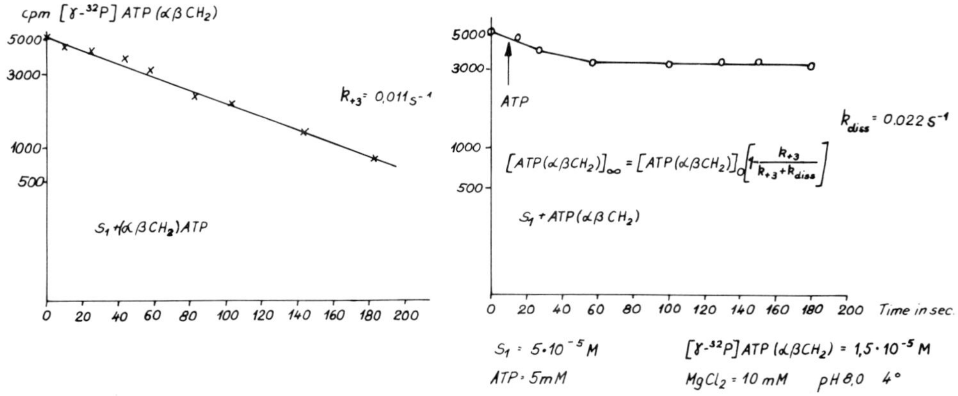

Fig. 3. Left, kinetics of ATP($\alpha\beta CH_2$) cleavage by S1 at 4°. Right, as on left, with addition of excess cold ATP after 5 seconds.

In addition, kass for this analog is significantly smaller than for ATP (13). Thus, for the two analogs for which it has been possible to measure the rate of release as intact analog molecules from myosin it is seen that they do this at a rate which is many orders of magnitude faster than ATP. For ATP(γS), however, this is not yet established, and analogous experiments to those described above for ATP-synthesis from ADP and Pi are probably best suited to resolve this problem.

Table 2

Analog	$kass=K_1k_{+2}(M^{-1}s^{-1})$	k_{+3} (or k'_{+3}) $=kcat(s^{-1})$	$kdiss=k_{-2}(s^{-1})$
ATP(γS)	10^6 (21°) (12)	0.24 (21°); 0.014 (4°)	0.014 (4°)
ATP(αβCH_2)	6.10^4 (21°) (13)	0.1 (21°); 0.011 (4°)	0.022 (4°)
ATP(βγNH)	6.10^4 (21°) (12)	not cleaved	0.02 (21°)

In the light of these results, it is of interest to consider whether the complex M^*·ATP really has ATP bound, in other words if K_1K_2 really describes the simple binding of ATP to myosin or perhaps some other process as well. It has recently been suggested (14) that the complexes M^*· ATP and M^{**}·ADP·Pi both contain bound nucleotide entities in which the terminal P-O-P linkage is intact, but in which water has already attacked to give a pentacovalent phosphorus atom. The step which we have until now regarded as the cleavage step (M^*·ATP $\rightleftarrows$ M^{**}·ADP·Pi) would then represent a pseudorotation about the phosphorus atom between two stereochemically distinct intermediates, in the first of which ADP is in an equatorial position and therefore unable to act as a leaving group, and in the second is in an apical position (i.e. with a longer bond length) from which it can leave. This explains the observation that the first complex yields ATP and the second ADP on chemical quenching. The expression K_1K_2 would then not represent the binding constant of ATP to myosin, but would reflect the energy change involved with the combined processes of binding, attack of water and stabilisation of the pentacovalent intermediate. In terms of this mechanism, the weak binding of ATP(βγNH) and ATP(αβCH_2) might be considered to stem partially from a less favourable equilibrium constant for formation of the pentacovalent intermediate. It should be pointed out that, additionally, the initial binding of these analogs (i.e. the step described by K_1 in scheme 2) is probably weaker (8).

At the moment it is difficult to decide between the mechanism outlined above and a mechanism involving an easily reversible cleavage step. It should be pointed out that it is not easy to make a correlation between step 2 in scheme 1 and the pseudorotation mechanism, since this step also appears to occur with ADP (8), where there is no attack of water on phosphorus. This would mean that step 2 cannot represent the attack of water on the terminal phosphorus of ATP, and must be a step preceeding this attack. The latter process could, of course, be very much faster than k_{+2} so that resolution of this step is impossible.

Effect of Actin on Individual Steps in the Hydrolysis of ATP by Myosin.

Lymn and Taylor (15) showed that the rate of product release from myosin is accelerated by actin. Since the maximum rate of ATP-turnover by myosin in the presence of actin is much faster than k_{+4} and k_{+6}, both processes must be accelerated by actin. It would be of interest to know which other steps are affected by actin, and since it seems fairly certain from the work of Lymn and Taylor (15) that k_{+3} and kass for ATP are not strongly affected, it is of interest to design experiments to test the effect on other processes, particularly on kdiss for ATP. Since k_{+4} and k_{+6} are both accelerated, one or more rate constants in the reverse direction must also be accelerated, among them possibly kdiss ($=k_{-2}$). This process is easier to study with ATP-analogs than with ATP itself.

ATP(γS), ATP($\alpha\beta CH_2$) and ATP($\beta\gamma$NH) are all able to dissociate actomyosin in the ultracentrifuge (13). A reciprocal effect, the dissociation of myosin-analog complex by addition of actin might also be expected. To test this, the effect of actin on the rate of hydrolysis of ATP(γS) and ATP($\alpha\beta CH_2$) was undertaken. In a simple steady-state experiment, the hydrolysis of the two analogs was found to be accelerated at low actin concentrations (although much less strongly than ATP-hydrolysis) and not further affected by higher concentrations in the case of ATP(γS) but inhibited in the case of ATP($\alpha\beta CH_2$) (Fig. 4).

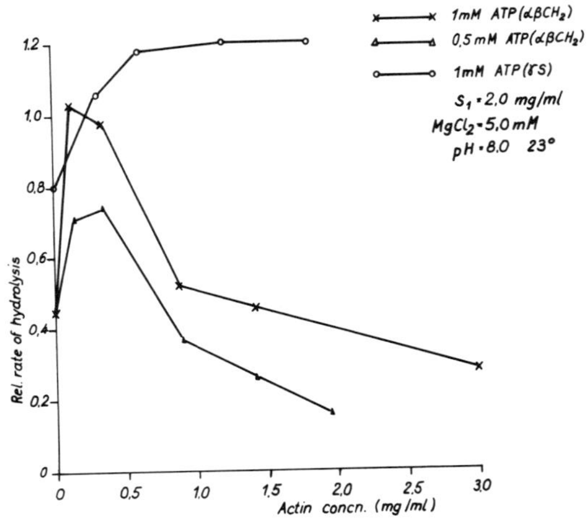

Fig. 4. Influence of actin on the steady-state rate of hydrolysis of ATP(γS) and ATP($\alpha\beta CH_2$) measured using a pH-Stat.

Furthermore the inhibition appears to be competitive, since the inhibition was stronger at lower ATP($\alpha\beta CH_2$) concentrations. Since k_{+3} is not affected by actin, this inhibition must be a competitive inhibition of analog binding, and since the association rate constant, at least for ATP, is not greatly affected by actin, this would indicate an acceleration of the dissociation rate. The small acceleration in the rate of hydrolysis of these two analogs can be explained if k_{-3} is greater than k_{+3} (in this case kcat $= \frac{k_{+3}k_{+4}}{k_{+3}+k_{-3}+k_{+4}}$, and increasing k_{+4} leads to an increase in kcat).

To investigate this effect more directly, the effect of actin on the rate of release of ATP-analogs is being studied. In Fig. 5, the effect of 1.2 mg/ml actin is shown on the rate of release of ATP(βγNH) from S1. kdiss is increased from $0.018s^{-1}$ under these conditions to a value which was too fast to measure using the coupled enzyme system described in the legend to Fig. 6. An experiment to demonstrate the slow release of ATP(βγNH) from S1 in the absence of actin is shown in Fig. 6.

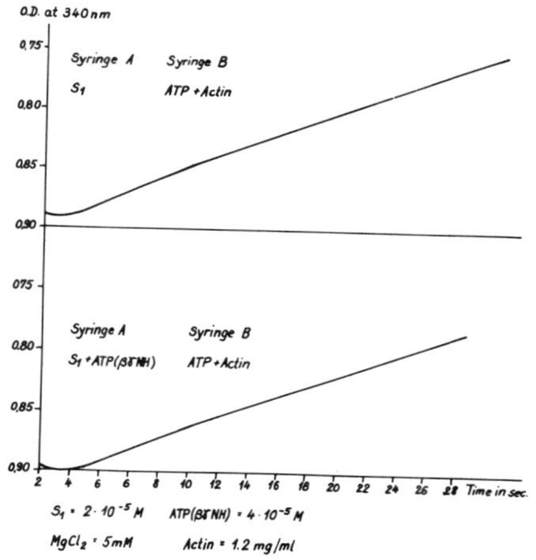

Fig. 5. Stopped-flow experiment under similar conditions to Fig. 6, except that 0.1 mg/ml pyruvate kinase and 0.1 mg/ml lactate dehydrogenase were used. Top, approach the steady-state on mixing S1 with ATP and actin. The lag-phase at the beginning comes from the linked assay system. Bottom, as before, but using S1-ATP(βγNH).

Experiments with ATP(γS) at 4° indicate that with this analog and at this temperature, the effect of actin on kdiss is not measurable at the actin concentrations (0.1 and 1.0 mg/ml) used. Further work is needed to determine whether this is due to a strong temperature dependence of kdiss, or because this analog binds more tightly than ATP(βγNH).

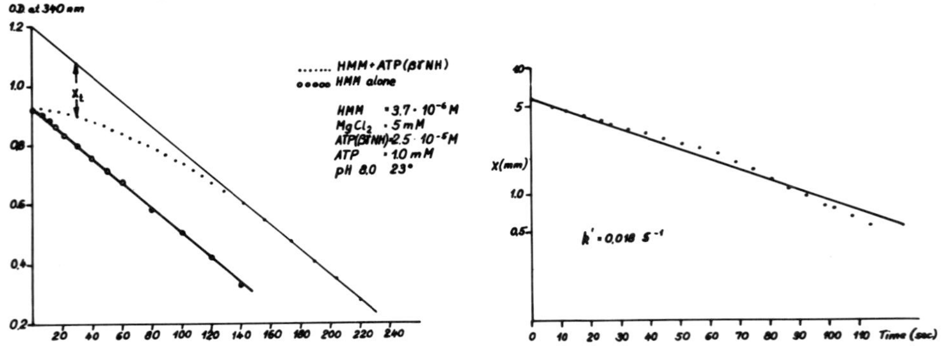

Fig. 6. Experiment to show the slow rate of release of ATP (βγNH) from HMM by measuring the course of approach to the steady-state rate of hydrolysis after addition of excess ATP to HMM.ATP(βγNH). The linked assay system contained pyruvate kinase (0.05 mg/ml), lactate dehydrogenase (0.05 mg/ml) phosphoenolpyruvate (1mM) and nicotaminide-adenine dinucleotide (reduced form, 0.2mM). On the left, a plot of the original data. On the right, a half-logarithmic plot to determine the first-order rate constant for release of ATP(βγNH).

Although the results on the effect of actin on kdiss for ATP or analog are incomplete, they do indicate that the binding of actin and ATP are competitive, although the sites are not identical. There would appear to be a type of allosteric interaction between the two sites, and the acceleration of the myosin ATPase could occur via the same mechanism.

Possible Models of Cross-Bridge Cycle

It is tempting to try to correlate the kinetic scheme of ATP-hydrolysis of ATP by myosin and actomyosin to the cross-bridge cycle, namely to assign defined intermediary steps of the kinetic pathway to certain cross-bridge orientations. In the relaxed state in insect flight muscle, the cross-bridge takes up the 90°-position detached from the actin filament, whereas in the rigor state it is attached to the actin filament, and angled at 45° relative to the filaments (4). From kinetic considerations, it follows that in the relaxed state, when there is an unmodified myosin ATPase, the predominant enzyme species of the kinetic scheme (M^{**}·ADP·Pi), corresponds to the rectangular cross-bridge orientation. Recently, Marston and Tregear (16) have succeeded in demonstrating that ADP is the predominant nucleotide species bound to myosin in relaxed muscle fibres from insect flight and rabbit psoas muscle. In the rigor state, i.e. in the absence of any nucleotide, the cross-bridges take up the angled position attached to the thin filament corresponding to the actomyosin-complex of the kinetic scheme of the ATP-hydrolysis by actomyosin. One can imagine a four state cycle of cross-bridge movement (Fig. 7).

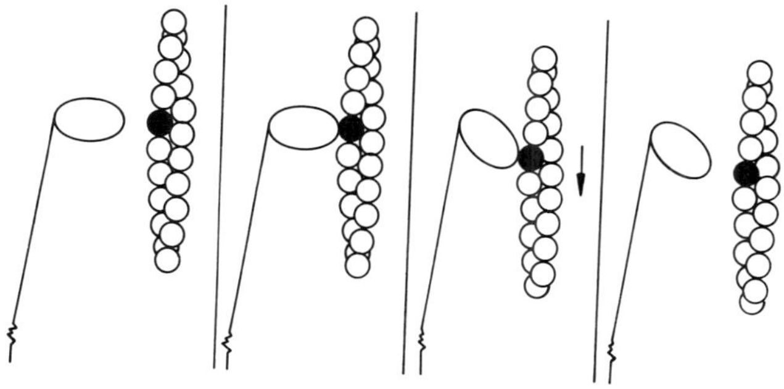

Fig. 7. **Four state model of cross-bridge attachment to and detachment from the actin filament producing a sliding motion of the filaments relative to each other.**

Starting from the situation of ATP-relaxed muscle, when the cross-bridge takes up a perpendicular position, this first state being characterized by the cross-bridge conformation T detached from the actin filament. In state 2, the cross-bridge still possesses the conformation T, but now is attached to the actin filament. The conformational change from T to R while attached to the thin filament is visualized to produce the power stroke. Detachment of the cross-bridge without change of angle leads to state 4. States 1 and 3 are equivalent to the relaxed and rigor state, respectively, and can be tentatively assigned to intermediary steps of the kinetic scheme. States 2 and 4 are still hypothetical. From the evidence mentioned above, it follows that state 1 corresponds to the myosin product complex ($M^{**}\cdot ADP\cdot Pi$) and state 3 to the

actomyosin complex AM.
One experimental approach towards identifying intermediary states in the analysis of the cross-bridge orientation taken up when the muscle is relaxed by ATP-analogs which are either not hydrolysed at all or which produce a different predominant steady-state complex. ATP-analogs used for this kind of investigation are ATP(βγNH), ATP(γS) and ATP(αβCH_2). The kinetic data of these analogs are given in Table 2. It can be seen that ATP(γS) and ATP(αβCH_2) are hydrolysed somewhat more rapidly than ATP by myosin, but the rate-limiting step is the cleavage step. Therefore, under steady-state conditions the predominant enzyme species is the myosin-substrate complex. The influence of actin on the interaction of these ATP-analogs with myosin has been discussed before. It was hoped that after relaxation of a muscle fibre in rigor the cross-bridge cycle could be "frozen" in a state not identified before and earlier reports from our laboratory gave the impression (13, 17), that state 4 could be produced by ATP(αβCH_2) and ATP(γS). Recent X-ray analysis, however, of insect flight muscle relaxed with these analogs reveals the cross-bridges in the 90°-position (18). With ATP(γS) the cross-bridges appear to be detached, with ATP(αβCH_2) predominantly detached, but with ATP(βγNH) apparently still attached to actin despite a change of angle to 90°. Taking into consideration the relative intensities of the equatorial reflections (1,0) and (2,0) arising from the hexagonal filament lattice and the 145 Å meridional reflection produced by the myosin cross-bridge repeat along the filament, the similarity of ATP(γS)- and ATP(αβCH_2)-relaxed X-ray diagram is striking. Both show a strong 145 Å layer line and a (1,0)/(2,0) ratio greater than 1, which is strong indication that the cross-bridges have swung back to the perpendicular position and are detached from the actin filament. Even though it has not been directly demonstrated that uncleaved ATP-analog is bound in this state, there is good reason to believe that the myosin-substrate complex in intact muscle fibres resembles the myosin-product complex, i.e. state

1 of the hypothetical cross-bridge cycle. Although these structural experiments do not add further evidence for the correctness of the four state cycle of cross-bridge movement, the correlation of state 1 to the myosin-substrate and product complex may simplify model-building when trying to assign the intermediary steps of the kinetic pathway to cross-bridge movement. Kinetically, there are seven distinguishable steps opposed by 4 states of the cross-bridge cycle, therefore, some cross-bridge states must correspond to more than one kinetic intermediate, which has been proved for state 1. Within the kinetic scheme there is evidence for at least three possibly different myosin conformations (M, M^{*} M^{**}). In the following we shall discuss two possibilities of correlation of the cross-bridge cycle and kinetic pathway, both of which are to our mind compatible with available evidence.

The first model which we want to discuss was first formulated by Mannherz and Schirmer (1970) (19) using preliminary evidence of X-ray analysis from ATP-analog-relaxed muscle fibres and later by Lymn and Taylor (15) using their kinetic scheme of ATP-hydrolysis by actomyosin.

This model is characterized by the power stroke of the cross-bridge being correlated to the product release processes. Starting from the rigor situation, state 3, ATP-binding induces cross-bridge detachment and formation of cross-bridge state 1. State 4 occurs transiently, and it was originally suggested that ATP-cleavage led from state 4 to state 1. However, the more recent X-ray results mentioned above (18) indicate that ATP-binding causes this transition. The large ΔG^{o} of ATP-binding to myosin could lead to efficient dissociation of the cross-bridge from actin. Cleavage occurs followed by recombination of the cross-bridge-product complex with actin to state 2 in the presence of high Ca^{2+}-ion concentration concomitant with or followed by product release and force generation. In this model, starting from the actomyosin complex (state 3), the kinetic steps 1 and 2 lead to state 1, the kinetic steps 4 to 7 lead to state 3.

One prominent feature of this model, is that the liberation of so-called chemical energy stored in ATP is not directly coupled to the step producing mechanical force. Furthermore, when comparing the ΔG^o-values for the ATP-binding and cleavage step, it can be seen that the largest free energy change occurs on binding of ATP, although it should be pointed out that the actual free energy change associated with each step in the steady state is dependent on the steady-state concentrations of intermediates. Due to the cyclical nature of the contraction mechanism these steady-state concentrations and therefore the ΔG^o-values for individual steps are interdependent.

A second conceivable model imagines the cross-bridge power stroke to take place after or during ATP-binding. In this model the step within the kinetic scheme associated with the largest change in ΔG^o is directly coupled with the mechanical force producing tilting of the cross-bridge. The cross-bridge product complex would still be the species with which actin interacts, but after recombination of actin and myosin the products would be displaced and ATP-binding would induce the conformation of the attached cross-bridge to change and in a second step to dissociate it from the actin filament. Such a sequence of events is thinkable in view of the two step binding mechanism of ATP. ATP-binding would cause a shift from cross-bridge state 2, via 3 to 4.

The first model strongly rests on the significance of the rigor state cross-bridge orientation. It is assumed that the rigor state is a natural intermediate state of cross-bridge movement during active contraction and of the kinetic pathway during ATP-hydrolysis. There is, however, no a-priori reason to believe this, especially since the angled attached cross-bridge orientation could only be demonstrated in insect flight muscle, whereas attempts using mammalian skeletal muscles have failed so far. Therefore, a nucleotide-free cross-bridge of the rigor type may be an artificial situation, with no context to the cross-bridge movement during active contraction. This does not mean that the cross-bridge does not produce mechanical force by a tilting movement, but the extreme orientation of the actomyosin-complex after washing out the nucleotide might not be reached during its active power stroke.

The second model is easier to imagine if state 4 does not exist as a discrete state, i.e. after detachment cross-bridges can only exist in state 1, which is probably composed of a distribution of cross-bridge angles. Tilting of the cross-bridges occurs only through simultaneous interaction of ATP and actin with the myosin head.

The X-ray results obtained from muscle fibres relaxed with ATP(γS) and ATP(αβCH_2) are compatible with both models. The results obtained with ATP(βγNH), however, are difficult to reconcile with either of the two models, unless it possibly represents state 2 in the second model.

References

(1) Huxley, H. E. and Hanson, J.: Nature 173, 973 (1954).

(2) Huxley, A. F. and Niederkerke, R.: Nature 173, 971 (1954).

(3) Huxley, H. E.: In Muscle. Proc. of a Symposium in University of Alberta, London: Pergamon Press (1965).

(4) Reedy, M. K., Holmes, K. C. and Tregear, R. T.: Nature 207, 1276 (1965).

(5) Huxley, H. E., Brown, W. and Holmes, K. C.: Nature 206, 1358 (1965).

(6) Weber, A. and Hasselbach, W.: Biochem. Biophys. Acta 15, 237 (1954).

(7) Lymn, R. W. and Taylor, E. W.: Biochem. 9, 2975 (1970).

(8) Bagshaw, C. R., Eccleston, J. F., Eckstein, F., Goody, R. S., Gutfreund, H. and Trentham, D. R.: in press.

(9) Bagshaw, C. R. and Trentham, D. R.: Biochem. J. 33, 323 (1973).

(10) Bagshaw, C. R. and Trentham, D. R.: Biochem. J. in press.
(11) Alberty, R. A.: J. Biol. Chem. 244, 3290 (1969).
(12) Bagshaw, C. R., Eccleston, J. F., Trentham, D. R., Yates, D. W. and Goody, R. S.: Cold Spring Harbor Symp. Quant. Biol. 37, 443 (1972).
(13) Mannherz, H. G., Barrington Leigh, J., Holmes, K. C. and Rosenbaum, G.: Nature New Biol. 241, 112, 226 (1973).
(14) Young, J. H., McLick, J. and Korman, E. F.: Nature 249, 474 (1974).
(15) Lymn, R. W. and Taylor, E. W.: Biochem. 10, 4617 (1971).
(16) Marston, S. B. and Tregear, R. T.: Nature New Biol. 235, 23 (1972).
(17) Barrington Leigh, J., Holmes, K. C., Mannherz, H. G., Rosenbaum, G., Eckstein, F. and Goody, R. S.: Cold Spring Harbor Symp. Quant. Biol. 37, 443 (1972).
(18) Barrington Leigh, J., Goody, R. S., Holmes, K. C., Mannherz, H. G. and Rosenbaum, G.: to be published.
(19) Mannherz, H. G. and Schirmer, H. R.: Die Molekularbiologie der Bewegung. Chemie in unserer Zeit, 165, (1970).
(20) Mannherz, H. G., Schenck, H. and Goody, R. S.: Eur. J. Biochem., in press.

Received July 22, 1974

DISCUSSION

Pohl: What is known about the temperature dependence of k_{+4} and k_{-4}, because this might shed some light on the kind of conformational change involved in this isomerization step.

Mannherz: At 4°C k_{+4} is approximately equal to k_{+5}, since k_{+5} is much more strongly temperature dependent than k_{+4}. From new results measuring release of ATP (ß,γ-NH) in the manner presented before it appears that k_{-2} has the same temperature dependence as k_{+5} with a Q_{10} of about 4.

Sund: In the right part of Fig. 6 you connected the experimental points by a straight line. It seems to me that the experimental points deviate characteristically from a straight line and should be connected by a sigmoidal curve. If you do so what would be the consequence to the postulated mechanism?

Mannherz: In Fig. 6 (right part) the experimental points might give the impression of sigmoidal curvature. But this might be due to low stability of the machine used for this experiment. More recent experiments indicate quite clearly that a straight line is obtained, when plotting the data obtained semilogarithmically.

Isolation of a Guanylnucleotide Binding Protein from Pigeon Erythrocyte Membranes

Thomas Pfeuffer and *Ernst J. M. Helmreich*

INTRODUCTION

Guanylnucleotides seem to play a quite universal role as activators of hormonally sensitive adenylate cyclases (Rodbell et al. 1971 a,b, Krishna et al. 1972, Puchwein et al. 1974). We have been working mainly with pigeon erythrocyte membrane preparations containing catecholamine stimulated adenylate cyclase. In this biological system GTP is rapidly degraded mainly to GMP, even in the presence of a regenerating system. We have therefore employed analogs of GTP in which the terminal P-O-P group is replaced by the $P-\overset{|}{N}-P$ or $P-\underset{|}{\overset{|}{C}}-P$ group and GTPγS, where one of the terminal oxygens is replaced by SH: $-\overset{SH}{\underset{O}{P}}-O^-$ (Goody and Eckstein, 1971).

ACTIVATION BY ANALOGS OF GTP (Fig 1)

In contrast to activation of isoproterenol-stimulated adenylate cyclase by GTP, activation by these analogs was not or only little dependent on substrate, ATP, concentrations (all experiments were carried out at 0.1mM ATP concentrations). The analogs in contrast to natural nucleoside triphosphates, activated the catecholamine-stimulated adenylate cyclase in pigeon erythrocyte membranes much more effectively than GTP. While

---

Abbreviations: GTPγS = guanosine-5′-0(3-thiotriphosphate); Gpp(NH)p = guanylylimidodiphosphate; Gpp(CH_2)p = guanylyl-ß,γ-methylenediphosphonate; MOPS = morpholino propane sulfonic acid.

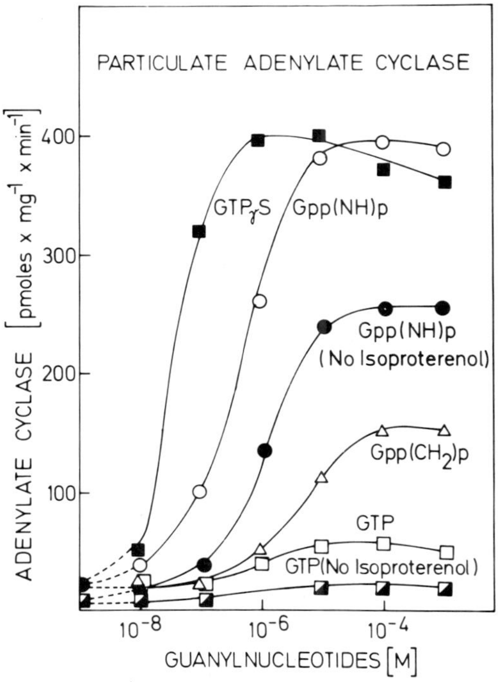

Fig. 1. Activation of membranous adenylate cyclase by GTP and analogs.

220μg of membrane protein were incubated in a buffer mixture containing 40 mM MOPS , pH 7.4, 0.1 mM [α-^{32}P] -ATP (10-15 μCi/μmole), 5 mM $MgCl_2$, 6 mM theophylline with and without 0.05 mM D, L-isoproterenol and with guanylnucleotides in the concentrations indicated. Incubations were at 37° for 10 min. [^{32}P] cAMP formed was measured according to Ramachandran (1971).

the activity of the isoproterenol-activated cyclase was enhanced maximally about 20-fold by Gpp(NH)p and GTPγS, it was increased only 2-fold by GTP. (In some preparations 0.1 mM GTP even inhibited). Among the analogs tested the most effective was GTPγS. The concentrations required for half maximal activation under the conditions of the experiments were 0.03 μM for GTPγS, 0.4 μM for Gpp(NH)p and 3 μM for Gpp(CH_2)p. It is of interest that basal activity - not stimulated by isoproterenol - was enhanced even more markedly - about 30 to 40 fold - by Gpp(NH)p, but only about 1.5 fold by GTP. Thus, in the case of the pigeon erythrocyte membrane adenylate cyclase the GTP analogs replaced, at least partly, the natural activator, the hormone.

SOLUBLE ADENYLATE CYCLASE (Table 1)

Lubrol PX (20 mM) made soluble about 15 to 20% of the protein of the erythrocyte membrane preparation and about all (90 to 95%) of its adenylate cyclase activity enriching the latter about 4-to 6-fold. Less than 10% of the soluble enzyme activity was precipitated by centrifugation at 105,000 x g for 60 min. Moreover, the solubilized adenylate cyclase did not appear in the void volume of various molecular sieve columns. The basal activity of adenylate cyclase solubilized from untreated membranes was low, 10 - 20 pmoles per mg of protein per min, but was stimulated - provided 0.5 mM Lubrol was present - 20-to 40-fold by Mg^{2+}/F^{-}. Soluble adenylate cyclase failed, however, to respond to catecholamines regardless whether Lubrol was present or not. Soluble adenylate cyclase prepared from untreated membranes was still activated by Gpp(NH)p but maximal activity was only about 1/10 of that solubilized from Gpp(NH)p-treated membranes. Activation of soluble adenylate cyclase by Gpp(NH)p was, like activation by magnesium fluoride, strictly dependent on the presence of small amounts of Lubrol. Adenylate cyclase could be released with Lubrol from membranes pretreated with Gpp(NH)p in an active and stable form which stripped of lipids and detergent, had an even (10 - 30%) higher specific activity then before. Its activity was not further increased on addition of Gpp(NH)p.

SYNERGISM BETWEEN HORMONE AND GUANYLNUCLEOTIDE (Table 2)

A SCATCHARD plot revealed two slopes indicating two sites differing about 100-fold in affinity for Gpp(CH_2)p. Because of the tight binding the question arose whether the nucleotide is bound reversibly (noncovalently) or irreversibly (covalently). Gpp(CH_2)p and Gpp(NH)p are nonphosphorylating because they lack an energy-rich phosphate bond and are not, or

Table 1: Solubilization of adenylate cyclase following treatment with Gpp(NH)p.

Membranes, 4.5 mg/ml, were incubated with or without 0.1 mM Gpp(NH)p and 50 μM D,L-isoproterenol at 37° for 20 min. After washing with 10 mM Tris HCl, 1 mM EDTA, 1 mM Mg Cl_2, 2 mM 2-mercaptoethanol, 0.25 M Sucrose buffer, pH 7.4, membranous adenylate cyclase was solubilized with 20 mM Lubrol PX in the same buffer by incubation for 30 min at 0°. Lipids were removed by treating the soluble preparation with snake venom phospholipase A (20 μg per ml) in the presence of 1 mM Ca^{2+}. After addition of EGTA at a final concentration of 1.5 mM (to stop lipase action) the preparation (3 mg of solubilized protein) was applied to a Biogel A 0.5 column (1.5 x 30 cm) equilibrated with 10 mM imidazole, 1 mM Mg EDTA, 5 mM 2-mercaptoethanol, pH 7.4. 95% or more of the phospholipids and more than 98% of Lubrol were separated from the protein peak containing adenylate cyclase activity. Adenylate cyclase activity was measured as indicated in the legend to Fig 1 with or without 0.1 mM Gpp(NH)p or 10 mM Mg^{2+}/F^-.

	Adenylate cyclase activity tested with		
	none	Gpp(NH)p	F^-
	pmoles x mg^{-1} x min^{-1}		
I Particulate Membranes (untreated)	10	270	130
Ia Adenylate Cyclase extracted with Lubrol from I	18	190	520
Ib Lubrol and Phospholipids removed	17	21	61
IIa Adenylate Cyclase extracted with Lubrol from Membranes treated with Gpp(NH)p	1.420	1.470	
IIb Lubrol and Phospholipids removed	1.680	1.690	

only very slowly, enzymatically hydrolyzed. GTPγS to the contrary, is hydrolyzed in erythrocyte membranes, but at a slower rate than GTP. We have studied the metabolism of $[^3H]$-Gpp(CH_2)p in membranes: The bound radioactivity could be extracted from the membrane with 0.5 M

Table 2: Synergistic activation by guanylnucleotides and catecholamines. Membranes (2 mg/ml) in isotonic phosphate/NaCl buffer pH 7.4 were preincubated with 5 μM Gpp(CH_2)p or 50 μM D, L-isoproterenol(alone or together) at 37° for 25 min. After washes with the same buffer adenylate, cyclase activity was determined. In parentheses is given the additive activation.

Membranes preincubated	Adenylate cyclase activity	
with	pmoles x mg^{-1} x min^{-1}	
D, L-isoproterenol +)	17.1	
Gpp(CH_2)p	54.0	(71.1)
Gpp(CH_2)p and D, L-isoproterenol	102.5	

+) and assayed in the presence of 50 μM D, L-isoproterenol

acetic acid at 20° and recovered to 95% or more as authentic Gpp(CH_2)p. Moreover, bound [^{3}H] -Gpp(CH_2)p was completely displaced by a 10,000-fold molar excess of unlabeled Gpp(CH_2)p or GTP, suggesting that the natural nucleoside triphosphate, GTP, and the GTP analogs share a common site in the membrane. Mutual exchange also rules out covalent attachment of the nonphosphorylating nucleotide analogs. The adenylate cyclase reaction in membranes which were preincubated with isoproterenol and Gpp(CH_2)p had reached a plateau with about 1.5 times greater activity than that of the same preparation preincubated with either Gpp(CH_2)p or isoproterenol alone. The hormone did not increase the amount of guanylnucleotide bound;it merely accelerated and potentiated synergistically its action. Interestingly, this activity level was maintained even after removing or blocking the catecholamine with propranolol.

Table 3: Separation of a guanylnucleotide-binding protein.
Membranes preincubated with 1 mM GMP and 50 µM D, L-isoproterenol at 37° for 15 min were solubilized with 20 mM Lubrol PX. Soluble adenylate cyclase, 200 µl, 0.9 mg protein/ml, was mixed with 200 µl of a suspension of Sepharose 4B or chemically modified Sepharose 4B and gently shaken for 30 min at room temperature. The concentration of coupled GTPγS was 2 mM. The slurry was centrifuged at 5,000 x g for 10 min at 4° and 50 µl of the supernatant were assayed for adenylate cyclase activity. The concentrations of Gpp(NH)p and F^- were 0.03 mM and 10 mM respectively. Relative activities [%] in parentheses.

	Conditions for affinity chromatography ("batch-method")	Adenylate cyclase activity in the supernatant assayed in the presence of: none	Gpp(NH)p	F^-
		nmoles x mg^{-1} x min^{-1}		
Ia	Sepharose 4B	0.08	1.95 [100%]	0.99
Ib	Seph-$(CH_2)_6$-$NHCOCH_2$-S-CH_2-CH_2OH	0.07	1.89	0.97
II	Seph-$(CH_2)_6$-$NHCOCH_2$-S-pppG (GTPγS-Sepharose)	0.08	0.49 [29%]	0.96
III	As in Ia and Ib but with 0.1 mM Gpp(NH)p		1.78$^+$ [91%]	
IV	As in II but with 0.1 mM Gpp(NH)p		1.77$^+$ [91%]	

+) Adenylate cyclase assayed without additional Gpp(NH)p.

SEPARATION OF A NUCLEOTIDE-BINDING FRACTION (Table 3)

We have been able to separate on Sepharose 4B a nucleotide-binding fraction from the adenylate cyclase activity. Separation was also achieved with soluble preparations from untreated membranes. This was exploited for affinity chromatography. Hexamethylenediamine was used as spacer and coupled to cyanogenbromide-activated Sepharose. The bromoacetamido derivative of the aminohexyl Sepharose was prepared with O-bromoacetyl-N-hydroxy succinimide (Cuatrecasas, 1970). Next, the bromoacetamido intermediate was reacted with GTPγS in glycerophosphate buffer pH 7.4 and the corresponding thioester with GTPγS was formed. Excess GTPγS was removed by repeated washes with buffer and the free bromoacetyl groups were reacted with 2-mercaptoethanol. For control purposes the S-carboxamidomethyl [8-^{3}H] guanosine-5'-0 [3-thiotriphosphate] was prepared. It was not hydrolyzed by erythrocyte membrane preparations or Lubrol-extracted soluble preparations. Thus, in contrast to free GTPγS, GTPγS coupled to a Sepharose derivative used for affinity chromatography was not hydrolyzed enzymatically. When a soluble adenylate cyclase preparation made from untreated membranes or from membranes weakly activated with GMP and D, L-isoproterenol was passed over the GTPγS Sepharose affinity column in the presence of 10 mM Lubrol it lost about 70% of its response to Gpp(NH)p whereas basal activity and Mg^{2+}/F^- stimulated activity were completely recovered. When affinity chromatography was carried out in the presence of 0.1 mM Gpp(NH)p which effectively competes with Sepharose bound GTPγS, the high activity characteristic for the Gpp(NH)p activated adenylate cyclase remained.

PURIFICATION OF THE REGULATORY AND THE CATALYTIC COMPONENTS OF ADENYLATE CYCLASE (Table 4)

Thus, we have detached from soluble adenylate cyclase a protein fraction which transmits the effect of the guanylnucleotide to the catalytic unit of adenylate cyclase. We have estimated an apparent molecular weight of about 230,000 for the nucleotide-binding fraction by gelfiltration on Sepharose 6B. Starting from the membrane preparation, the nucleotide-binding fraction was about 40-to 80-fold purified. Although the fraction is still far from being homogenous a considerable purification has been achieved (Fig 2). The same applies to the soluble adenylate cyclase fraction which was extracted with Lubrol from membranes incubated with

Table 4: Purification of adenylate cyclase.

Membranes, 4 mg/ml, preincubated with 0.1 mM Gpp(NH)p and 50 μM D, L-isoproterenol for 20 min at 37° were solubilized with Lubrol PX. Soluble adenylate cyclase was treated with Phospholipase A and chromatographed on a Sepharose 4B column (1.5 x 40 cm) equilibrated with 10 mM imidazole, 1 mM Mg EDTA, 1 mM 2-mercaptoethanol, 50 mM NaCl pH 7.4. Note, that in the adenylate cyclase assay the ATP concentration was 1 mM (rather than 0.1 mM).

Preparations	Activity	Purification	Yield
	nmoles x mg^{-1} x min^{-1}	(x times)	%
Untreated Membranes	0.021		
Membranes treated with Gpp(NH)p	0.75		100
Lubrol extracted	4.45	5.9	79
Phospholipase A treatment and chromatography on Sepharose 4B	28.5	38	58

Gpp(NH)p and D, L-isoproterenol and subsequently treated with phospholipase A (Crot. terr. terr.) and fractionated and separated from Lubrol by chromatography on Sepharose 4B. The specific activity rose to 28.5 nmoles x mg^{-1} x min^{-1} in the course of purification, a value which may be compared with the specific activities of the untreated membrane and the Gpp(NH)p treated membranes which were 0.021 and 0.75 nmoles x mg^{-1} x min^{-1}, respectively. The specific activity of the purified preparation obtained from membranes activated with Gpp(NH)p may be compared with that of an otherwise identical preparation obtained from

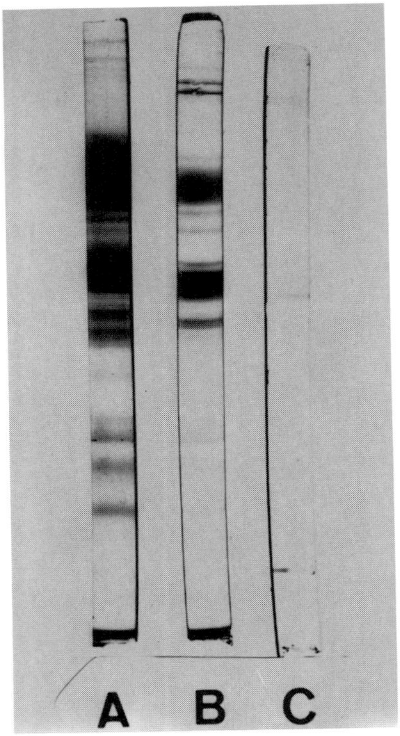

Fig. 2. Sodium dodecylsulfate - acrylamidegelelectrophoresis of solubilized membranes. Total membrane protein solubilized by Lubrol PX (20 mM) (A); following Sepharose 4B chromatography; fraction containing adenylate cyclase activity (B); and the binding protein (C). Concentration of acrylamide was 12%.

membranes activated with 10 mM Mg^{2+}/F^- which was 10 nmoles per mg per min. Although the soluble adenylate cyclase preparation is likewise by no means homogeneous, a considerable part of the membrane-bound Ca^{2+}, Mg^{2+} and K^+ and Na^+ activated ATPase activities was removed on Sepharose chromatography. The ratio of specific activities of Mg^{2+} activated ATPase to Mg^{2+}/F^- activated adenylate cyclase decreased from 50 to 1 in the soluble preparation to 10 to 1 after chromatography on Sepharose.

SUMMARY

In summary, we would like to suggest that the quasi-irreversible activation of the adenylate cyclase system by the GTP analogs is an unphysiological consequence of the metabolic stability and the tight binding of these artificial compounds. We assume that these analogs which bind to the same site as the natural activator GTP merely mimic the action of metabolizable guanosine triphosphates and further speculate that adenylate cyclase in the untreated membrane is inhibited and that inhibition is released by interaction of nucleotides with a regulatory subunit analogous to the mechanism of activation of protein kinase by cyclic AMP (Brostrom et al. 1970). Thus, one would like to know whether association-dissociation of the nucleotide-binding protein and adenylate cyclase actually plays a physiological role in activity control of this membrane-bound enzyme system. Another question which needs to be answered relates to the enzymatic activity of the guanylnucleotide-binding protein. For example, one could visualize the stable GTP analogs as transition-state intermediates of a hypothetical GTPase, what would explain the tight binding and the quasi-irreversible action of these analogs. Preliminary experiments have indicated that the protein fraction which binds nucleotides hydrolyses GTP with a specific activity of about 10 nmoles per mg per min at 37°. Moreover, the ratio of the specific activities of Mg^{2+} activated GTPase to Mg^{2+} activated ATPase rose from 0.8 in the unfractionated Lubrol-extracted membrane proteins to 2.0 in the nucleotide-binding fraction. But we cannot yet decide whether binding and hydrolysis are properties of one and the same protein because this fraction is still too inhomogenous (cf. Fig 2). Another problem which is currently investigated in our laboratory concerns the functional relationship of the guanylnucleotide-binding moiety to other GTP-requiring processes. Recently, Dr. H.P. Zenner could show that vinblastine inhibits catecholamine-sensitive adenylate cyclase much more than the Mg^{2+}/F^{-} stimulated or the Gpp(NH)p-treated enzyme. This effect is strictly dependent on the order of addition. For example, nucleotide

Table 5: Inhibition of adenylate cyclase activity by vinblastine.

In experiments Ia - Ic, adenylate cyclase activity, stimulated as described, was measured in particulate membranes (3.6 mg protein/ml) with or without 62 μM vinblastine for 20 min at 37°. In experiments IIa - IIc, soluble adenylate cyclase was prepared with 20 mM Lubrol PX from membranes preincubated with 0.3 mM GMP and 50 μM D,L-isoproterenol for 20 min at 37°. In IIa, soluble adenylate cyclase (1mg protein/ml), stimulated with Mg^{2+}/F^-, was assayed accordingly with or without vinblastine. In IIb, soluble adenylate cyclase was first incubated with or without vinblastine for 10 min at 37° and the assay was started by addition of substrate and Gpp(NH)p. In IIc, soluble adenylate cyclase was first incubated with Gpp(NH)p and the assay was started by addition of substrate with and without vinblastine. ATP concentration was 0.1 mM in each experiment.

Adenylate cyclase preparations		Adenylate cyclase activity with vinblastine	none	Inhibition
		pmoles cAMP x min^{-1} x mg^{-1}		%
Ia	Particulate; stimulated with Isoproterenol (0.05 mM)	28	64	56
Ib	Particulate; stimulated with Mg^{2+}/F^- (10 mM)	227	251	10
Ic	Particulate; stimulated with Gpp(NH)p (0.1 mM)	259	272	5
IIa	Soluble; stimulated with Mg^{2+}/F^- (10 mM)	428	415	—
IIb	Soluble; stimulated with Gpp(NH)p (0.1 mM)	165	244	32
IIc	Soluble; preincubated with Gpp(NH)p (0.1 mM)	493	532	7

activation is likewise impaired when the soluble preparation is treated with vinblastine prior to the addition of guanylnucleotides. (See Table 5) These preliminary results are of interest with respect to a dual role of nucleotide control of adenylate cyclase activity and structure and function of contractile systems such as microtubules. (cf: Wilson et al. 1974)

ACKNOWLEDGEMENTS

This work was supported by the Deutsche Forschungsgemeinschaft and Fonds der Chemie. A detailed account of this work is in press in J. Biol. Chem. (1974). We gratefully acknowledge the expert technical assistance of Mrs. E. Pfeuffer and Miss H. Dietrich. We also thank Mr. R. Thomas for assisting us in some of the affinity chromatography experiments and Dr. H. P. Zenner for making preliminary results on vinblastine effects on adenylate cyclase available prior to publication.

REFERENCES

RODBELL, M., BIRNBAUMER, L., POHL, S.L. and KRANS, H.M.J., (1971) J. Biol. Chem. 246, 1877

RODBELL, M., KRANS, H.M.J., POHL, S.L. and BIRNBAUMER, L. (1971) J. Biol. Chem. 246, 1872

KRISHNA, G., HARWOOD, J.P., BARBER, A.J. and JAMIESON, G.A. (1972) J. Biol. Chem. 247, 2253

PUCHWEIN, G., PFEUFFER, Th. and HELMREICH, E.J.M. (1974) J. Biol. Chem. 249, 3232

GOODY, R.S. and ECKSTEIN, F. (1971) J. Amer. Chem. Soc. 93, 6252

RAMACHANDRAN, J. (1971) Anal. Biochem. 43, 227

BROSTROM, M.A., REIMANN, E.M., WALSH, D.A. and KREBS, E.G. (1970) Advances in Enzyme Regulation, Pergamon Press, Vol. 8, 191

WILSON, L., BAMBURG, J.R., MIZEL, S.B., GRISHAM, L.M. and CRESWELL, K.M. (1974) Fed. Proc. 33, 158

Received August 12, 1974

DISCUSSION

Changeux: What do you know of the kinetics of the various reactions you postulate?

Helmreich: Nothing. I can only speculate that dissociation of the guanylnucleotide binding protein from adenylate cyclase may be rather slow and rate limiting. The catecholamine receptor interaction accelerates because the catecholamine might help to pull the catalytic unit out of the dissociation-association equilibrium:

$$[G \cdot S] \cdot [C] \underset{}{\overset{\text{slow}}{\rightleftharpoons}} [C^*] + [G \cdot S]$$

Liganded guanyl nucleotide-binding protein, [G · S]

Active catalytic unit

$$[C^*] + [R] \xrightarrow{\text{Lipids}} [C^*] \cdot [R]$$

Liganded catecholamine receptor, [R]

von Hippel: How do you visualize the activation of the cyclase systems by MgF_2? Do you have a hypothesis for how the two ions participate?

Helmreich: Recently, Constantopoulos and Najjar (Biochem. Biophys. Res. Commun. 53, 794 (1973)) have tried to explain adenylate cyclase activation by F^- by assuming that F^- removes covalently bound phosphate; the phospho-form of adenylate cyclase being the inactive form. Moreover pyrophosphate, one of the reaction products of the adenylate cyclase, inhibits. But may that be as it may, activation by GTP analogs does not involve phosphorylation or guanylation. Hence I do not have a plausible explanation for the general activation of adenylate cyclases by Mg^{2+}/F^-, although membrane bound adenylate cyclase activated by guanyl analogs and by Mg^{2+}/F^- have some interesting properties in common, among them the irreversibility of activation (that is: the enzyme remains active even after the activator is removed).

Boos: Can you restore the fully regulated system from the purified components?

Helmreich: I wish I could. No, this we have not yet achieved but we are trying.

Mannherz: Do you have any ideas about the binding constant of isoproterenol to the receptor site and is it changed by the non-hydrolysable analogs of GTP?

Helmreich: The K_{diss} for D,L-isoproterenol and the membrane bound receptor in the pigeon erythrocyte membrane is about 8×10^{-6} M. Only about 25 - 30 % of the bound D,L-isoproterenol may be replaced by propranolol. Because of the low affinity of the catecholamine to its receptor and because of the ambiguities with respect to the specificity of the receptor site for catecholamine it is difficult to test the effect of GTP analogs on the catecholamine receptor interaction in our system. But I agree that it would be of interest to know whether or not GTP analogs modulate catecholamine-receptor interaction.

Sund: Your Scatchard plot shows that the number of binding sites with the low dissociation constant is much lower than those with the higher dissociation constant. Can you correlate these data with e.g. kinetic data and are both kinds of binding sites specific? If they are specific do you know the different functions?

Helmreich: I do not know whether the low- and the high-affinity binding sites have different functions although it seems that the high-affinity sites are mainly involved in activation. But there is certainly more work to be done to decide whether both sites interact and show negative cooperativity or whether they are independent.

Wagner: Are the binding activities for the GTP analogues and the GTP hydrolysis activity located in the same particle?

Helmreich: I cannot yet answer your important question. As you may have noticed from one of the slides, the guanylnucleotide-binding fraction is still too inhomogeneous to allow to decide whether binding and GTP hydrolysis are properties of one and the same protein.

Kirschner: Following up on the question of Dr. Wagner with regard to the identity of GTP hydrolysis and GMPPNP binding sites, have you found competitive inhibition of the hydrolysis by the analog? If inhibition were competitive and the inhibition constant were the same as you find by equilibrium dialysis, this would constitute evidence for the same site being involved.

Helmreich: Thank you for this suggestion. I agree that this

should be done. We have only measured competition between GTP and guanylnucleotide analogs for activation of adenylate cyclase but not as you suggest for GTPase activity which would require that the binding protein is free of ATPase which also hydrolyzes GTP.

McNamee: In the preprint you briefly mention the effect of vinblastine on your system. Would you like to comment further on those effects?

Helmreich: Better not, these are preliminary experiments and their significance is still rather dubious. Obviously we are intrigued by a possible functional and structural relationship of the guanylnucleotide-binding protein which exhibits GTPase activity to other GTP-requiring processes such as microtubular function. But at present, further speculation seems premature, we just have to wait and see.

Kempfle: Have you any suggestion about the influence of the phosphodiesterase?

Helmreich: The phosphodiesterase activity did not concern us. In contrast to adenylate cyclase it is readily detached from membranes on solubilization. Moreover, in the assay we add theophylline to inhibit phosphodiesterase and to overcome ATPase we add an ATP-regenerating system.

Veeger: Can you exclude completely that the activating effect is caused by the presence of lipids?

Helmreich: No, because activation in the membrane by guanylnucleotides is strongly affected by perturbation of the lipid matrix. But what I wanted to say is that once activation has been achieved by deinhibition perhaps by removal of an inhibitory guanylnucleotide-binding protein, the adenylate cyclase activity now becomes independent of P-lipids and may be solubilized with Lubrol in a stable active form and stripped of lipids with phospholipase A without change in activity. Although we cannot exclude that 20-30 molecules of Lubrol still remain bound per 100,000 g, this would not be enough to form micelles.

On the Binding of Isoleucine and Related Amino Acids to Isoleucyl-tRNA Synthetase from Escherichia Coli MRE 600

J. Flossdorf, H.-J. Prätorius and *M.-R. Kula*

Investigations of Loftfield and Vanderjagt [1] have shown that under physiological conditions L-valine substitutes less than 3 times per 100 000 for L-isoleucine in the 10th and 17th position of the α-chain of hemoglobin of rabbits. This result gives an idea of the high fidelity of genetic translation, which becomes even more mysterious if we especially remember the close structural similarities between the naturally occurring aliphatic amino acids.

Key substances of the protein biosynthesis are the aminoacyl-tRNA synthetases. Each of these enzymes is specific for only one amino acid and catalyzes the activation and the attachment of this amino acid to either one or only a few cognate tRNA's. In the following, we shall be concerned with the isoleucine-specific synthetase from E.coli MRE 600. It was our aim to measure the dissociation constants of various protein-amino acid complexes and to determine the structural criteria which serve the enzyme to identify a small molecule as the true substrate L-isoleucine.

MATERIALS

Isoleucyl-tRNA synthetase was isolated from E.coli MRE 600 as described earlier [2]. The higher homologues of L-isoleucine

(C_7 and C_8-amino acids) were synthesized by the method of Schöllkopf [3] and separated into the D- and L-isomers by enzymatic hydrolysis of the N-chloro-acetyl derivatives [4]. Separation of the threo- and erythro-forms of the β-branched amino acids was achieved by high-voltage paper electrophoresis [5]. Amino alcohols were obtained by reduction of the corresponding amino acid methyl esters [6]. All other reagents were purchased from various firms mentioned elsewhere [7] and were of the highest purity available.

METHOD

Dissociation constants were determined by an ultracentrifugal method. It is based on the difference between the sedimentation coefficients of the enzyme and the amino acid and allows the separation and subsequent estimation of the unbound free ligand from both free and complexed protein. In this respect, it is comparable with an equilibrium dialysis. The binding of L-isoleucine was directly followed by counting radioactivity, whereas the binding of other ligands was measured by their competition with L-isoleucine. Details of this procedure, together with a discussion of possible sources of error, can be found in an earlier publication [7].

RESULTS AND DISCUSSION

Our results, as far as they have already been published [7], are summarized in Table 1. It shows that the binding of L-isoleucine is stronger than that of any other amino acid listed therein. Binding of the smallest amino acids glycine and L-alanine is presumably so weak that it could not be detected under our experimental conditions. Similarly the D-isomers and the α-hydroxy analogues of the amino acids listed in Table 1 fail to compete with L-isoleucine. The alcohols corresponding

Table 1

Dissociation constants of various amino acids to isoleucyl-tRNA synthetase from E.coli MRE 600 at 20°C in 0.05 M potassium phosphate buffer pH 7.5

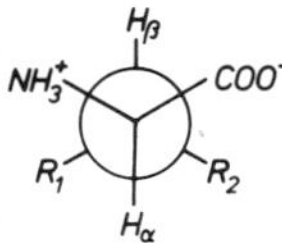	R_1	R_2	K_d [M]	ΔG^o $\left[\frac{kcal}{mole}\right]$
glycine	-	-	no binding detected	
L-alanine	-H	-H	no binding detected	
L-α-amino-n-butyric acid	$-CH_3$	-H	$2 \cdot 10^{-2}$	2.3
L-norvaline	$-CH_2CH_3$	-H	$9.0 \cdot 10^{-4}$	4.1
L-valine	$-CH_3$	$-CH_3$	$5.7 \cdot 10^{-4}$	4.4
L-norleucine	$-CH_2CH_2CH_3$	-H	$2.3 \cdot 10^{-3}$	3.5
L-isoleucine	$-CH_2CH_3$	$-CH_3$	$4.4 \cdot 10^{-6}$	7.2
L-allo-isoleucine	$-CH_3$	$-CH_2CH_3$	$1.3 \cdot 10^{-3}$	3.9
L-leucine	$-CH(CH_3)_2$	-H	$2.8 \cdot 10^{-3}$	3.4
D,L-tert.-leucine ($-H_\beta$ replaced by $-CH_3$)	$-CH_3$	$-CH_3$	no binding detected	

to L-isoleucine, L-valine and L-norvaline do, in contrast, bind to the enzyme, but more weakly by a factor of 2 000 as compared to the corresponding acids.

Thus, we should first note that some structural conditions must obviously be fulfilled by a molecule in order to be recognized as similar to L-isoleucine and to be subsequently bound to the enzyme. Those are the presence of a protonated amino group, of not too small a side chain and, preferably, of a carboxyl group. Their combination forms the S-configurated environment of a carbon atom. These requirements being fulfilled, it is only the nature of the side chain, which determines the strength of binding.

Let us imagine, as pointed out by the Newman-projection at the top of Table 1, that L-isoleucine is an L-alanine derivative with one methyl and one ethyl substituent at the β-carbon atom. Replacing ethyl by methyl would yield L-valine which binds to the enzyme more weakly than L-isoleucine by 3 kcal·mole^{-1}, whereas its replacement by larger groups, e.g. n-propyl or i-propyl, lowers the binding energy by 3.5 kcal·mole^{-1}. Omitting the methyl lowers the binding energy by 3 kcal·mole^{-1}; omitting the ethyl by as much as 5 kcal·mole^{-1}. Finally, conversion of the S-configuration to the R-configuration lowers the binding energy by more than 3 kcal·mole^{-1}. The presence of a hydrogen atom as the third substituent of the β-carbon atom seems to be essential, since no binding could be detected in the case of tert.-leucine. These results support the assumption that the presence of one methyl and one ethyl group in the S-configuration with one hydrogen atom and the glycine moiety are just those features that enable the enzyme to identify an amino acid as L-isoleucine, when offered for binding.

For this reason, we have investigated the binding of those amino acids in which either the methyl or the ethyl substituent was replaced by other aliphatic substituents. Table 2 shows the effect of replacing the methyl group. As can be seen, the enzyme sensitively discriminates between the methyl normally

Table 2

Dissociation constants of various 2-amino-3-alkyl-pentanoic acids to isoleucyl-tRNA synthetase from E.coli MRE 600 at 20 °C in 0.05 M potassium phosphate buffer pH 7.5

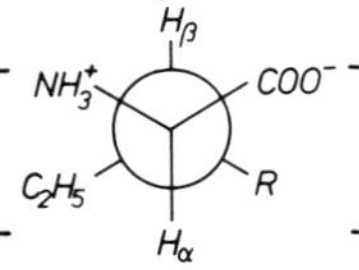	R	K_d [M]	ΔG^0 [kcal/mole]
L-norvaline	-H	$9.0\cdot10^{-4}$	4.1
L-isoleucine	$-CH_3$	$4.4\cdot10^{-6}$	7.2
L-2-amino-3-ethyl-pentanoic acid	$-CH_2CH_3$	$5.6\cdot10^{-4}$	4.4
L-2-amino-3-ethyl-hexanoic acid	$-CH_2CH_2CH_3$	$\sim 5\cdot10^{-3}$	$\sim$ 3

present and other substituents, either smaller or larger. The difference in binding energy is at least as high as 3 kcal·mole^{-1}.

Table 3

Dissociation constants of various 2-amino-3-alkyl-butyric acids to isoleucyl-tRNA synthetase from E.coli MRE 600 at 20 °C in 0.05 M potassium phosphate buffer pH 7.5

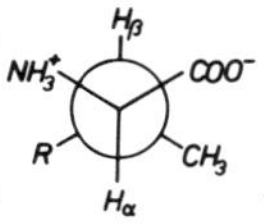	R	K_d [M]	ΔG^o [kcal/mole]
L-α-amino-n-butyric acid	-H	$2 \cdot 10^{-2}$	2.3
L-valine	$-CH_3$	$5.7 \cdot 10^{-4}$	4.4
L-isoleucine	$-CH_2CH_3$	$4.4 \cdot 10^{-6}$	7.2
L-2-amino-3-methyl-hexanoic acid	$-CH_2CH_2CH_3$	$2.7 \cdot 10^{-5}$	6.1
L-2-amino-3,4-dimethyl-pentanoic acid	$-CH(CH_3)_2$	$4.7 \cdot 10^{-5}$	5.8
L-2-amino-3-methyl-heptanoic acid	$-CH_2CH_2CH_2CH_3$	$2.7 \cdot 10^{-3}$	3.4
D,L-2-amino-3,3-dimethyl-pentanoic acid ($-H_\beta$ replaced by $-CH_3$)	$-CH_2CH_3$	$1.3 \cdot 10^{-3}$	3.9

Similarly, Table 3 shows the influence of the second β-substituent. Replacement of ethyl by other groups generally lowers the binding energy, which is particularly pronounced in the case of smaller groups: for example, replacement of ethyl by the smaller methyl lowers the binding energy by 3 kcal·mole^{-1}, whereas replacement by the larger propyl group reduces the binding energy only by slightly more than 1 kcal·mole^{-1}. The last line also shows the next higher homologue of tert.-leucine that exhibits a weak but significant binding to the enzyme. Thus, the presence of a β-hydrogen atom favours the binding of an amino acid, as stated above, but is by no means as decisive as the protonated amino group or the L-configuration.

In summary: The binding ability for amino acids is an intrinsic quality of the synthetase and is not basically dependent upon the presence of ATP and/or cognate tRNA's. The enzyme is triggered by a protonated amino group correctly positioned relative to other substituents of the α-carbon atom. The gain of binding energy is determined exclusively by the size and shape of the aliphatic side chain and is highest with a (2,S)-n-butyl side chain, that is, the enzyme is taylored to its natural substrate L-isoleucine.

This work was supported by the Ministry of Research and Technology (BMFT) of the Federal Republic of Germany within the Technology Program.

[1] Loftfield,R.B. and Vanderjagt,D.: The frequency of errors in protein biosynthesis. Biochem.J. 128, 1353-1356 (1972)

[2] Dureković,A., Flossdorf,J. and Kula,M.-R.: Isolation and properties of isoleucyl-tRNA synthetase from Escherichia coli MRE 600. Eur.J.Biochem. 36, 528-533 (1973)

[3] Schöllkopf,U., Gerhart,F. and Schröder,R.: α-Formylaminoacrylsäureester aus α-metallierten Isocyanessigsäureestern und Carbonylverbindungen. Angew.Chemie 81, 701 (1969)
Hoppe,D.: Aminosäure-Synthesen mit metallierten α-Isocyan-carbonsäureestern. Thesis, Göttingen 1970

[4] Greenstein,J.P.: Resolution of DL mixtures of α-amino acids. in: Colowick,S.P. and Kaplan,N.O.: Methods of Enzymology III, Academic Press, New York 1957, pp. 554-570

[5] Prätorius,H.-J., Flossdorf,J. and Kula,M.-R.: in preparation

[6] Karrer,P., Portmann,P. and Suter,M.: Die Reduktion von α-Aminocarbonsäureestern zu Aminoalkoholen mittels Lithiumaluminiumhydrid. Helv. Chim.Acta 31, 1617-1623 (1948)

[7] Flossdorf,J. and Kula,M.-R.: Ultracentrifuge studies on binding of aliphatic amino acids to isoleucyl-tRNA synthetase from Escherichia coli MRE 600. Eur.J.Biochem. 36, 534-540 (1973)

Received July 22, 1974

DISCUSSION

Hinz: I would like to add some preliminary results of an investigation which directly pertain to the talk of Dr. Flossdorf. He has demonstrated, that the specificity of recognition of isoleucyl-tRNA synthetase is a characteristic intrinsic property of the enzyme which stems from considerable differences in Gibbs free energy of binding for the different amino acids. The general validity of these free energy values constitutes the forcefulness and the elegance of the thermodynamic approach. However, these quantities completely lack the quality of being interpretable in terms of molecular models. One first further step towards this goal which still is thermodynamic in nature but which might facilitate an interpretation of the binding free energies, is to separate the Gibbs free energies into enthalpic and entropic terms. Therefore, I performed direct calorimetric determinations of the enthalpies involved in the binding of some amino acids to isoleucyl-tRNA synthetase in collaboration with Dr. Flossdorf and Frau Dr. Kula.

Although the results are preliminary, two conclusions might be drawn from them:

1. In the vicinity of 20 °C (i.e. between 15 and 25 °C) there is a linear dependence of the binding enthalpy of isoleucine to the enzyme on temperature corresponding to a temperature coefficient of approximately -700 ± 100 cal per mole per degree with the binding enthalpy vanishing around 18 °C.
2. The second result obtained from the calorimetric experiments is the following. Reaction enthalpy measurements were carried out at 25 °C under saturating conditions of the amino acids with valine and leucine and it was found that, within the limits of experimental error, there does not seem to be a significant difference in binding enthalpy for isoleucine, valine, and leucine. Thus the large differences in binding free energies have to arise from entropic changes of the system in the course of the binding reaction. It should be emphasized, however, that this statement can be applied only to a temperature of 25 °C with the results presently available because of the lack of data on the temperature dependence of the enthalpies of binding of the different amino acids. I do not want to attempt any molecular interpretations of the thermodynamic quantities since without additional non-thermodynamic information it is difficult to do anything else but to draw analogy conclusions.

Steinhardt: In order to fix the sign of the thermodynamic parameters, please give the values of ΔH and $T\Delta S$ at any temperature. It would appear from the tables that the entropic contribution is rather small for isoleucine at about 25° but at least as large as ΔH at some other temperatures.

Hinz: Enthalpy of binding of isoleucine to the enzyme amounts to approximately -5.5 kcal per mole of enzyme at 25 °C, thus the entropic contribution to the Gibbs free energy value of -7.2 kcal per mole is -1.7 kcal per mole. This number depends on the reasonable assumption, that the Gibbs free energy exhibits a small temperature dependence in the temperature range between 15 and 25 °C, since the value of -7.2 kcal per mole refers to 20 °C. However, as you mention yourself, the ratio of enthalpic and entropic contributions to the Gibbs free energy is bound to vary considerably with temperature due to the high temperature coefficient of the reaction enthalpy.

von Hippel: Have you made measurements at various temperatures with the various analogs? Is the specificity better or worse at higher (or lower) temperatures?

Hinz: Temperature dependence of the equilibrium constant and of the binding enthalpy has been determined up to now only for the reaction between isoleucine and the synthetase, the measurements involving valine and leucine refer to a temperature of 25 °C. Therefore, we cannot make any statements about the dependence of specificity on temperature yet.

Maass: According to the data of Loftfield there is a frequency of errors in the substitution of valine relative to isoleucine of 3 to 100 000. The difference in the dissociation constants of about two orders of magnitude, however, indicates an error frequency of about 1 %. Can it be concluded from these data that the presence of the cognate $tRNA^{ile}$ makes up for this difference?

Flossdorf: This is correct. There are at least two models of influence to be regarded: first, simultaneous binding of yet uncharged $tRNA^{ile}$ enhances binding of the true substrate L-isoleucine by a factor of two. If it would be possible to show that the binding ability for L-valine is not affected or even lowered, then it would come out that the error frequency is lower than 1 %. We are in the progress of investigating this problem. Secondly, it is a well known phenomenon that isoleucyl-tRNA synthetase not only catalyzes the ATP-PP_i[^{32}P]-exchange reaction in the presence of L-isoleucine but also in the presence of L-valine. Nevertheless, it has not yet been observed that L-valine has been transferred to $tRNA^{ile}$ to yield val-$tRNA^{ile}$. In contrast, it has been found that L-valyl-adenylate, though undoubtedly formed from L-valine and ATP by enzymatic action, is hydrolyzed in the presence of $tRNA^{ile}$. Most of the substrate specificity is due to that hydrolyzing activity.

The Effect of Small Molecules on the Affinity of Individual Chains in Hemoglobin and on its Cooperativity

K. H. Winterhalter, A. Mansouri and *E. E. Di Iorio*

INTRODUCTION

Adult human hemoglobin (Hb A) consists of two α and two β chains. Each of these chains carries one heme which is able, as long as its iron is in the divalent state, to reversibly bind oxygen. Detailed structural data on both the ligand free (deoxy or T state) and the liganded form have come from the laboratory of Max Perutz (1,2). Classically, the four heme groups of hemoglobin have been considered identical with respect to their functional properties (3). In recent years, however, the evidence that this is not the case has accumulated. ^{19}F-NMR spectra obtained on increasingly liganded trifluoroacetylated hemoglobin implied the possible existence of inequality of the two chains and the existence of states intermediate between T and R (4). Evidence to the same effect was obtained by NMR on progressively more liganded hemoglobin (5).

Heme in aqueous solution is very rapidly converted into hemin, a process that is enormously slowed down by the binding to the hydrophobic environment in the protein. In view of the fact that the protein plays such an important role in the rate of oxidation of the heme iron, it was to be suspected that the two different chains would have a different rate of oxidation. The argument was strengthened by the considerable differences in heme-protein contacts between α and β chains (1). Experimental data are consistent with postulate (6). In fact, at 37°C α chains in phosphate buffer at pH 6.5 have a rate of autooxidation almost ten times as large as β chains. Both previous (7) and presently conducted work demonstrated that deoxyhemoglobin oxidized much more rapidly than liganded hemoglobin. Thus, lowering the oxygen pressure in the tonometer used for the autooxidation experiments increased the rate of autooxidation, but more so for the β chains than for the α chains. This implies that the β chains have a lower affinity for O_2 than the α chains when incorporated into an oxyhemoglobin A molecule.

2,3-Diphosphoglycerate (2,3-DPG) binds in the central cavity of deoxyhemoglobin (8). The following residues have been implicated in the binding on the basis of crystallographic evidence: H_2N- of the N-terminal of both β chains; His 143 (H_{21}) on both β chains, His β_2 (NA_2) and Lys β_{82} (EF_6). The N-terminal amino groups and His β_{143} seem to be the two most important residues in this binding, since Hb F_1 which has the N-terminal blocked by an acetyl residue and in which on the γ chains His 143 is replaced by serine while residues 2 and 82 are as in β chains, exhibits no effect of 2,3-DPG (9,10). The central cavity of oxyhemoglobin, on the other hand, is too narrow to accommodate 2,3-DPG. However, even oxyhemoglobin is in a T $\rightleftharpoons$ R equilibrium with most of its molecules in R. The addition of 2,3-DPG will shift the equilibrium somewhat towards T, thus increasing the proportion of molecules in the low-affinity state. In the following we will demonstrate which chains have their affinity most affected by 2,3-DPG and how the cooperative behaviour of hemoglobin is influenced by orthophosphate ions.

MATERIALS AND METHODS

Hemoglobin A, A_{1C}, F and F_1 were prepared as described elsewhere (10). 2,3-DPG was purchased from Calbiochem as ammonium salt and converted to the acid form by recycling 30 ml of a 10 mM solution of the commercial product in water on a column (10x20 mm) packed with Amberlite IR 120. Hemoglobin was stripped free of bound 2,3-DPG by chromatography on Sephadex G 25 equilibrated with 0.1 M NaCl (11). The resulting Hb solutions were tested for the absence of phosphates (12).

Autooxidation was measured as previously described (6). At intervals aliquots of the partially autooxidized sample were withdrawn, CN^- ions were added and the remaining bound O_2 replaced by CO. The separation of the α and β chains was then carried out in a CO atmosphere by the rapid separation method described elsewhere (6).

Oxygen equilibria were obtained by the spectrophotometric method (13). The value n of the Hill equation was calculated by least square fitting of the Hill plot between 10 and 90% saturation. The kinetics of oxidation of deoxyhemoglobin by various concentrations of ferricyanide were performed in a Durrum Gibson stopped-flow apparatus Model D 110, equipped with a 2 cm observation tube. The progress curves were recorded at 630 nm on a transient recorder Datalab Model DL 905.

RESULTS

Interaction of 2,3-DPG with Hemoglobin

Fig. 1 shows progress curves of autooxidation of hemoglobin with and without 2,3-DPG at three different pH values. Clearly, in this situation (oxyhemoglobin) the effect of 2,3-DPG is marked at pH 6.8 and minimal at pH 9. Further increase of the 2,3-DPG/Hb ratio at pH 6.8 did not further influence the rate of autooxidation.

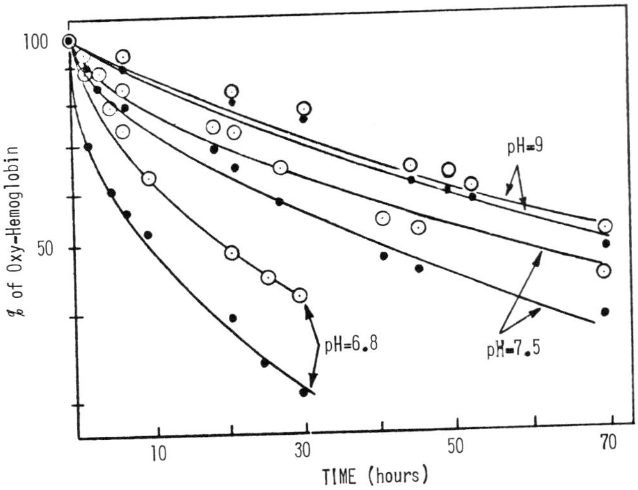

Fig. 1 Autooxidation of Hb A at 37°C, pH 6.8, 7.5 and 9, in the presence of 2 molar excess of 2,3-DPG (●—●) and in its absence (o—o). Buffer 0.05 M Tris-HCl, [Hb] ≃ 1.5%.

Table I Rates of autooxidation of α and β chains in Hb A in the absence and presence of 2,3-DPG

	- 2,3-DPG	+ 2,3-DPG	Ratio
α	0.073 hr^{-1}	0.195 hr^{-1}	~2.6
β	0.023 hr^{-1}	0.032 hr^{-1}	~1.5

Rapid chain separation of partially autooxidized hemoglobin (40%) under the conditions of Fig.1, pH 6.8, gave an α/β ratio of 1.24 in the absence and 2.7 in the presence of 2,3-DPG. In the presence of 2,3-DPG, α chains thus experienced a greater increase in the rate of autooxidation than did the β chains. Table I gives the rate constants of the rapidly oxidizeable (α) chains (6) and slowly oxidizeable (β) chains (6) in the presence and absence of 2,3-DPG under these conditions.

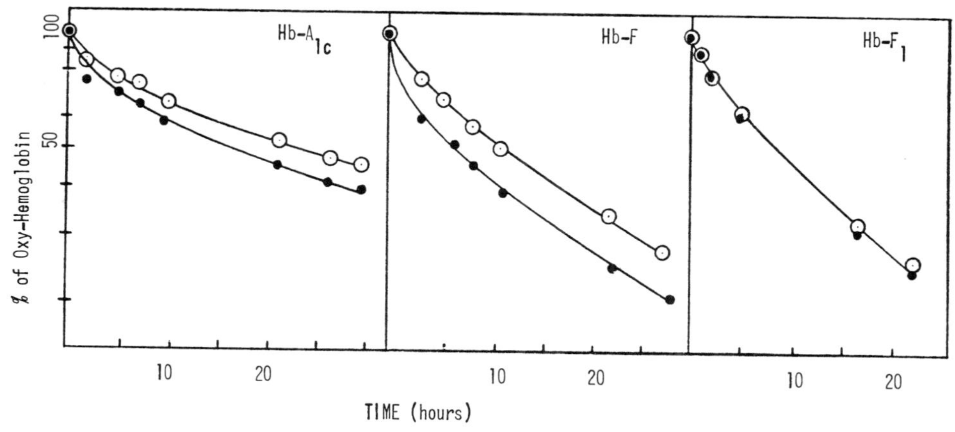

Fig. 2 Autooxidation of Hb A_{1C}, F and F_1 at 37°C, pH 6.8, in the presence (●—●) and absence (o—o) of 2,3-DPG -[2,3-DPG], [Hb] and buffers as for Fig. 1.

Fig. 2 shows the influence of 2,3-DPG on Hb A_{1C} which has its β chain N-terminal blocked, hemoglobin F in which β His 143 is replaced by γ Ser 143 and finally Hb F_1 which has its γ N-terminal blocked by an acetyl residue and His 143 replaced by Ser. It is clear that the 2,3-DPG effect is much smaller than in Hb A for both Hb A_{1C} and Hb F and absent in Hb F_1.

Fig. 3a illustrates the progress curves of the oxidation of deoxyhemoglobin with ferricyanide in 0.1 M citrate buffer pH 6 in the presence and absence of 2,3-DPG. There is no detectable effect of the organic phosphate. The progress curve is heterogeneous without clear biphasicity. These properties were not influenced by varying the ferricyanide to hemoglobin ratio from 10 to 50. Fig. 3b illustrates the same progress curves in 0.1 M TRIS-HCl buffer pH 9.0. Here in addition to the heterogeneity there is a clear biphasicity and the size of the rapidly reacting compartment is markedly increased by the presence of 2,3-DPG. Separation of chains of deoxyhemoglobin oxi-

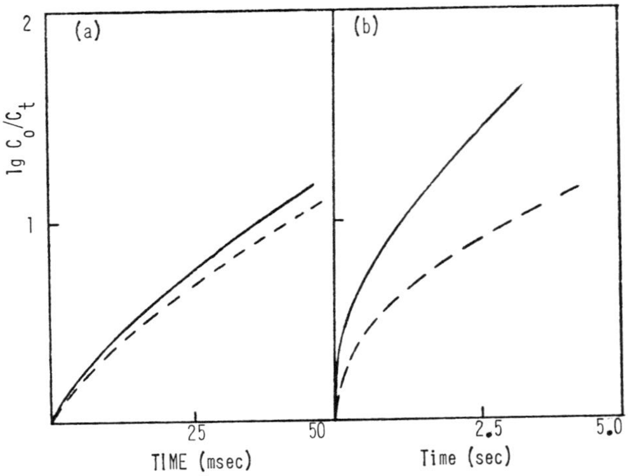

Fig. 3 Time course of chemical oxidation of Hb A at 20°C, (a) pH 6.0 and (b) pH 9.0, in the presence of 10 molar excess of 2,3-DPG (——) and in its absence (---) -[Hb] ≃ 0.1 mM, [K_4[Fe(CN)$_6$]] ≃ 5 mM, both before mixing. Buffers: citrate 0.1 M, pH 6.0, and TRIS-HCl 0.1 M, pH 9.0.

Table II Percentage of oxidation of α and β chains in deoxy Hb A in the absence and presence of 2,3-DPG.

pH	% oxid.	- 2,3-DPG			+ 2,3-DPG		
		α	A	β	α	A	β
6	~ 40	32	42	50	36	40	40
9	~ 40	40	52	61	29	40	58

dized to about 40% with ferricyanide under the four conditions described above gave the following results: Both at pH 6 and 9 in the absence of 2,3-DPG about 1.5 times more β chains were oxidized than α chains. In the presence of 2,3-DPG this ratio increased to about 2 at pH 9 but remained unchanged at pH 6. Some of the actual data are represented in Table II.

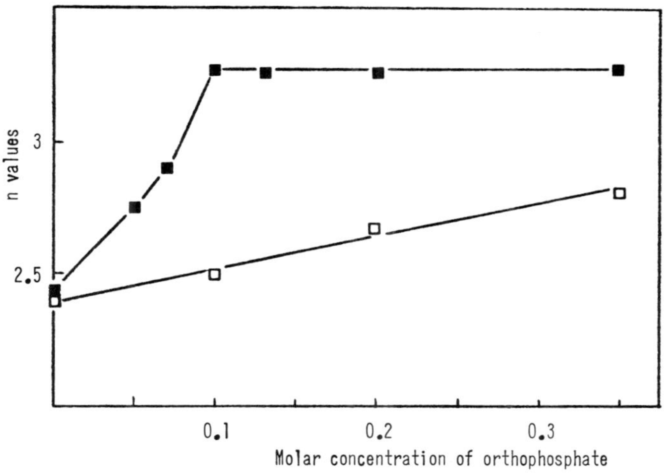

Fig. 4 Dependence of n of Hb A on orthophosphate concentration at pH 8.2 (□) in the presence of 2 moles of 2,3-DPG per mole of tetramer and (■) in its absence. Hb concentration ∼1 mg/ml, T = 20°C.

Interaction of Orthophosphate with Hemoglobin and its Influence on Cooperativity

The n values of stripped Hb A obtained at various pH in the presence of 0.13 M orthophosphate and in borate buffer alone are reported in Table III.

Clearly, in borate buffer alone no significant variations of n are observed between pH 7.5 and 9. In the presence of 0.13 M orthophosphate there is a relatively sharp rise in the n values, with increasing pH, giving a maximum cooperativity of about 3.3 at pH 8.2 with a return to about 2.6 at pH 8.9.

Fig. 4 illustrates the variability of n of Hb A at pH 8.2 dependent on the absence or presence of 2,3-DPG.

In the absence of organic phosphates the maximum effect is reached at concentrations of 0.1 M orthophosphate and the n

Table III pH dependence of n of Hb A on pH in the presence of 0.13 M orthophosphate and in borate buffer alone. Hb concentration ~1 mg/ml. T = 20 °C.

pH	borate phosphate n	borate n
8.9	2.65	2.50
8.5	3.01	2.51
8.2	3.27	2.48
8.1	3.07	2.55
7.5	2.65	2.49

Table IV Influence of orthophosphate on the cooperativity of Hb F. Hb concentration ~1 mg/ml, T = 20 °C.

Borate n	Borate + 0.13 M phosphate n
2.64	3.18

value cannot be further influenced by increasing the orthophosphate concentration. In the presence of a twofold molar excess of 2,3-DPG the n values at pH 8.2 are also influenced by the presence of the orthophosphate ions, however, much larger concentrations of these ions are needed to reach the same effect as in the absence of organic phosphates. In fact, at 0.35 M orthophosphate concentration, the maximum possible without increase of conductivity and loss of buffering capacity by lowering of borate concentration, n only reached 2.8. Its value appeared, however, to increase linearly with the concentration of the inorganic phosphate. In Table IV the n values of stripped Hb F, both in the presence and absence of orthophosphate at pH 8.2, are reported. As can be seen, the same phenomenon observed for Hb A is found here.

Table V Influence of orthophosphate on the cooperativity of Hb A_{1C}. Hb concentration ~1 mg/ml, T = 20°C.

pH	Borate n	Borate + 0.13 M phosphate n
7.5	---	1.66
8.2	1.65	1.65
8.8	---	1.64

Table V reports the n values of stripped Hb A_{1C} at different pH values, in the presence of 0.13 M orthophosphate and at pH ~8.2 in borate buffer alone. In this instance no change of cooperativity can be observed in the same pH range where Hb A and Hb F have been shown to be influenced by orthophosphate ions.

DISCUSSION

The effect of 2,3-DPG is completely manifest only in Hb A which has both β N-terminals free and a histidine residue in position β 143. If the N-terminal is blocked as in Hb A_{1C}, the 2,3-DPG effect is roughly halfed. If His β 143 is replaced by serine as in the γ chains of Hb F, the 2,3-DPG effect is also roughly one-half. In contrast, the orthophosphate effect observed in Hb A is fully present in Hb F (His 143 → Ser) and completely absent in Hb A_{1C} (N term blocked). In hemoglobin F_1 where the N-terminal is blocked and His 143 replaced by Ser, 2,3-DPG has no effect. From this we conclude that the N-terminal amino group and His 143 on each β chain are responsible for the overwhelming part of the binding of 2,3-DPG and no important role is played by His 2 and Lys 82 which have also been implicated on topological grounds (8). The sole residue responsible for the binding of orthophosphate is the N-terminal of β chains (8).

2,3-DPG enhances the autooxidation. This is achieved by lowering the oxygen affinity of hemoglobin and thus increasing the minute amounts of deoxy sites present in an oxyhemoglobin solution. The situation in this respect is analogous to the observation that lowering the ambient oxygen pressure (thus also increasing the concentration of deoxy sites) increased the rate of autooxidation (6). Since the topology permits binding 2,3-DPG in the central cavity of hemoglobin only in the T structure, the observed effect of 2,3-DPG must involve the shift of the equilibrium towards T.

In view of the pK values of the residues involved in the binding, it is not surprising that in oxyhemoglobin no effect of 2,3-DPG in 2 x molar excess is observed at pH 9.0, in contrast to the marked effect at pH 6.0 (Fig. 1).

The enhancement of autooxidation by 2,3-DPG is more marked for the α chains than for the β chains (Table I), although direct interaction of 2,3-DPG is limited to residues on the β chains. We therefore conclude that the difference in affinity between T and R structure is much more marked for the α chains than for the β chains. Similar conclusions have been reached by Olson and Gibson (15) for n-butyl isocyanide and methemoglobin.

In the experiments on the oxidation of deoxyhemoglobin by ferricyanide, the situation is different. At pH 6.0 the salt bridges stabilize the molecule in the T structure and 2,3-DPG does not shift the equilibrium appreciably further. In fact, the allosteric constant given for this situation by Ogata and McConnell (16) is $\frac{T}{R} = L = 10{,}000$. However, as the oxidation goes on, aquomethemoglobin is formed with a concomitant shift in the T - R equilibrium. This appears a plausible explanation of the observed heterogeneity of the progress curves. One could further argue that the minor differences observed at the end of the curves in presence of 2,3-DPG reflect the kinetics of 2,3-DPG unbinding, the speed of the reaction between hemoglobin and 2,3-DPG being in the milliseconds range (17).

At pH 9 where in view of the pK values of the residues involved even in deoxyhemoglobin the salt bridges are weak, 2,3-DPG, which does not bind strongly - not even to deoxyhemoglobin - at this pH, still exhibits some effect on the T - R equilibrium. The differences in the rates of oxidation between α and β chains (Table II) at this pH are not big enough to explain the marked biphasicicity of the curves in either presence or absence of 2,3-DPG. It therefore appears that the shift from prevalently T to prevalently R form is responsible for the impressive decrease of the rate of oxidation after about 50% of conversion of deoxy- to methemoglobin. In fact, 2,3-DPG increases the rate(s) of oxidation markedly, particularly in the second part of the reaction.

Orthophosphate ions increase the cooperativity of Hb A and Hb F between pH 7.5 and 9. The results reported in Fig. 4 show that orthophosphate ions have a much lower affinity constant for hemoglobin than 2,3-DPG.

The binding of orthophosphate ions to hemoglobin, like the binding of 2,3-DPG, stabilizes the T form, but in a way different from 2,3-DPG. The most visible difference is that only one residue on each β chain instead of two is involved.

The pH region in which the increase of n values by orthophosphate is observed suggests that

a) only the HPO_4^{2-} ions are capable of affecting the cooperativity of hemoglobin. In fact, the phenomenon appears around pH 7.5 and the pK of the reaction $H_2PO_4^- \leftrightarrows HPO_4^{2-} + H^+$ is about 7.3.

b) The binding of the orthophosphate is regulated, in analogy with the binding of organic phosphates, by the T $\leftrightarrows$ R equilibrium. In fact, above pH 8.2, when the groups involved in the conformational changes (1,2) become prevalently deprotonated, the T $\leftrightarrows$ R equilibrium will shift towards R as a function of pH. The already low affinity constant of orthophosphate for hemoglobin is thus decreasing further and the phenomenon progressively disappears.

The two phenomena described here illustrate that different mechanisms influencing the functional behaviour of hemoglobin all appear to be operative by their effect on the conformational equilibrium.

REFERENCES

1. Perutz, M.F., Muirhead, H., Cox, J.M., Goaman, L.C.G.: Nature 219,131-139(1968).
2. Muirhead, H., Greer, J.: Nature 228,516-519(1970).
3. Monod, J., Wyman, J., Changeux, J.P.:J.Mol.Biol. 12,88-118(1965).
4. Huestis, W.H., Raftery, M.A.: Biochem.Biophys.Res.Commun. 49,1358-1365(1972).
5. Lindstrom, T.R., Noren, I.B.E., Charache, S., Lehmann, H., Ho, C.: Biochemistry 11,1577-1681(1972).
6. Mansouri, A., Winterhalter, K.H.: Biochemistry 12,4946-4949(1973).
7. Antonini, E., Brunori, M., Wyman, J.: Biochemistry 4, 545-551(1965).
8. Arnone, A.: Nature 237,146-149(1972).
9. Bunn, H.F., Briehl, R.W.: J.Clin.Invest. 49,1088-1095(1970).
10. Mansouri, A., Winterhalter, K.H.: Biochemistry (1974) in press.
11. Benesch, R., Benesch, R.E., Yu, C.I.: Proc.Natl.Acad.Sci. U.S. 59,526-532(1968).

12. Jørgensen, S., Chen, P.S.: Scand.J.Clin.& Lab.Invest. 8,145-154(1956).

13. Rossi Fanelli, A., Antonini, E.: Arch.Biophys. and Biochem. 77,478-492(1958).

14. Garby, L., Gerber, G., DeVerdier, C.M.: Europ.J.Biochem. 10,110-115(1969).

15. Olson, J.S., Gibson, Q.H.:J.Biol.Chem. 247,1713-1725(1972).

16. Ogata, R.T., McConnell, H.M.: Cold Spring Harbor Symposia on Quantitative Biology, Vol.XXXVI,325-335(1971).

17. Gibson, Q.H.: Biochem.Biophys.Res.Commun.40,1319-1324(1970).

Received July 4, 1974

DISCUSSION

Steinhardt: Why was 40 % oxidation chosen for this work?

Di Iorio: The spectral differences between α and ß chains are optimal at 40 % oxidation due to the technical limits of the method of chain separation.

Steinhardt: Both DPG and $HPO_4^=$ stabilize liganded and unliganded hemoglobins against unfolding by dilute acid. Tris does also but only if octanol has destabilized.

The concentration effect on ferricyanide oxidation suggests the existence of a second-order part of the reaction path, which becomes rate-limiting only at low concentrations.

Veeger: Is the present information about the differences between the T- and R-states large enough to explain on a molecular basis the effects you describe by differences in conformations?

Di Iorio: In my opinion - yes. There are, in fact, X-ray analysis and NMR spectroscopy data providing good evidence for considerably big differences between T- and R-structures.

Veeger: It is clear from data in the literature that halogenide and phosphate ions have in general a competing effect on protein? From these data it is clear that the effects induced by one ion are counteracted by the other. Is it thus possible that the effect of phosphate, as you describe, is due to a replacement of Cl^- rather than a shift of the equilibrium between the T- and R-states?

Di Iorio: It is very well possible. Evidence in favor of this hypothesis came from data published by Dr. E. Chiancone in 1972 concerning the binding of Cl^- ions.

Kempfle: Did you try to investigate some mutants like hemoglobin Bart's where you have only ß chains or some hemoglobin M mutants?

Di Iorio: No.

G. Weber: How do you account for the fact the two processes of ligand binding and chemical oxidation have similar physical basis?

Di Iorio: It is a matter of the model one chooses for explaining data. Of course, other explanations are possible and suggestions in this sense are welcome.

G. Weber: Do the separate chains have intrinsically different chemical reactivity to oxygen and ferricyanide?

Di Iorio: There is a difference in autooxidation between isolated α and ß chains. I do not know of any experiment done on the chemical oxidation of isolated chains.

Helmreich: Did I get it right that the phosphate anion with pK_2 of about 7.0 is responsible for the effect of phosphate anions on O_2 binding to hemoglobin? It might interest you that we prepared the pyridoxal-5'-P monomethylester and that Dr. Benesch in New York has used this analog and could prove that point directly. In the monomethylester of pyridoxal phosphate the phosphate group of pyridoxal phosphate, pK_2 6.2, is blocked and unprotonable at physiological pH. Dr. Benesch has published these results recently in the Proc. Natl. Acad. Sci. (USA).

Di Iorio: Many thanks for the suggestion. It is very interesting for me to know of the existence of such a compound.

Hinz: Would you dare generalizing your result that interactions postulated by X-ray analysis on the mere basis of distances need not always exist?

Di Iorio: No.

Cooperative Interactions without Symmetry or Subunits

Jacinto Steinhardt

The purpose of this communication is a modest one: it is an attempt to demonstrate how much - or how little - may be learned about the detailed mechanism of the binding of oxygen by hemoglobin by attempts to fit the equilibrium oxygen dissociation curve by means of model equations. The goal, although modest, is not trivial, since the literature abounds with painstaking efforts to derive the "cooperativity" of the oxygen-binding process by means of various models applied to more and more accurate data. It also abounds with determinations from the data, and detailed hypotheses of calculated "interaction energies" ascribed to the effect of occupation of each site of the tetrameric deoxy molecule on hypothetical changes in the binding constants of the other sites. An excellent review of most of the proposed models has been given by Antonini and Brunori (1971).

The Adair equation, with its four independent binding constants (K_1, K_2, K_3, and K_4) has been used to describe the oxygen dissociation curve of hemoglobin for many years. The fact that the constants can be manipulated to give a very good fit of the best experimental data does not necessarily demonstrate that the equation is unique or that the physical model which was the basis of its original derivation is correct. Thus, it has been found necessary to assume that the values of the K_i depends on the state of occupancy by ligand of the other sites, and this dependence has been commonly stated in terms of an interaction energy which is formally related to $RT \ln \frac{K_{i+1}}{K_i}$ where the latter fraction expresses the change in any K_i brought about by the occupancy of an other specified site in the tetramer. A number of qualitatively different theories have attempted to specify how the occupancy of one site affects the K of other sites, or just particular other sites. Some of these theories invoke the existence of a tetramer-dimer equilibrium as a mechanism of mediating the interaction (Guidotti,

1967). Others explicitly distinguish between two hemoglobin conformations (commonly designated by R and T) which have different sets of Adair K_i values, and propose conditions for the transformation between R and T conformations on the basis of the occupancies of particular sites or numbers of sites (Monod, et al, 1965). Some of these theories have appealed to fits of the experimental dissociation data for support of their applicability.

Since it is known from both X-ray and pmr studies that R and T conformers of each subunit exist, the present paper takes the simplest possible case of a two-conformer theory, and explores how predictions of the oxygen dissociation curve depend on attributed numerical values of the parameters, or ratios of parameters.

Our model assumes that after combination of a specified exposed site or sites with oxygen, the binding of additional oxygen requires a conformational transition which permits all four sites to combine with the ligand (Reynolds, et al, 1967; Steinhardt and Reynolds, 1969). There is an equilibrium between the two forms expressed by the constant U = R/T. The value of U used here refers to equilibrium between the deoxy forms of R and T. The binding constant for oxygen of the "native" site or sites is K and the intrinsic binding constant of each of the four sites of the R form is J. The cases of interest are those in which J is equal to, or larger than, K since otherwise no appearance of "cooperativity" would result. In the interest of simple exposition, we consider first the case where only one site has the binding constant K. We shall see that the results are not very different if there are two native sites (K) rather than one.

All the intrinsic values of the K_i are taken to be the same. All of the J_i are also assumed equal. There are thus no $K_1 \ldots K_4$ or $J_1 \ldots J_4$ but only U, K, and J. The following equations hold:

$$\frac{\text{moles } O_2}{\text{moles Hb}} = \frac{pO_2}{1 + F(pO_2)} \left[\frac{nK}{1 + F(pO_2)} + \frac{F(pO_2)mJ}{1 + J(pO_2)} \right] \tag{1}$$

$$\text{where } F(pO_2) = \frac{U\,[1 + J(pO_2)]^m}{[1 + K(pO_2)]^n} \tag{2}$$

Here n is the number of sites per tetramer that bind with the constant K, and m is the number of sites in the transformed conformation R (presumably 4) that bind with the constant J.

We shall see from the accompanying figures that the value of U exerts an important controlling effect on the results. It is therefore important to understand that the equilibrium between R and T is not necessarily brought about through the unliganded forms. With appropriate trivial changes eqs. (1) and (2) prevail regardless of whether Hb°, HbO_2, or $Hb(O_2)_x$ are specified as the form in terms of which the $T \rightleftharpoons R$ transition equilibrium is described.

It should be noted that this model does not assume that there is any interaction whatsoever between the hemes. It is simply assumed that there is an equilibrium between R and T forms. It is also assumed in the main part of the paper (Figs. 1 and 2) that the T form has fewer sites than the other. All the sites combine with ligands in a simple fashion determined by K or J, respectively, according to the law of mass action.

It is readily apparent that Eqs. 1 and 2 would represent a simplified "symmetrical" Monod-Wyman-Changeux model for the oxygen-hemoglobin equilibrium if n and m were equal, and both set to 4. Our model is more general than the latter theory because, as the Figures will show, it allows for "cooperative" effects which are in part the result of $m > n$, as well as for similar effects which may result from $J \gg K$ or for combinations of both effects. It is shown below that the actual results predicted in both cases are very dependent on the numerical magnitude assigned to the transition constant U.

Fig. 1 includes an illustration of the effect of changes in J on the extent of oxygenation at fixed U. The calculated extents of oxygenation are plotted logarithmically to give Hill plots. Curves A, B, and C all have U equal to 1 (representing relatively large amounts of R in equilibrium with T). In Curve A the affinities (K and J) of both forms are equal. In curves B and C, J is tripled and quintupled without changes in K, which remains 10^4. With all three curves the Hill coefficient is close to 1, but as J increases oxygenation occurs at progressively lower values of (pO_2).[1]

---

1. The abscissa of Figs. 1-3, given as log [A] represent log $[pO_2]$.

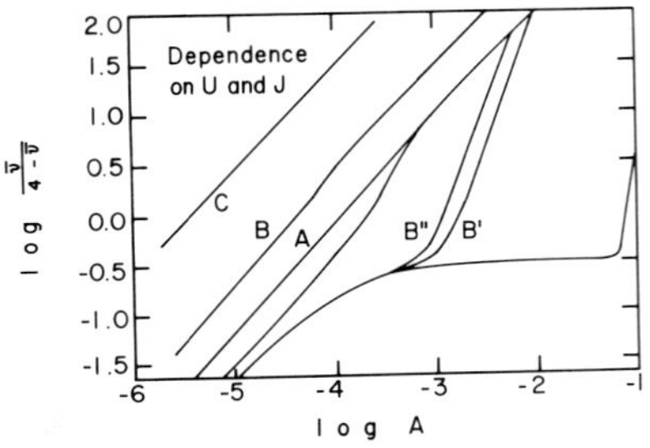

Fig. 1. Logarithmic Hill plots of dependence of oxygen binding on pO_2 (A on abscissa). Numerical values: n = 1, m = 4, K = 10^4; the values of U and J are given in the text.

When U is smaller (0.1), as in the curve to the right of A, the top and bottom portions of the curve are parallel to curve A, but the bottom is displaced relative to the top. As a result the central portion has a steeper slope (Hill coefficient greater than 1).

Curve B' shows how diminishing U further to 10^{-6} affects the results obtained with U = 1 at J = 3 x 10^4 (curve B). The titration with oxygen of the single sites postulated to be available on the T form is clearly separated from the remaining 3 sites. The Hill coefficient for the latter is very high, approaching 2.5. Relatively high oxygen tensions are required for oxygenation of these sites to occur. The higher Hill coefficient is due wholly to the diminished value of U. If there were no conformational transition, (and J were smaller than K) the slope of the upper portion would hardly exceed 1.

Curves B'' and B''' are more extreme examples of the foregoing. In B'' J has been slightly increased to 4 x 10^4, and in B''' J has been diminished to 2 x 10^4, but U has been greatly reduced to 10^{-10}. The Hill coefficients of the upper portion gradually approach 2.9. It is clear that the effect of the sharp diminution in U (from 10^{-6} to 10^{-10}) has outweighed the smaller effect of diminishing J.

In Fig. 2 the results of calculations are presented which are more directly applicable to the experimental results, represented as a broken curve (Roughton, et al, 1955). Curves A' and C differ only in the values of J which are 10^6 and 10^8 respectively. Their common parameters are

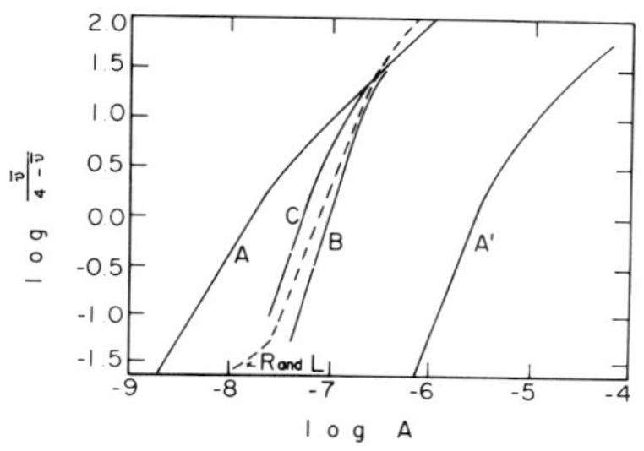

Fig. 2. Logarithmic Hill plots of dependence of oxygen-binding on pO_2. Numerical values: n = 1, m = 4, K = 10^4. The values of U and J are given in the text.

U = 10^{-3} and K = 10^4. The increase in J from 10^6 to 10^8 not only makes the oxygen binding occur at much lower values of pO_2 but also increases the Hill coefficient from 2.76 to 3.02. The linear region of the Hill plot also prevails to higher oxygen saturation with the higher value of J. If U is reduced to 10^{-4}, the effect is small - the curve is moved to slightly higher oxygen tensions, and becomes steeper (Hill coefficient near 3.4).

The importance of having relatively small values of U is further shown by curve A which has a relatively large U, 10^{-1}, combined with large J, 10^8. The large J value makes oxygen bind at very low oxygen pressure, but the Hill coefficient is nevertheless small, only 1.9, and the linear portion of the Hill plot fails beyond about 75 percent saturation.

The actual experimental data of Roughton, et al (1955) fall between curves B and C.[2] More painstaking curve fitting would yeild a value of U of about 10^{-2} if the ratio of J to K is taken as about 1000. There is no particular point in such careful curve-fitting since the model is admittedly crude. All that one can learn from such a comparison with good data is that J must exceed K by a substantial amount, and that the conformational transition constant must be small enough ($< 10^{-1}$) to cause deoxy sites of R to combine oxygen at substantially higher pressures than those which cause corresponding T sites to be oxygenated in spite of their higher affinities for oxygen. Obviously, therefore, a more fully specified model (with a diversity of K_i and J_i values for different sites) cannot be

2. The Roughton-Otis-Lyster data permit the establishment of a scale factor between A and pO_2. An oxygen pressure of unity represents 10^8 on the A concentration scale.

established by comparison with such experimental data.

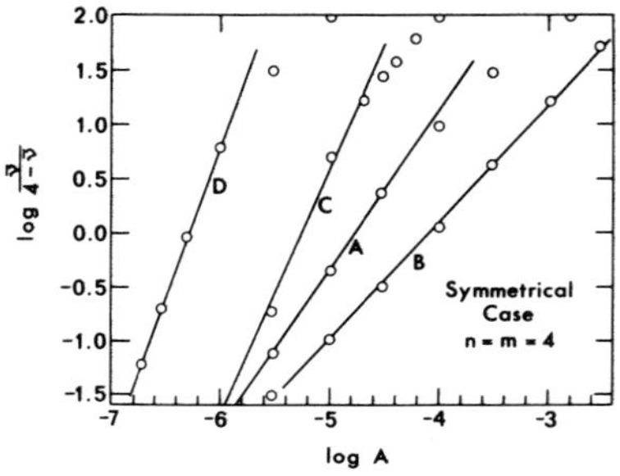

Fig. 3. Logarithmic Hill plots of the dependence of oxygen binding on pO_2 (A), when n=m=4; K = 10^4. The values of U and J are given in the text.

Monod, et al (1965) chose to specify in their essentially similar model that both T and R forms had the same number of binding sites, 4. Then the entire burden of "cooperativity" had to fall on differences in the binding constants A_i and J_i. We have therefore calculated as a special case the results given by such a "symmetrical" model with n=m=4. The results are given in Fig. 3.

All except one (A) of the lines in Fig. 3 have been calculated with K = 10^4 and U = 10^{-3}. While all of the calculations yield excellent linear Hill plots, the Hill coefficients exceed 2 only in lines C and D, representing respectively J = 10^6 and 10^7. The latter gives a Hill coefficient of nearly 3. With U = 10^{-1} a higher Hill coefficient results than with U = 10^{-3} but the linearity does not prevail to as high values of saturation.

Clearly the "symmetrical" Monod-Wyman-Changeux model, with identical sets of intrinsic K_i and identical sets of intrinsic J_i, will generate Hill plots with coefficients between 2.5 and 3.0 as long as J/K is in the vicinity of 10-1000. Calculations not shown on any of the figures have also shown that assumptions of n = 2, m = 4 give results, under conditions of high J/K and low U, which are indistinguishable from those shown in Figs. 1 and 2 (n = 1, m = 4), and which give satisfactory fits to experimental data.

By way of summary of the results of the calculations the following conclusions are stated:

1. If no conformational transition occurs ($U \rightarrow 0$) only one set of binding sites is titrated. The Hill plot slope is 1. K alone affects the position on the concentration axis.

2. If unfolding is easy ($U \geq 1$), a single continuous Hill plot with a slope of 1 is obtained. However the position on the concentration axis is determined by J as well as K. Variations in J and K are about equally effective.

3. The falling off in slope at high saturation is less evident when $J \gg K$ but the slopes of the main (lower and central) regions are not affected.

4. Reducing the value of U at ratios of J/K not far above unity introduces a rapid rise near 50 percent saturation. This is caused by a shift in the bottom section of the curves to higher concentrations without an accompanying shift in the upper section. The shift in the bottom portion is due to a lower contribution of the J groups to the ordinate. This can be seen at $U = 10^{-6}$ to 10^{-10} where no further shift occurs. It is also shown at U = 1 where changes in J cause large shifts in the line of slope 1.

5. The important generalization emerges that cooperativity occurs when J groups are prevented from participating in binding at concentrations at which they would bind if freely exposed. The higher Hill slope (> 1) results in the first instance from small U. It is contributed to by high ratios of J/K.

6. When U is very small, the J groups and K groups are clearly separated. However only the J groups show cooperativity with Hill slopes of > 2.5. Sufficiently high J can raise these slopes to 2.9 and higher.

7. Insofar as this model applies a Hill coefficient between 2.5 and 2.9 is practically inevitable. Such values show that $U \ll 1$ and, if near 2.8 - 2.9, that $J \gg K$.

8. Under conditions which lead to a Hill slope of 2.8 - 2.9 it is practically impossible to distinguish the results calculated for n = 2 from those calculated for n = 1. The symmetrical case (n=m=4) also gives a Hill coefficient of 2.5 to 3.0 within closely similar ranges of J/K and U.

9. The conditions of symmetry or of polymeric subunit structure thus appear to be inessential to the explanations of cooperativity in oxygen binding phenomena. Such a conclusion confirms what has long been apparent in cases of cooperative denaturation of single chain proteins (e.g. serum albumins, myoglobins) by interactions with such ligands as hydrogen-ions and detergent ions. The oxidation of hydroquinones to quinones also represent examples of small scale cooperative phenomena.

10. Since no differences in the K_i or J_i have been made use of, the present model is independent of assumed differences in interaction energies between sites.

11. Three-state systems have been studied by extension of this model, and have been applied to the binding of certain detergents by serum albumins. They have not been applied to hemoglobin equilibria, since there would be no hope of distinguishing their oxygen binding equilibria from those of the two-state system described by the model in this paper.

1. Antonini, E. and Brunori, M. (1971) "Hemoglobin and Myoglobin in their Reactions with Ligands", Chapter 14, American Elsevier, New York.
2. Guidotti, G. (1967). J. Biol. Chem. 242, 3673, 3685, 3694, and 3704.
3. Monod, J., Wyman, J. Jr., and Changeux, J. P. (1965). J. Mol. Biol. 12, 88.
4. Reynolds, J. A., Herberts, S., Polet, H., and Steinhardt, J. (1967). Biochemistry 6, 937.
5. Roughton, F. J. W., Otis, A. B., and Lyster, R. J. J. (1955). Proc. Roy. Soc. Loundon, B144, 29.
6. Steinhardt, J. and Reynolds, J. A. (1969) "Multiple Equilibria in Proteins" Academic Press, New York.

Received September 21, 1974

DISCUSSION

Changeux: The most salient feature of the theory which was proposed by Monod, Wyman and myself was not to derive mathematical equations which give an adequate fit of the oxygen binding energy. It was to propose a physical interpretation of the cooperative interactions between hemes. We postulated that these interactions are mediated by transitions between a small number of conformational states of the oligomeric molecule. The question is whether or not physical techniques might lead to the identification and characterization of these states.

Steinhardt: I completely agree.

G. Weber: One must distinguish between the formalism of two states, and the existence of actual unique protein conformation. The very large number of degrees of freedom of proteins would make it virtually impossible to prove that only two molecular forms are present, although of course one could come to prove that more than two forms are required to be postulated to explain the experimental facts.

Steinhardt: No quarrel, except as to relevance. The point is that an excessively simple formal model suffices to express a certain kind of data. The data will not, therefore, serve to verify or disprove more sophisticated variants.

von Hippel: One can often prove a multistate system exists by monitoring different aspects of a transition. As you say, on the other hand, one cannot "prove" 2-statedness specifically.

In actual fact, are you not using the equivalent of several (2 or 3) binding constants in your argument, and doesn't that in itself build in Hill coefficients > 2?

Steinhardt: In general, no. The result depends, in general, on the relative values of the constants. If far enough apart, and with no conformational transition, the Hill coefficient would be ONE. Actually the transition constant and $J > K$ are essential to a Hill coefficient much over 2.2.

Veeger: What is the difference between your model on which your equation is based and the Monod et al.-model?

Steinhardt: There is no difference, except that the model is numerically specified. No mechanism for the conformation change is postulated; no effort is made to interpret in terms of control mechanisms.

Veeger: I don't understand why in the first figure you showed, assuming $U = 0$, the Hill coefficient becomes zero?

Steinhardt: The Hill coefficient does not become zero. There is a two-step association. The Hill coefficient of the second is quite large.

A Nuclear Magnetic Resonance Study of Activator Binding by Carboxypeptidase A

Tamar Kushnir and *Gil Navon*

The applicability of nuclear magnetic resonance (NMR) to the study of binding small molecules to macromolecules provides us with quantitative data on the structure and dynamics of the binding process. Many experiments dealing with the NMR spectrum only detect this binding, or probe various changes occurred in the proteins and bound molecules. However, it is felt that in the quantitative data provided by the NMR method lies its most significant value. NMR studies of diamagnetic systems, i.e. measurements of relaxation times and nuclear Overhouser effect, give us information concerning the freedom of motion of some parts of the bound molecule (1), its conformation in the free and bound state and even the nature of its adjacent groups in the macromolecule (2).

NMR studies of paramagnetic systems provide us with different quantitative information, both structural and kinetic. For example, interatomic distances can be estimated in complexes of small molecules and macromolecules in the presence of paramagnetic metal ions or organic free radicals attached to the macromolecules (3,4). The obtained structural information can be compared only to X-ray crystallography and although it is not as detailed, it has the obvious advantage that it is measured in solution.

In this paper we shall illustrate the use of NMR in structural determination of protein-ligand binding, using the example of the complex of manganese carboxypeptidase A (Mn-CPA) and one of its activators, carbobenzoxyglycine

(CbzGly).

Carboxypeptidase A (CPA) is a zinc metalloenzyme with a molecular weight of 34,500. The zinc atom, which is essential to the activity of the enzyme, can be replaced by a manganese atom, retaining about 35 percent of its original activity. The enzyme exhibits both peptidase and esterase activities.

It undergoes both substrate and product inhibition and activation. In addition, various small molecules modify the activity of the enzyme, by acting as activators for its peptidase activity towards dipeptides, and as inhibitors when either longer peptides, or esters were used as substrates (5).

This mode of modification of activity was interpreted (5) by multiple substrate binding sites. Some of the sites which are productive for the esterase activity are non-productive for the peptidase activity, and vice versa. However, a kinetic proof for this model is still lacking (6).

Although there is no X-ray diffraction data on complexes of CPA with activators, Lipscomb et al.(1968) (7) have suggested that the activator is bound in the substrate secondary binding site, in a region containing the Arg 71, Tyr 198 and Phe 279 amino acid residues.

In order to check Lipscomb's model we decided to investigate the location of the binding site of activators. As a first example we chose the activator carbobenzoxyglycine (CbzGly), which is also the product of the substrate carbobenzoxyglycine-L-phenylalanine (CGP).

THEORETICAL BACKGROUND

The theory and application of NMR relaxation rates to the study of geometries of enzyme-ligand complexes has been summarized in the literature

(3,4), therefore only the essential relationships and some comments will be given here.

The net relaxation rates caused by an exchange of the ligand with a small fraction (f) of it bound to an enzyme is given in the absence of chemical shift by:

$$\frac{1}{T_{ip}} = \frac{f}{T_{iM} + \tau_M} \qquad i = 1,2 \tag{1}$$

T_{1M}, T_{2M} and τ_M are the longitudinal and transversal relaxation times and the exchange lifetime of the bound ligand, respectively. For ligand concentrations which are in excess over that of the enzymes, f is given by

$$f = \frac{[EL]}{[L]} = \frac{[E_o]}{K + [L]} \tag{2}$$

where K is the dissociation equilibrium constant.

The relaxation times of the bound ligand T_{1M} and T_{2M} are determined by nuclear - nuclear dipolar interaction, electron-nuclear dipolar interaction and scalar interaction due to finite unpaired spin density on the nucleus.

In paramagnetic complexes the first interaction is often negligible and its contribution can be deduced by running an experiment with a control solution containing a diamagnetic enzyme under similar conditions. In our case the control solution contained the zinc enzyme.

The scalar interaction is transferred along chemical bonds and it is a very strong decreasing function of the number of bonds. Therefore it is assumed to be negligible in our system. It should be noted that for other nuclei like C^{13} and P^{31} this contribution may have a higher value and more care should be paid to take it into account.

Relaxation rates caused by electron nuclear dipolar interaction are given by the following formula: (Solomon) (8).

$$\frac{1}{T_{1M}} = \frac{6}{15}\,\frac{S(S+1)\,\gamma_I^2\,g^2\,\beta^2}{r^6}\left[\frac{\tau_c}{1+\omega_I^2\tau_c^2}\right]$$

$$\frac{1}{T_{2M}} = \frac{1}{15}\,\frac{S(S+1)\,\gamma_I^2\,g^2\,\beta^2}{r^6}\left[4\tau_c + \frac{3\tau_c}{1+\omega_I^2\tau_c^2}\right] \qquad (3)$$

where τ_c is the correlation time for the modulation of the dipolar interaction and is given by :

$$\frac{1}{\tau_c} = \frac{1}{\tau_r} + \frac{1}{\tau_s} + \frac{1}{\tau_M} \qquad (4)$$

τ_r is the rotational correlation time, τ_s the electronic longitudinal relaxation time and τ_M the exchange lifetime. Eq. (3) is a simplified expression for $\omega_s\tau_c >> 1$ which is the case for manganese enzymes. Further assumptions which are included in the derivation of Eq.(3) are:

(a) negligible effect of the zero field splitting of the manganese electronic levels.

(b) An isotropic rotation of the complex.

Eq. (3) and Eq. (1) are used for the determination of interatomic distances r between the nuclei and the paramagnetic metal ion. However, τ_c and τ_M are also unknown and should be calculated before the value of r can be determined. This can be done if both T_{1P} and T_{2P} are measured for at least two groups of nuclei (9).

The procedure is based on the fact that while r is different for each group, both τ_M and τ_c are common to groups of nuclei on the same ligand.

If we subtract the relaxation times belonging to the two groups, from Eq.(1) we get:

$$\Delta T_{iP} = \frac{\Delta T_{iM}}{f} \qquad i = 1,2$$

$$\text{or : } \frac{\Delta T_{1P}}{\Delta T_{2P}} = \frac{\Delta T_{1M}}{\Delta T_{2M}} \tag{5}$$

Now, from Eq. (3):

$$\frac{T_{1M}}{T_{2M}} = \frac{7}{6} + \frac{2}{3} \quad \omega_I^{\ 2} \ \tau_c^{\ 2} \tag{6}$$

Thus T_{1M}/T_{2M} is common to all proton groups in the same ligand. Therefore:

$$\frac{\Delta T_{1P}}{\Delta T_{2P}} = \frac{T_{1M}}{T_{2M}} \tag{7}$$

$$\text{or:} \qquad \tau_c = \frac{1}{\omega_I} \left(\frac{3}{2} \frac{\Delta T_{1P}}{\Delta T_{2P}} - \frac{7}{4} \right)^{1/2} \tag{8}$$

and since $\tau_M = f\ T_{iP} - T_{iM}$

$$\frac{\Delta T_{1P}}{\Delta T_{2P}} \tau_M - \tau_M = \frac{\Delta T_{1P}}{\Delta T_{2P}} [f\ T_{2P} - T_{2M}] - f\ T_{1P} + T_{1M} = \frac{\Delta T_{1P}}{\Delta T_{2P}} f\ T_{2P} - f\ T_{1P}$$

$$\tau_M = \frac{f\left(\frac{\Delta T_{1P}}{\Delta T_{2P}} T_{2P} - T_{1P}\right)}{\frac{\Delta T_{1P}}{\Delta T_{2P}} - 1} \tag{9}$$

Thus Eqs. (8) and (9) enable us to calculate τ_c and τ_M directly from the experimental T_{1P} and T_{2P} .

EXPERIMENTAL SECTION

Bovine pancreatic carboxypeptidase A_α was prepared by the method of Cox et al.(10) and in later experiments the enzyme supplied by Sigma was used. The replacement of the zinc atom by manganese was done following Vallee et al.(11).

Carbobenzoxyglycine (CbzGly) was supplied by Sigma. For the high resolution NMR experiments, where it was important to reduce the amount of protons in the solution it was dissolved in D_2O and NaOD, precipitated by DCl solution, filtered and washed with D_2O.

Ultrapure Tris chloride was used as a buffer throughout, except for the high resolution NMR experiments, where no buffer was used, due to the overlap of the signal of the tris with that of the glycyl signal of the CbzGly.

Enzymatic Activity was determined spectrophotometrically (12) by following the hydrolysis of hippuryl-L-phenylalanine at 280 nm. It was not possible to follow the hydrolysis at 254 nm due to the strong absorption of the CbzGly at that wavelength.

Concentration of Mn-CPA was determined by the increase of the longitudinal relaxation rates at 100 MHz assuming molar relaxivity at that frequency of $(NT_{1P})^{-1} = 3\cdot10^4 M^{-1}sec^{-1}$ (13).

NMR spectra (100 MHz) were obtained with a Varian HA-100 spectrometer. The probe temperature was measured by the peak separation of dry methanol and was 28.5 $\pm$ 1° unless otherwise stated. Hexamethyldisiloxane (HMS) was used as an external lock. Values of $1/T_2$ were obtained from the spectral line width using the expression $1/T_2 = \pi\Delta\nu_{1/2}$, where $\Delta\nu_{1/2}$ is the full line width at half-maximum peak height. T_1 values were obtained using the progressive saturation method. H_1 was calibrated by progressive saturation of a standard manganese solution. T_1 and T_2 of the standard solution were measured independently by pulse method using a pulse NMR attachment to a Varian HA-100 high-resolution spectrometer (14).

RESULTS

EFFECTS ON ENZYME ACTIVITY. The effect of CbzGly on the activity of the Mn-CPA was checked in order to see whether it is an activator of the manganese enzyme like the case of the zinc enzyme and to measure the binding constant for the activation site for comparison with the NMR results. This effect is shown in Fig.1. It is seen that both the manganese and the zinc enzymes behave similarly except that the activation of the manganese enzyme is slightly higher. The maximum enhancement was by factors of 2.4 and 2.8 for the zinc and manganese enzymes respectively. The determination of the activator binding constant by activity measurements was done by plotting $V_o/(V-V_o)$ vs. $1/[L]$ (15).

The negative x-axis intercept is equal to 1/K, where K is the dissociation constant, at the limit of small substrate concentrations. However, in a similar study using the same substrate the negative x-axis intercept was not found to be dependent on substrate concentration (15). Therefore we assume the x-axis intercept to be close to 1/K in our case as well.

The result for the dissociation equilibrium constants of CbzGly to the zinc enzyme was 8.3±0.8 mM. at 20.5°c. This result may be compared with the dissociation constant obtained for the inhibition by CbzGly of the esterase activity of Zn-CPA towards the hydrolysis of cinnamoyl-β-phenyllac-

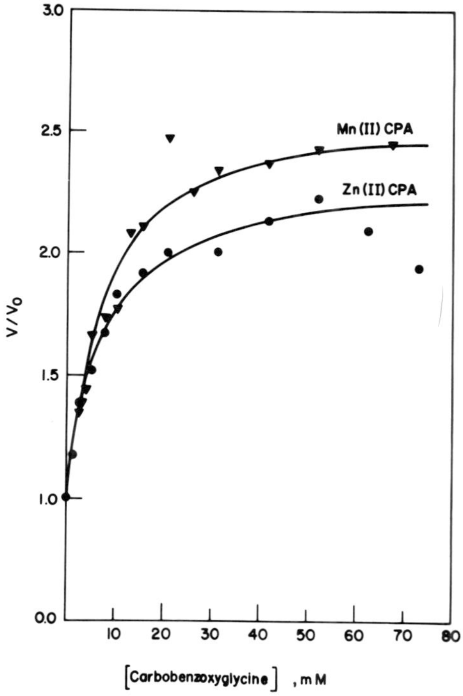

Fig. 1. Effect of carbobenzoxyglycine on the activities of manganese and zinc carboxypeptidase A.

tic acid (16 mM) (16) and of the peptidase activity for the hydrolysis of the tripeptides Bz-Gly-Gly-L-Phe and Cbz-Gly-Gly-L-Phe at 25°c (25 and 29 mM respectively) (17). The discrepency may be due to the different temperature used and to the effect of the substrates.

In the measurement of the dissociation constant of CbzGly to the manganese enzyme a correction due to the presence of small quantities of residual zinc enzyme should be applied. Since very small concentrations of the enzyme are used in the activity measurement, there is always a danger that some of the manganese will be replaced by extraneous zinc.

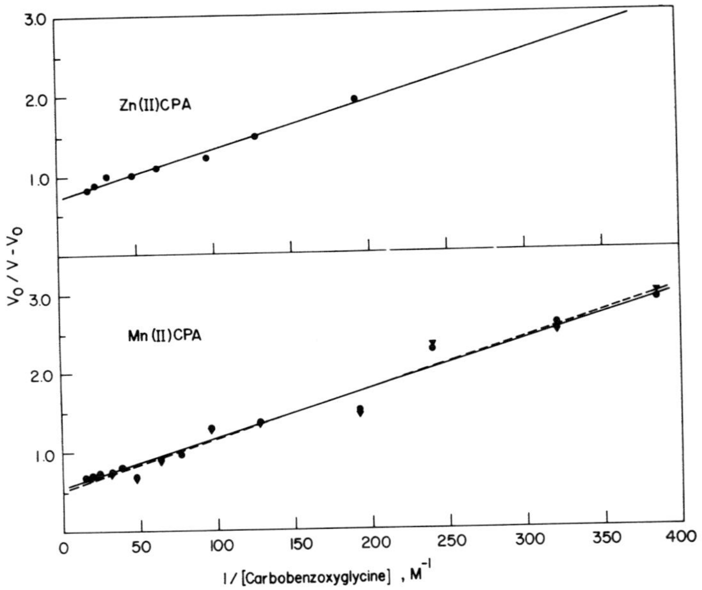

Fig. 2. A linear plot for the determination of the binding constant of an activator.

Considering that the specific activity in the solution used was 35 percent of that of the zinc enzyme under identical conditions we concluded that more than 94 percent of the enzyme was bound to manganese and less than 6 percent to zinc. In Fig. 2 the continuous line was drawn with the assumption that no zinc enzyme is present and the results corrected for a presence of 6 percent zinc enzyme are shown by the dashed line. The calculated binding constants based on these two assumptions are 11 and 13 mM, respectively.

WATER RELAXATION RATES. The contribution of the manganese enzyme to the water relaxation rates can be affected by the ligands. This is shown for the case of Mn-CPA and CbzGly in Fig. 3. There are two possible causes to the effect: The ligands can change the hydration number of the manganese

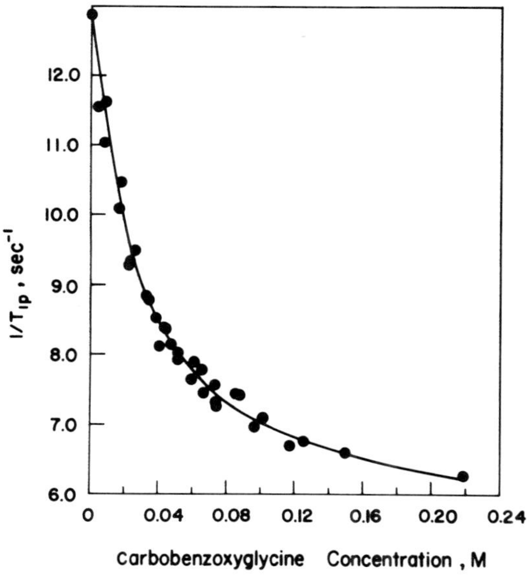

Fig. 3. Effect of CbzGly on the water longitudinal relaxation at 28.5°c.

atom on the enzyme, or can change the various correlation times which govern the relaxation. In any case, it is possible to utilize the variation in the water proton relaxation rates upon the binding of ligands for the determination of their binding constants, without any reference to the mechanism of the effect. From the results of Fig. 3 a value of K=27±3 mM was derived, and the relaxivity of the water due to the enzyme-Cbz-Gly complex was 40 percent of that of the enzyme alone.

RELAXATION TIMES OF THE CbzGly PROTONS. The NMR spectrum of CbzGly consists of three singlets at 4.01, 5.46 and 7.77 ppm from external HMS, belonging to the glycyl methylene (Gly), carbobenzoxy methylene (Cbz) and phenyl (Phe) protons respectively. All three signals were broadened by small amounts of added Mn-CPA, with the broadening decreasing in the order Gly > Cbz > Phe. This shows immediately that the system is not in the slow exchange limit (since otherwise all the signals would be equally broadened), and therefore structural information can be obtained. Furthermore it can be concluded that distances of the above three groups from the manganese are in the order Gly < Cbz < Phe.

The signal of the phenyl residue is composed of three proton groups, ortho, meta and para, which although they happen to have the same chemical shift, are expected to undergo different broadening effects. In fact, preliminary T_1 measurements using 180°- 90° sequence in Fourier transform NMR did indicate differential T_1 relaxation within the phenyl signal. Owing to this heterogeneity of the relaxation effect on the phenyl signal its relaxation was not further pursued. An example of the broadening effect on the two methylene groups is shown in Fig. 4.

In order to ensure ourselves that the effect of the Mn-CPA on the NMR relaxation times of the CbzGly protons is due to a binding to the same site which is responsible for the activation effects and changes of the water relaxivity, we have repeated the determination of the equilibrium constant following the line broadening as a function of the CbzGly concentration. The equilibrium constant can be extracted from this data utilizing the following relation : (18)

$$\Delta\nu_p^{-1} = \frac{\pi(T_{2M} + \tau_M)}{[E_o]} ([L] + K) \qquad (10)$$

In this plot the equilibrium constant is obtained from the x-axis intercept in the plot of $1/\Delta\nu_p$ vs. [L]. This procedure was repeated at several temperatures. Such plots are given in Fig. 5 and the results of the equilibrium constants obtained by the three methods are summarized in Fig. 6.

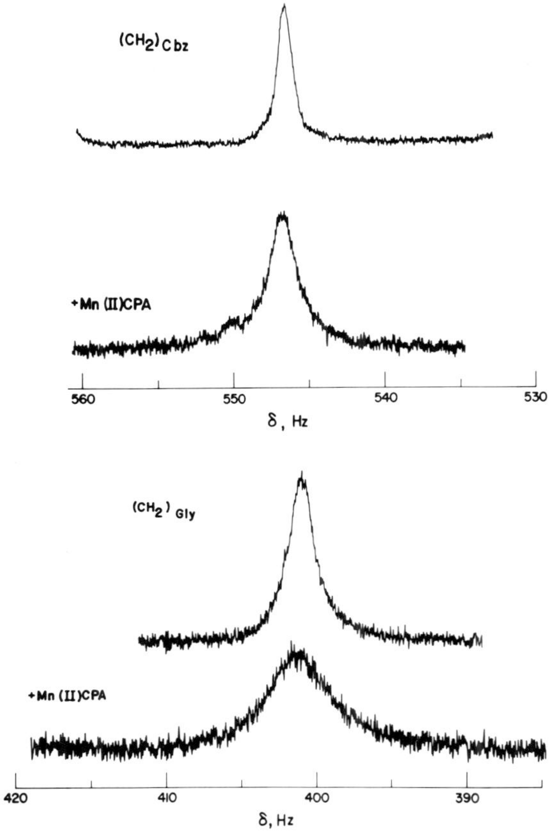

Fig. 4. 100 MHz NMR spectra of the glycyl, $(CH_2)_{Gly}$, and carbobenzoxy, $(CH_2)_{Cbz}$, methylene protons of 0.06 M CbzGly in the absence and presence of 1.1×10^{-5} M Mn-CPA in D_2O and 28.5°c.

From the line broadening at various temperatures, the function $T_{2M}+\tau_M$ was calculated using Eq.1 (see Fig.7). f- the fraction of bound CbzGly was calculated for each temperature using the values of K interpolated from Fig.6. In order to extract from the experimental data values for exchange times τ_M, correlation times τ_c and interatomic distances r it is necessary to know the values of both transversal and longitudinal

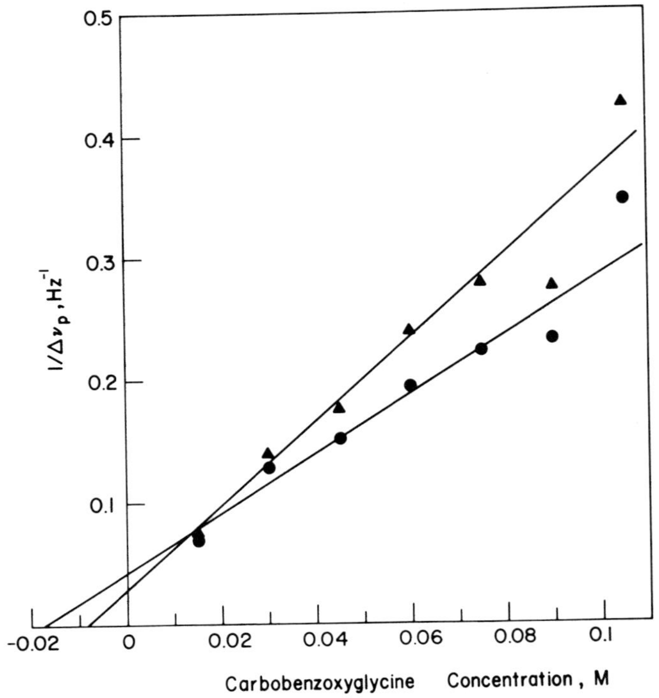

Fig. 5. Linear plot for the determination of the equilibrium constant, from NMR line broadening data. ▲ 13.5°c, ● 22.5°c.

relaxation times for at least two groups of protons on the same ligand. These are shown for the glycyl and carbobenzoxy methylene groups in Table 1. It is seen from the table that r values calculated at the various temperatures are fairly constant and their variations give us an estimate of the experimental error. The average values (with the average error) for the distances thus obtained for the Gly and Cbz methylene groups are 5.2±0.3 and 7.0±0.2 Å, respectively.

DISCUSSION

From our results for the distances of the methylene groups of carbobenzoxyglycine 5.2±0.3 and 7.0±0.2 for the glycyl and the carbobenzoxy

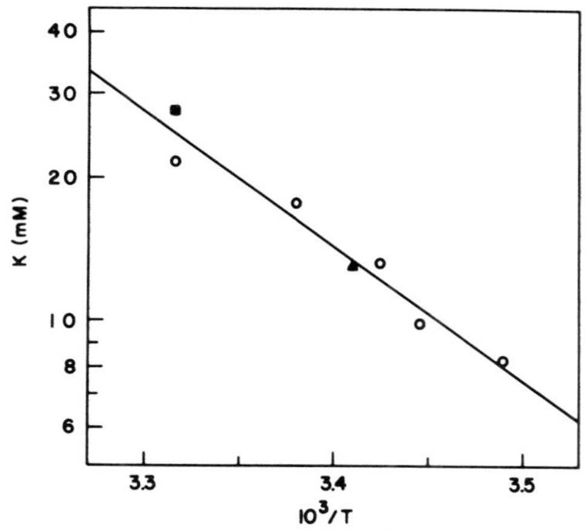

Fig. 6. Semilogarithmic plot of the Mn-CPA-CbzGly equilibrium constant vs. reciprocal absolute temperature. o NMR line broadening data. ■ Water relaxation rates. ▲ Enzymatic activity.

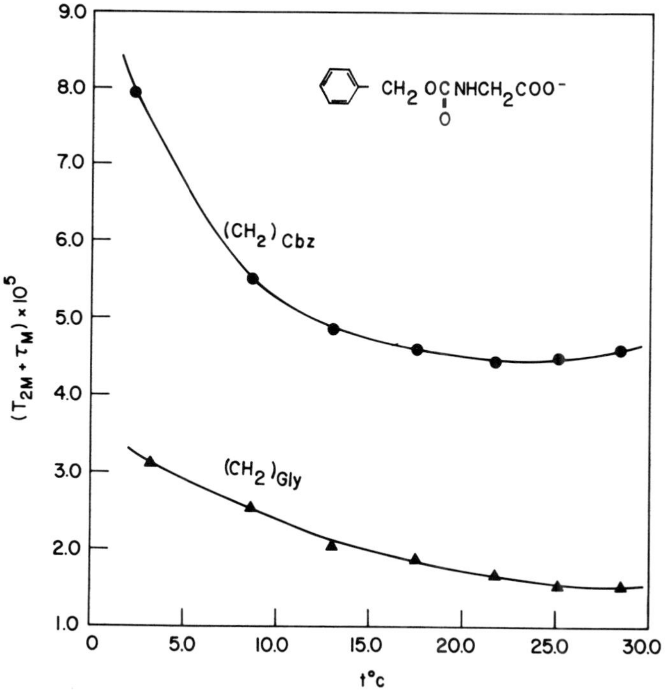

Fig. 7. Temperature dependence of $T_{2M} + \tau_M$ for CbzGly bound to Mn-CPA.

Table 1.

t°c	fx10^4	T_{1p} (msec)		T_{2p} (msec)		τ_Mx10^5	τ_c	r (Å)	
		(CH_2) Gly	(CH_2) Cbz	(CH_2) Gly	(CH_2) Cbz	(sec)	nsec	(CH_2) Gly	(CH_2) Cbz
28.5	5.16	125.5	632	31.8	91.8	1.0	5.3	5.3	7.1
24.5	4.33	144	716	36.1	96.2	1.0	5.6	5.2	6.9
20.5	4.5	159	1280	37.5	92.5	1.4	8.5	4.9	7.2
15	2.02	517	2230	116	224	1.8	7.5	5.4	7

methylenes, respectively, it appears that this molecule is bound with the carboxylate group close to the metal ion, but is not bound directly in the first coordination sphere of the metal ion.

For comparison, inhibitors of CPA were found to bind closer to the metal ion (19). The methylene group of methoxyacetic acid, for example, was found to be only 4.3 Å away from the metal ion and was assumed, therefore, to be bound through its carboxylate group to the metal ion. It is reasonable to assume that the activator, in contrast to the inhibitor, does not replace the water of hydration on the metal ion. This is now in a process of investigation by the measurement of the relaxation times of the water protons (13). Unfortunately, only two distances cannot allow us to draw a complete picture for the activator mode of binding. They may be of a great help, however, for eliminating other possibilities, in the process of finding a plausible structure using space-filling models of the protein and the activator. By such a procedure, Quiocho and Lipscomb (20) proposed a model in which the distances of the methylene groups of carbobenzoxyglycine seem to be higher than those found in the present work.

Since r appears in Eq.3 in the sixth power, its value is not very sensitive to experimental errors. This is not the case with the values of τ_M and τ_c given in Table 1. Therefore, until a more accurate measurement of T_{1p} will be made, using the Fourier-transform technique, rather than the progressive saturation method, they will be discussed only from a qualitative point of view.

For a one-step association reaction of enzyme with the ligand, τ_M, the residence time of the ligand on the enzyme is equal to k_{off}^{-1}, where k_{off} is the first order dissociation rate constant. Having measured the equilibrium constant which equals to k_{off}/k_{on}, the association rate constant k_{on} can be also estimated and it is of the order of 10^6-10^7. It is interesting to compare these results with the kinetic data obtained for inhibitors which were also carboxylic acids (19). In that case a correlation was found between the binding constants and the on and off rate constants, so that the rate constants were smaller for the inhibitors with the higher binding constants. Comparing the activator carbobenzoxyglycine with inhibitors of similar binding constant reveals that both k_{on} and k_{off} are larger for the activator. This result is consistent with the larger distances found for the activator proton-manganese ion. Thus, it seems possible that the activator is bound not as deep in the active site crevice as the inhibitors.

Although the results given here are not as definite as they may be and more experimental work is planned on this system, they may serve as a demonstration of structure determination of ligand-protein complexes containing paramagnetic metal ions, by NMR.

REFERENCES

1. Navon, G., Lanir, A.: J. Mag. Res. 8, 144-151 (1972).

2. Bothner-By, A.A., Gassend, R.: Ann. New York Acad. Sci. 222, 668-676 (1973).

3. Mildvan, A.S., Cohn, M.: Advan. Enzymol. 33, 1-70 (1970).

4. Dwek, R.A.,"Nuclear Magnetic Resonance in Biochemistry: Application to Enzyme Systems." Clarendon Press. Oxford (1973).

5. a) Davies, R.C., Auld, D.S., Vallee, B.L.: Biochem. Biophys. Res. Comm. 31, 628-633 (1968).

   b) Vallee, B.L., Riordan, J.F., Betune, J.L., Coombs, T.L., Auld, D.S., Sokolovsky, M.: Biochemistry 7, 3547-3555 (1968).

6. Shechter, I., Zazepizki, E.: Eur. J. Biochem. 18, 469-473 (1971).

7. Lipscomb, W.N., Hartsuck, J.A., Reeke Jr., G.N., Quiocho, F.A., Bethge, P.H., Ludwig, M.L., Steitz, T.A., Muirhead, H., Coppola, J.C.: Brookhaven Symposia in Biology 21, 24-90 (1969).

8. Solomon, I.: Phys. Rev. 99, 559-565 (1955).

9. Lanir, A., Navon, G.: Biochemistry 11, 3536-3544 (1972).

10. Cox, D.J. Bovard, F.C., Bargetzi, J.P., Walsh, K.A., Neurath, H.: Biochemistry 3, 44-47 (1964).

11. Vallee B.L., Rupley, J.A., Coombs, T.L., Neurath, H.: J. Biol. Chem. 235, 64-69 (1960).

12. Folk, J.E., Schirmer, E.Q.: J. Biol. Chem. 238, 3884-3894 (1963), (as modified in Wortington Biochemical Corp. Catalogue).

13. Navon, G.: Chem. Phys. Letters 7, 390-393 (1970).

14. Ginsburg, A., Lipman, A., Navon, G.: J. Sci. Inst. 3, 699-701 (1970).

15. Epstein, M., Navon, G.: Biochem. Biophys. Res. Commun. 36, 126-130 (1969).

16. Awazu, S., Carson, R.W., Hall, P.L., Kaiser, E.T.: J. Amer. Chem. Soc. 89, 3627-3631 (1967).

17. Auld, D.S., Vallee, B.L.: Biochemistry 9, 602-609 (1970).

18. Navon, G., Shulman, R.G., Wyluda, B.J., Yamane, T.: J. Mol. Biol. 51, 15-30 (1970).

19. Navon, G., Shulman, R.G., Wylada, B.J., Yamane, T.: Proc. Nat. Aca. Sci. 60, 86-91 (1968).

20. Quiocho, F.A., Lipscomb, W.N.: Adv. Protein Chem. 25, 1-78 (1971).

Received August 10, 1974

DISCUSSION

Kirschner: How can you exclude that two molecules of inhibitor are bound?

Navon: The equality of the binding constants obtained by either NMR or activity measurements indicate that we are looking in the NMR experiment at the same activator molecule which is also responsible for the activation effect. It is possible, however, that more carbobenzoxyglycine molecules are bound to other sites on the protein, but neither affect the activity of the enzyme nor the NMR results.

Perkins: My question is related to what Dr. Kirschner has just said. Is it not possible that you are seeing the binding of carbobenzoxyglycine to the part of the catalytic site that would be occupied by that part of the substrate if substrate were present? Would not the distances from the Mn ion then match more closely the X-ray picture of Lipscomb?

Navon: As was pointed out in my talk this is possible and it is being checked now by repeating some of the measurements in the presence of a slowly hydrolysed substrate.

Perkins: Can you suggest any mechanism by which the carbobenzoxyglycine can inhibit esterase activity while activating peptidase activity? How much inhibition of esterase activity do you get?

Navon: We do not have enough new data that will allow me to speculate on this matter. It may be better to refer you again to the article of Vallee et al. (ref. 5). In their working hypothesis part of the productive site for the esterase activity is a non-productive site for the peptidase activity. In this way, a bound ligand can both inhibit the esterase activity and interfere with the binding of peptides to the non-productive site and thereby increase their hydrolysis rates.

Veeger: I admire your courage of extending your work by studying substrate binding, but I wonder whether the presence of product will not interfere with your interpretation. Would it not be more logical to use a substrate analogue?

Navon: The substrate we are planning to use is glycyl-tyrosine which is a very slowly hydrolysed substrate. Also, its hydrolysis products - glycine and tyrosine - do not strongly bind to the enzyme. Therefore, I do not expect serious problems in this respect.

Veeger: Concerning the binding of Mn^{2+}, have you any information by which protein-residues it is liganded? Have you looked at the protein resonances, presumably by difference Fourier analysis?

Navon: Although we have no data concerning this part (except that there is only one manganese ion bound per protein molecule), it seems plausible to assume that the manganese is bound to the same site as the zinc ion (i.e. to His-69, Glu-72 and His-196). This is since: (a) The binding of these two ions is competitive, and (b) the activity characteristics of the manganese enzyme are very similar to those of the zinc enzyme. We have not looked for the protein resonances since we do not have the necessary high-field equipment.

Maass: Could you elaborate on the experimental reasons for your conclusion that the inhibitor is bound in the inner hydration sphere of the metal ion whereas the activator is being bound in the relatively labile position of the outer hydration shell?

Navon: The evidence for the conclusions you have mentioned is the following: (a) The results of water relaxation rates indicate that while inhibitors replace the one water molecule in the first coordination sphere of the manganese in Mn-CPA (R.G. Shulman et al., Proc. Natl. Acad. Sci. **56**, 39 (1966)), the activator carbobenzoxyglycine does not. (b) The manganese-proton distance obtained for the α-methylene group of the inhibitor methoxyacetic acid was 4.3 Å (ref. 19), in agreement with a direct binding of this inhibitor to the metal ion. On the other hand, the distance 5.2 Å which was obtained for the activator presently investigated is too large for an inner sphere binding. (c) The exchange rate of carbobenzoxyglycine is relatively greater than that of inhibitors of comparable binding constants in spite of its larger size.

Also it may be pointed out that carbobenzoxyglycine binds equally well to manganese and zinc carboxypeptidase while the inhibitors bind preferentially to the zinc enzyme (ref. 18) indicating that the metal ion plays a larger role in the inhibitor's binding as compared with the activator.

Pecht: Could the lecturer comment in some more detail on the mechanism by which the activator is operating? The given scheme implies a rigid insertion of the activator into the combining site and I would expect to observe an induced structural transition following the binding step.

Navon: There is no question that Vallee's model mentioned before (ref. 5) may be an oversimplification and induced changes in the protein molecule may occur. In fact, Quiocho and Lipscomb (ref. 20) do mention that they have data indicating such changes upon binding of carbobenzoxyglycine. However, we do not have any data to support this conclusion.

Section III. Repressors

Chairmen: *G. Blauer* and *E. Helmreich*

Lactose Operator Sequences and the Action of Lac Repressor

Walter Gilbert, Jay Gralla, John Majors and *Allan Maxam*

How does a repressor recognize a specific site on the DNA chromosome of E. coli and act there to prevent transcription of genes? We know that the lactose repressor, a 150,000 dalton protein of four identical subunits, searches out and binds to the control site, the lactose operator, with high precision and affinity. When the repressor is bound to the operator, transcription is prevented; when that binding is weakened by the interaction of the repressor with inducing ligands such as IPTG (isopropyl-thio-galactoside) transcription of the operon into RNA can proceed. During the last few years we have been able to isolate and to sequence the protein-DNA interaction site that constitutes the lac operator, identified, in those experiments, as a DNA fragment protected by the lac repressor against digestion with DNase. Now we have been able to confirm that identification genetically by sequencing a large number of mutant operators (O^c, operator constitutive, mutations that weaken the interaction of the repressor with DNA in vivo and in vitro). These mutant changes lie in our previously determined sequence and identify the position of base pairs relevant to the protein-DNA interaction.

Here we shall discuss our current knowledge of the DNA sequences recognized by the lac repressor protein and explore the implications that these sequences have for the mechanisms by which the repressor recognizes DNA and blocks transcription.

THE SEQUENCE OF THE LAC OPERATOR:

The lac repressor protects a 25-30 base-pair long fragment of DNA against digestion with DNase (1,2). A twenty-four base-pair sequence, covering most of this fragment, was determined by transcribing strands of this DNA with the RNA polymerase and sequencing the resulting RNA (2). That sequence is:

```
              1           5               10              15              20
5' ... T G G A A T T G T G A G C G G A T A A C A A T T  3'
3' ... A C C T T A A C A C T C G C C T A T T G T T A A  5'
```

There are a few bases still undetermined at the left end of the protected fragment; however, the right end of the minimal fragment is probably correct as given; we did not find certain pyrimidine tracts that should appear if the protected fragment extended further to the right. A striking property of this sequence is that sixteen base pairs (marked with arrows) are arranged in 2-fold symmetry centered on the base pair at position 11. Furthermore, in the longer DNA sequence containing this protected sequence eight more base pairs fall into this dyad symmetry:

```
...GTTGTGGAATTGTGAGCGGATAACAATTTCACACAGG....
...CAACACCTTAACACTCGCCTATTGTTAAAGTGTGTCC....
```

(The sequence at the left has been determined by Barnes, Reznikoff, Dickson and Abelson, private communication). Overall 24 base pairs fall in an exact 2-fold symmetry, and 2 more base pairs (between the longer blocks at positions 0 and 22) are purine-purine symmetric. There is no evidence, however, that this extended symmetry is involved in repressor binding (3).

LAC MESSENGER RNA:

Nancy Maizels (4) showed that the operator sequence appears at the beginning of the lac mRNA transcribed in vitro from the UV-5 lac promoter mutant. This mutant promoter bypasses the CAP factor requirement of the wild-type promoter and functions in *vivo* and *in vitro* to initiate the RNA polymerase under the negative control of *lac* repressor alone. (The wild-type promoter is under a further pleiotropic positive control mediated by the CAP factor, which senses the cyclic AMP level in the cell and monitors the availability of glucose and the need for alternative carbon sources). The UV-5 promoter is a double mutant, containing one mutation, L8, in the CAP interaction site and a further mutation that increases the low level promotion left in the absence of CAP function to a level comparable to that of the wild type. UV-5 DNA can be easily transcribed *in vitro* and provided a model system to study *lac* messenger synthesis.

In vitro, *lac* messenger RNA begins with a triphosphate at the adenosine marked 1 on the operator fragment, and the first 21 bases carry a sequence derived from the *lac* operator. The messenger sequence is:

```
         1        5         10        15        20
     pppA A U U G U G A G C G G A U A A C A A U U U C

 25       30        35        40        45        50
A C A C A G G A A A C A G C U A U G A C C A U G A U U
                             fmet - thr - met - ile -

      55        60
A C G G A U U C A C U G G . . .
 thr - asp - ser - leu - ala - -
```

There is a 38 base long leader sequence before the first AUG, which initiates *z*-gene protein synthesis; the first 7 amino acids of the β-galactosidase sequence are defined by the mRNA sequence. *In vitro* this messenger can start at either of two neighboring bases, beginning with pppAAUUGU... 85% of the time and with pppGAAUUGU... 15% of the time.

Clearly, *lac* mRNA synthesis begins on a region of DNA covered by the repressor and thus unavailable to the RNA polymerase when the repressor is in place. We are led to a simple picture for the control of the UV-5 promoter by the repressor: steric hindrance. What about the wild-type promoter? Chen et al. (5) suggested that control at the wild-type promoter differed significantly from that shown by a CAP independent mutant (the p^S partial revertant promoter). They argued that the repressor acts by direct competition with the polymerase on the mutant promoter, as we found for UV-5, while in CAP-dependent situations, the repressor blocked the passage of the polymerase out of the initiation site. Naturally we have sought to sequence transcripts from the wild-type promoter both to resolve this problem and to develop the capability to sequence O^c mutations that were available in the wild-type background. Technically, we were unable to transcribe successfully the wild-type promoter until we developed new ways to isolate specific DNA fragments from the *lac* promoter-operator region.

DISSECTION OF THE PROMOTER-OPERATOR REGION BY RESTRICTION ENZYMES:

Several restriction enzymes, isolated from various strains of *Haemophilus* (*H. influenzae*, *aegyptius*, and *aphrophilus*; referred to as Hin, Hae, or Hap enzymes respectively), have been of great use in isolating small specific DNA fragments that contain the *lac* promoter-operator region. We took thousand-base-pair long fragments of DNA and purified the region that contains the lactose operator by adding lactose repressor protein, passing that mixture through a cellulose nitrate filter to which repressor binds only those DNA fragments carrying the lactose operator, and then eluting those DNA fragments bound specifically to the repressor by washing the filter with IPTG (6, 1, 2). We digested these purified thousand-base long pieces with restriction enzymes and electrophoresed the digestion products on 4 or 5% polyacrylamide gels in a 0.10 *M* Tris-borate buffer, pH 8.3, 1m*M* Mg^{++}, to display fragments from the operator region. Furthermore, by rebinding the digested fragments to filters, using the *lac* repressor, and electrophoresing material that had been eluted from the filters with IPTG, we determined which of the fragments binds to the repressor. By examining partial digests, and redigesting isolated fragments with other enzymes, we worked out a restriction enzyme map of the neighborhood of the lactose operator (Gilbert, Gralla, and Maxam to be published). The pattern of cutting is:

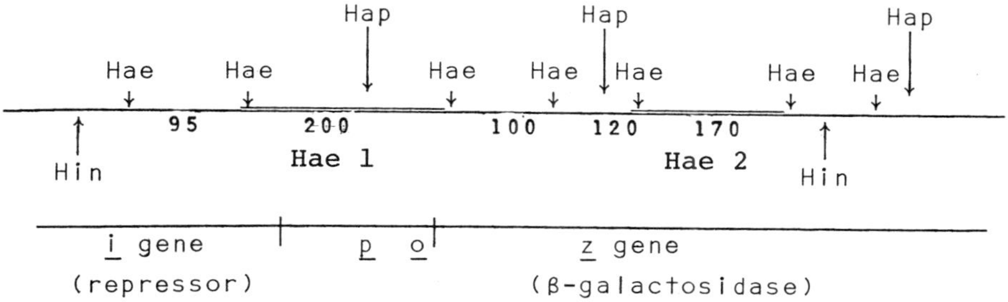

The sizes of the Hae fragments are given. The most useful enzyme has been the Hae III enzyme, which cuts at the center of a GGCC sequence (K. Murray, private communication), and yields a fragment, the Hae 1 fragment, that contains the end of the i-gene, the complete promoter-operator region, and the beginning of the z-gene. This fact was established by the following lines of evidence. In the pattern from a strain carrying the L_1 deletion mutant, a deletion that cuts into the C-terminus of the lac repressor and into the CAP factor interaction site, only the Hae 1 fragment is changed, becoming 60 to 70 bases shorter, all of the other fragments remaining unchanged. This places the end of the i gene on this fragment. Furthermore, the lac repressor binding site is on this fragment. In addition, when this fragment was isolated from the UV-5 mutant strain, it could be transcribed in vitro by the RNA polymerase to yield a messenger RNA fragment identical in sequence to those obtained from longer lac DNA fragments by Nancy Maizels. Since the messenger fragment is 60-70 bases long, the RNA initiation point must be 60-70 bases from one end of the fragment. The messenger fragment also contains the initial codons of the z gene polypeptide, starting at position

39. Therefore, this restriction fragment spans the promoter-operator region between the i and the z genes. The operator and the beginning of the z gene lie on a still smaller fragment, 80 bases long, cut by the Hap enzyme from the Hae 1 piece. This Hap cleavage was used to orient the fragment on the map.

The Hin enzyme, which cuts promoter sequences on SV40 and on phage λ (7,8) does not cut at the lac promoter. The closest Hin cuts are several hundred bases inside the i and z genes.

There is a second binding site for the lac repressor inside the β-galactosidase gene. To our surprise, when Hae fragments of lac DNA were bound to filters with repressor, not only did the Hae 1 fragment bind but so also did Hae 2. At first we worried that both these fragments contained the same operator sequence or that there was not a single controlling site for the operon, but an examination of partial products showed that these fragments were not overlapping, or adjacent, but were separated by some 200 base-pairs. Furthermore, when we eluted these fragments from the polyacrylamide gel and examined the half-life of the repressor-fragment complex, the lifetime of the repressor on the Hae 2 fragment was approximately one tenth that of the repressor-Hae 1 complex; the affinity of the repressor for the site on Hae 2 is at least an order of magnitude less than that for the Hae 1 fragment. That the Hae 1 fragment contains the "true" operator was further shown by isolating these fragments from an O^c mutation, UV-5 O^c2, that had been sequenced and was known to be in our operator sequence (3). This O^c mutation weakened the binding of the repressor to the Hae 1 restriction fragment only. Thus, we concluded that the repressor binding site on the Hae 2 fragment represents a sequence of weaker affinity than the true operator and is located 200-400 bases from the amino terminus of the z gene (5-10% of the length of this gene). Reznikoff and his co-workers (9) were the first to infer the existence of a secondary repressor-binding site inside the z gene; our knowledge of their work made easier the unravelling of the restriction fragment story. As yet there is no role known for this secondary binding site. There is no genetic evidence that it is involved in the control of the lac operon.

mRNA FROM THE WILD-TYPE PROMOTER:

The Hae 1 fragment, the 200 base long fragment containing the promoter-operator region, resolved our difficulties. A series of efforts to transcribe the wild-type promoter using the 1000 base-long sonicated fragments had not been successful; there were other sites on the sonic fragments at which CAP will stimulate the RNA polymerase to initiate, sites not related to lac control. On the Hae 1 fragment we finally were able to get CAP dependent, repressor sensitive RNA synthesis. This mRNA synthesis initiates at exactly the same base and has the same sequence as the mRNA from the UV-5 mutant. Order of addition experiments have confirmed the result predicted from the position of the initiation site: that repressor binding and the formation of an RNA polymerase initiation complex are mutually exclusive events, for both UV-5 and the wild-type promoter (J. Majors, to be published). We conclude that the repressor acts by preventing the formation of the initiation complex for RNA synthesis, by binding to and covering part of the ultimate attachment site for the RNA polymerase. The UV-5 change, which creates a new sequence that can substitute for the CAP function in aiding or permitting the RNA polymerase to get to the initiation site, must lie to the left of the beginning of the messenger, in the untranscribed region.

SEQUENCES OF OPERATOR CONSTITUTIVE MUTANTS:

The essential bases in the operator sequence define the shape that the repressor recognizes. Not all of the protected sequence is relevant, but some bases will make contacts with the protein while others may maintain specific aspects of DNA structure necessary to the interaction. These critical bases should be those that can be mutated---in which changes perturb the interaction of the repressor with DNA. To identify these positions, we have sequenced a number of operator constitutive mutations.

We began by isolating new O^c mutations in the UV-5 background. Five independent spontaneous mutations were sequenced, but only two sites in the operator were identified (3). A G·C to A·T transition at position 5 and a G·C to T·A transversion at position 9 produced O^c operators. However, the mutation at position 9 reoccurred four times---suggesting a rather discouraging prospect. Nonetheless, that 5 independent mutations fell in our sequence was reassuring.

Several years ago, Temple Smith and Jack Sadler (10, 11) isolated about a thousand lac O^c mutants. They argued that their collection contained many repeats at a few sites, and they assigned most of their mutants to seven different classes, each class being defined by a different ratio of the basal (constitutive) to the fully induced (maximal) β-galactosidase level. (This induction ratio was named the P value). Each of these classes was further subdivided into two subclasses, a and b, on the basis of a measurement of the maximal enzyme activity, which differed by up to a factor of two for some subclasses. Recombination studies suggested that each subclass represented a single site, that the two subclasses recombined with each other, and that the arrangement of the P-valued classes on a genetic map showed a 2-fold symmetrical structure whose order was:

$$\underline{i}\ \text{--}VII_a\text{-}VI_b\text{-}V_b\text{-}II_b\text{-}III_a\text{-}IV_a\text{-}IV_b\text{-}III_b\text{-}II_a\text{-}V_a\text{-}VI_a\text{--}\ \underline{z}$$

(They attributed class I, the highest level O^c mutations, to double mutations and did not map them).

Jack Sadler very kindly sent us 11 representatives of the different subclasses. We have examined each of these for changes in the operator sequence. In each case, we grew the defective phage carrying the mutation in a 150 liter fermentor, purified the phage and the phage DNA, isolated 1000 base long sonicated fragments carrying the lac operator by binding the O^c DNA to filters with the lac repressor in a buffer containing low (3 mM) Mg^{++} in order to tighten the binding, cut those purified fragments with the Hae enzyme and isolated the Hae 1 fragment from a polyacrylamide gel. We transcribed that Hae 1 template using CAP and RNA polymerase in four reaction mixtures each with a different α-^{32}P-labeled triphosphate, ran each reaction product on polyacrylamide gels in urea to separate the stutter products (4), then isolated single RNA species and ran T_1 and Pancreatic RNase fingerprints for each label. (This will be described more fully by Gilbert and Maxam, to be published). In most cases we examined the sequence extending from the beginning of the messenger through the beginning of the z gene; however, in each case we found only a single base change. Although the mutational change could often be read directly from the fingerprints, the critical spots were eluted and their sequence confirmed by further digestions. These eleven mutant strains, however, yielded only eight mutant loci, including repeats of the UV-5 O^c positions. These changes are:

```
               1         5         10        15        20
...T G G A A T T G T G A G C G G A T A A C A A T T
...A C C T T A A C A C T C G C C T A T T G T T A A
                 ↓   ↓ ↓ ↓ ↓ ↓     ↓     ↓
                 A   T G T T A     C     T
                 T   A C A A T     G     A
```

It is pleasing that these operator changes do lie in the sequence of the protected fragment. These mutations, however, lie at the center of the sequence and do not mimic the widely spaced symmetry regions.

Five of these changes are in symmetrically placed base pairs, and reduce the symmetry. However, two changes, transversions at positions 7 and 9, increase the symmetry while damaging the operator. One change, the G·C to A·T transition at position 11, is at the axis of the dyadic symmetry and was, in our estimation, a very unlikely place for a protein-DNA contact. Six out of the eight changes are from G·C to A·T or T·A base pairs. These changes may alter contacts directly or may change the interaction in more subtle ways by influencing the pitch and structure of the DNA helix.

The positions of the O^c mutations do not correspond to the reported fine-structure genetic map. Table 1 lists the mutations in more detail. Not each of the 11 O^c mutant strains contained a different sequence change. Although the representatives of class II and class III produced changes at different locations for the subclasses a and b, those mutations were not related symmetrically. The next three classes, IV, V and VI, each correspond to a single base change: IV_a and IV_b are both the transversion at position 7, V_a and V_b are the C·G to T·A change at position 17, and VI_a and VI_b are the G·C to T·A change at position 9. Furthermore, the order that we find does not resemble the order predicted. We think it likely that the genetic mapping arguments were not sufficient to order mutations in neighboring bases and to overcome marker effects.

The "a" and "b" designations refer to differing maximal levels of β-galactosidase activity; some, two-fold above the wild type. With only a single change in the operator for three such pairs, how can we explain this variation? There must be secondary mutations elsewhere that raise the enzyme activity. These could be mutations on the DNA to the left of the mRNA initiation point or could be changes in the structural gene for β-galactosidase itself. The latter is more likely, since the shift in β-galactosidase level is not fully reflected in a change in transacetylase level (10). The original selection for the O^c mutants involved the slow growth of the colony on phenyl-galactose (a non-inducing metabolite for the *lac*

Table 1

	Change 5'3'	Mutant	Class	P Value %	Lifetime i^Q	Lifetime i^{X86}
1	A T					
	A T					
	T A					
	T A					
5	G C → A T	RV1, O^c2	II_a	4.3	0.95'	30'
	T A					
	G C → T A	RV10, 112	IV_a & $_b$	1.9, 1.7	1.0', 1.2'	60', 75'
	A T → G C	RV17	III_b	2.8	1.1'	24'
	G C → T A	RV51, 52, O^c5	VI_a & $_b$	0.8	2.5', 2.6'	92', 95'
10	C G → T A	RV116	III_a	2.9	1.0'	19'
	G C → A T	RV120	II_b	4.8	0.95'	16'
	G C					
	A T					
	T A → C G	RV24	VII_a	0.4	3.8'	180'
15	A T					
	A T					
	C G → T A	RV29, 12	V_a & $_b$	1.0, 1.1	2.4', 2.3'	63', 63'
	A T					
	A T					
20	T A					
	T A					
	For the wild type: O^+			0.1	75'	480'

O^c2 and 5 are the UV-5 O^c's (3). The P values (the % constutivity; the ratio of the basal to the induced enzyme level) and the _in vitro_ lifetimes of the repressor-DNA complex are from Jobe, Sadler, and Bourgeois (12). The i^{X86} mutant repressor binds DNA more strongly.

operon) as the sole carbon source. This is a long period of constant selection pressure for further mutations to better growth.

The interaction of the repressor with DNA is not wholly governed by symmetry. Table 1 lists the *in vivo* P value (induction ratio) for each of these O^c mutations, and values for the *in vitro* lifetimes of the repressor-DNA complex, taken from Jobe, Sadler, and Bourgeois (12) (also Bourgeois *et al* this volume). The P values (the degree of constitutivity) and the lifetimes show the same trends: the stronger O^c's have shorter lifetimes, corresponding to weaker affinity of the repressor for the altered operators. The shortest lifetimes probably do not reflect the binding of the repressor to the mutant operator but reveal its interaction with the secondary binding site under the *z* gene. In general, O^c mutations on the left half of the operator are stronger than those on the right. Specifically, the symmetric changes in the sequence, the pair (5, 17) and the pair (8, 14) do not have the same P values but have (4.3%, 0.8%) and (2.8%, 0.4%), respectively.

Is this difference due to the fact that the RNA polymerase is trying to enter from the left? Probably not, because the repressor-operator lifetimes also show the same bias: the repressor binding is affected less by the changes on the right-hand side. However, it is still possible to argue that symmetric regions on the repressor multimer are trying to interact with symmetric regions on the DNA. The O^c mutations at positions 7 and 9 are mutations to a more symmetric sequence that shows a weaker interaction. If the symmetry is relevant, the original base pairs at positions 13 and 15 must be bad contacts. The effects of these bad contacts on the right-hand side could be to make difficult the formation of the rest of the symmetric contacts on the right. Thus the repressor would interact well on the left side, badly on the right, and the effects of mutations in disrupting that interaction would be drastic on the left side, small on the right. This model makes an explicit prediction: mutations at base pairs 13 and 15 to increase the symmetry should generate a better operator sequence, one that binds the repressor far more strongly than does the original.

CONCLUSION:

The sequences of operator constitutive mutants do not support the notion that the repressor simply recognizes a symmetric region on DNA. Rather, they show that the protein senses the detail of an irregular sequence of bases. The interaction may still have a symmetric component, reflecting a symmetry (assumed) of the protein tetramer, but this symmetry is not dominant. On recognizing and binding to the operator, the lac repressor covers a sequence transcribed into the initial portion of the lac messenger. Thus the repressor functions by blocking access of the RNA polymerase to its initiation site.

ACKNOWLEDGEMENTS

We thank J. R. Sadler for the O^c strains; Clyde Hutchison, Tom Maniatis, and Dean Hamer for providing Hemophilus strains and samples of enzymes; Edith Butler, Joanna Knobler and Annette MacKay for technical assistance; and the NIGMS, grant # GM09541, for their support of this work. W.G. is an American Cancer Society Professor of Molecular Biology. J.G. was supported by a Jane Coffin Childs Fellowship.

REFERENCES

1. Gilbert, W.: The lac repressor and the lac operator. In Polymerization in Biological Systems, the Ciba Foundation Symposium 7 (new series), ASP, Amsterdam, pp. 245-256 (1972).
2. Gilbert, W. and Maxam, A.: The Nucleotide Sequence of the lac operator. P.N.A.S. 70, 3581-3584 (1973).
3. Gilbert, W., Maizels, N., and Maxam, A.: Sequences of Controlling Regions of the Lactose Operon. Cold Spring

Harbor Symposium, 38, 845-856 (1973).

4. Maizels, N.: The Nucleotide Sequence of the Lactose Messenger Ribonucleic Acid Transcribed from the U V 5 Promoter Mutant of Escherichia coli. P.N.A.S. 70, 3585-3589 (1973).

5. Chen, B., de Crombrugghe, B., Anderson, W.B., Gottesman, M.E., Pastan, I., Perlman, R.L.: On the Mechanism of Action of lac Repressor. Nature New Biology 233, 67-70 (1971).

6. Bourgeois, S. and Riggs, A.: The Lac Repressor Operator Interaction IV: Aassay and Purification of Operator DNA. Biochem. Biophys. Res. Commun. 38, 348-354 (1970).

7. Allet, B., Roberts, R.J., Gesteland, R.F., and Solem, R.: Class of Promoter Sites for Escherichia coli DNA-dependent RNA Polymerase. Nature 249, 217-221 (1974).

8. Maurer, R., Maniatis, T. and Ptashne, M.: Promoters are in the Operators in Phage Lambda. Nature 249, 221-223 (1974).

9. Reznikoff, W.S., Winter, R.B., Hurley, C.K.: The Location of the Repressor Binding Sites in the lac Operon. P.N.A.S. 71, 2314-2318 (1974).

10. Smith, T.F. and Sadler, J.R.: The Nature of Lactose Operator Constitutive Mutations. J. Mol. Biol. 59, 273-305 (1971).

11. Sadler, J.R. and Smith, T.F.: Mapping of the Lactose Operator. J. Mol. Biol. 62, 139-169 (1971).

12. Jobe, A., Sadler, J.R. and Bourgeois, S.: Lac Repressor-Operator Interaction IX. The Binding of Lac Repressor to Operators Containing O^c Mutations. J. Mol. Biol. 85, 231-248 (1974).

Received August 15, 1974

DISCUSSION

Wagner: Is it feasable to replace some of the thymidines in the operator by 5-fluoro- or 5-bromo-deoxyuridine and to test how the repressor recognizes this change?

Gilbert: Yes. Lin and Riggs made 5-bromo-deoxyuridine-substituted operator and found a 10-fold tighter binding of the lac repressor.

G. Weber: Is there evidence as to a "unit of recognition" by the repressor, that is base pair: single base or molecular group?

Gilbert: There is no evidence yet. As Suzanne Bourgeois will describe, it is very unlikely that there is any melting of the operator on binding, therefore, it is not likely that the repressor recognises the open bases. The current hypotheses are that groups on the protein make hydrogen bonding contacts to groups on the outside of the base pairs, the specific placement of such bonds being sufficient to recognize the base pair. However, the final structural solution may show that specific sets of amino acids may make a series of contacts with a single base pair.

Changeux: Is it well established that four repressor subunits bind per operator?

Gilbert: No, but there is no suggestion in any experiment that any structure smaller than the tetramer binds. The lac repressor remains tetrameric in dilutions down to at least $10^{-10} - 10^{-11}$ M. At these concentrations the molecule that is released from the operator by IPTG is still a tetramer.

Steinhardt: Would it be fair to say that there are two elements of the recognition problem: 1. establishing the necessary elements of the recognition pattern; and 2. determining the stereochemical requirements for being able to use the pattern? If so, it would appear that the approach so splendidly presented by you cannot distinguish between these two elements, and may therefore suggest a more complicated recognition pattern than is actually required.

Gilbert: The pattern has to be complicated enough to recognize a rather long sequence - a sequence long enough to generate the required specificity (i.e. one that does not reoccur - or reoccur too often - on the DNA of E. coli). This means that at least 12 base pairs are recognized specificically (and possibly a greater number are recognized with partial specificity).

Sund: The repressor consists of four identical polypeptide chains and also four active sites. According to your model only two polypeptides chains are necessary for the contact with the DNA. Therefore, it seems that the repressor belongs to that group of "enzymes" which exhibit half-site-reactivity. What is your idea about the physiological significance that only half of the sites are used?

Gilbert: The significance probably lies in that two or four IPTG molecules are ultimately involved in pulling the repressor off the DNA. This spreads the energy requirement around. The binding of repressor is weakend by a factor of 20,000, i. e. by 6 kcal by the binding of four IPTG's, each binding with a $K_D = 10^{-6}$ M. This means that each IPTG needs to distort the protein by a few kcal. The sigmoidal induction curve, furthermore, may be of physiological use in the cell.

One should recall that the DNA binding site is very large - a groove 20 Å in diameter and 90 Å long - it may take contacts from all four subunits to create the binding site.

Kirschner: Is it feasible to use electron microscopy to determine whether the isolated operator exists - at least partially in a Gierer-type double hair pin?

Gilbert: No. The resolution of the electron microscope is not good enough to resolve details at this level.

Kirschner: Have you done difference-ultraviolet spectra between the O-R complex and the separated components?

Gilbert: No. The amounts of operator available have been in the 1/100'ths of an O.D. range, a small fraction of a µg. This has only recently changed, larger amounts are now available.

Knippers: How does the CAP protein fit into your "competitive binding" model?

Gilbert: The CAP factor binds to a DNA site, the RNAP site (J. Majors). We do not know how the CAP factor acts in detail: either it touches the polymerase protein and creates more contacts and hence a better binding site or it melts slightly the DNA or changes the DNA structure so that the polymerase can bind. In either case, the repressor still competes directly with the polymerase for the initiation site.

von Hippel: Evolutionary notion of either approaching symmetry or diverging from symmetry at operator site to "titrate" to the intermediate (and "appropriate") binding constants for R to O. Could you comment on how you might approach the analysis of this evolutionary problem from mutant sequence analysis?

Gilbert: If we can obtain mutant operators to which the repressor binds more strongly, that will show that the current operator is not at a maximum for the binding. If that "superoperator" is actually lac negative, then it will be clear that it is not evolving from a more symmetrical operator. An evolutionary picture might follow the line that a primordial repressor consisted of a molecule with a DNA binding site recognizing a sequence of no particular symmetry. A mutation in the protein to produce a multimer then doubles the size of the DNA binding site - and increases the binding by the non-specific component. Further mutation of the sequence under the second part of the protein would then let the operator evolve toward the much tighter and more specific operator that we now observe.

Steinhardt: What is the principal evidence that more than two sites affect the strength of the repressor-DNA binding? Might not the mutations at other sites have indirect effects, as by changing angles and distances by small amounts, reducing the time-averaged probability of the true sites being enganged simultaneously? This question has a bearing on what are the elements of the recognition pattern as distinguished from the steric requirements.

Gilbert: Genetics can only show the underlying informational structure - we learn that the exact base pairs at 8 sites control, in some way, the intention, but we do not know anything yet about the steric considerations that really are involved. Each of these mutations that we have studied might interfere with the binding by an indirect change in the structure - or might interfere by the loss of a hydrogen bond or by the interposition of a methyl group. Indirect effects, such as a change in pitch created by the change in the central base-pair, throwing all of the contacts slightly out of register, are attractive.

Brahms: In order to explain how the mutations which you described are affecting the binding of the repressor, I would like to indicate that recently we have determined in polydeoxynucleotides the orientation of the phosphate when you have alternating and alternating homopolynucleotide sequences rich in A and T versus various DNA phosphate orientation (Pilet, Blicharski and Brahms, Biochemistry 1974). It appears that the orientation in these A,T - rich polymers of the phosphate group O$\hat{P}$O bisector is about 5° lower than in DNA. This may explain the differences in your mutants when you replace G or C to T and A and this may influence the binding of protein to the phosphate group of DNA.

Bourgeois: How do you explain that Chen et al. did not observe competition between RNA polymerase and repressor, a result which suggested at the time that there was no overlap between the wild type promoter and the operator?

Gilbert: Possibly, the CAP and RNAP interaction on the wild-type promoter does not yield as stable a complex as the polymerase + UV-5, a less stable complex could have produced a misleading result in their experiment; our experiments were done at higher polymerase and repressor concentrations.

Pohl: Referring to the possible symmetry of DNA binding, is there anything known whether one repressor molecule binds to two operators?

Gilbert: So far, no experiment has successfully shown two DNA binding sites.

The Active Sites of Lac Repressor

B. Müller-Hill, T. Fanning, N. Geisler, D. Gho, J. Kania, P. Kathmann, H. Meissner, M. Schlotmann, A. Schmitz, I. Triesch and *K. Beyreuther*

How do proteins bind specifically to DNA? Are there rules which regulate the recognition of DNA sequences by proteins, and if so, what are they? These are questions we have asked ourselves during the last few years. We have analysed the lac repressor-lac operator-system of E.coli (1). It is well characterized chemically and physically. Lac repressor is a tetrameric protein with two different active sites: one to recognize the operator, and the other to recognize the inducer, a small sugar molecule. Each of its identical subunits is composed of 347 amino acids of known sequence (2). The operator is a stretch of 27 base pairs. Its sequence is also known (3,4,5). It shows a twofold symmetry (3,4,5).

MUTANTS THAT MAKE MORE LAC REPRESSOR

For the biochemical analysis it became necessary to work with mutants that made more lac repressor than wild-type strains. We used two such mutants. One (i^q) makes ten times more (6) and another (i^{q1}) makes one hundred times more lac repressor than wild type. These stable mutations were crossed onto a special lac-transducing prophage to increase the dosage of the i gene per cell. Table 1 shows that under optimal conditions 7% of the protein of E.coli can be made lac repressor.

THE THREE MAJOR TYPES OF POINT MUTANTS OF THE i GENE.

We have tried to characterize the two active sites of lac repressor genetically and chemically (8,9,10). For this purpose we made use of mutants of the i gene which produce altered lac repressor. A priori, all point mutants of the

Table 1. Specific activity of lac repressor in crude extracts.

i alelle	A	B
i^+	0.1	1
i^q	1.0	10
i^{q1}	10	100

A: Strain $[lacpro]_\Delta$F'lacpro was grown in rich medium at 33°.

B: Strain $[lacpro]_\Delta \lambda h80c_{I857}S7dlac$ (or a similar strain) was heat-induced as in ref.6. Specific activity of lac repressor was determined as in ref.7. The specific activity of pure lac repressor is 1300 (7).

structural part of the i gene might be classified as belonging to one of the following groups:

1) Mutants in which lac repressor is unable to bind operator, but able to bind inducer. We will call them i^{-op}.
2) Mutants in which lac repressor is unable to bind inducer, but able to bind operator (i^s).
3) Mutants in which lac repressor is unable to bind inducer and operator. We will call them $i^{-op,s}$.

The selection, detection and mapping procedures for such mutants have been described elsewhere (9,10). Since binding sites do not simply disappear, the mutant classes might be subdivided further according to the change in the binding constant of the defective binding site.

MUTATIONS WHICH MAKE A DEFECTIVE OPERATOR BINDING SITE.

We want to identify those amino acids of lac repressor which make direct contacts with the base pairs or the backbone of the operator. Since the same amino acids will presumably not make direct contacts with inducer we concentrated our effort on mutants which did not bind operator, but which still bound inducer. Mutations, however, which alter (A) the aggregation, (B) freeze the lac repressor in the inducer-binding conformation, or (C) distort the general structure of lac repressor might appear to have the same properties. Mutants in which the aggregation of lac repressor is altered have been looked for, but so far not found among i^{-op} mutants (11,12 and our own observations). Mutations which freeze lac repressor in the inducer binding conformation might map near the subclass of i^{s} mutations which freeze repressor in the operator binding form. We will see later that such mutants have probably been found.

The most troublesome group of mutants are those in which the general structure of repressor is distorted: they could map anywhere. We need a criterion to recognize such mutants and to exclude them from our analysis. We propose here that the strength of an i^{-op} mutation (i.e. the factor by which the dissociation constant of repressor-operator complex is increased) is adequate for this purpose. We argue that an exchange directly in or very near a contact site might sometimes reduce the operator-repressor binding drastically, whereas the chance for such a strong effect through long-distance interaction will be smaller.

How much will an exchange at or very near a contact point reduce the binding to operator? Lac operator has a twofold symmetry. It is reasonable to assume that if one amino acid binds to a particular base pair on one side of the operator, the corresponding amino acid of another subunit binds to the corresponding base pair on the other side of the operator. Since operator constitutive mutations (o^{c}) reduce the binding of lac repressor to operator 5- to 200-fold, the corresponding repressor mutants will reduce the binding by at least 25- to 40000fold. This is a minimal estimate, because some amino acid exchanges are sterically more drastic than the exchanges of base pairs. Furthermore, the exchange of one base pair should not influence too much the secondary structure of neighbouring base pairs. An amino acid with a larger side chain, however, might change the positions of amino acids nearby. This effect will be particularly large, if, what we think is quite reasonable, amino acids which are located near each other in the sequence are involved in the recognition of neighbouring base pairs.

An i gene mutant which is fully constitutive and which carries a normal i^+ promoter produces repressor which has its affinity to operator decreased by a factor of at least 10^4, a fully constitutive i^q by a factor of at least 10^5 and a fully constitutive i^{ql} by a factor of at least 10^6. Considering this, we decided deliberately to concentrate our analysis upon strong i^q and i^{ql} constitutives which still bound inducer in vitro.

42 $i^{ql-op,d}$ MUTANTS ARE FOUND AMONG 62 i^{ql} CONSTITUTIVES.

62 i^{ql} lac constitutives were isolated after mutagenesis with N-methyl-N'-nitro-N-nitrosoguanidine (NG). We did not use any selection techniques, but simply picked deep blue colonies from plates containing bromo-chloro-indolyl galactoside (13). Extracts of 42 of the mutants still bound the inducer isopropyl-thio-ß-D-galactoside (IPTG) in vitro. These mutants were by definition i^{ql-op}. The same 42 mutants were all transdominant (i^{ql-d}, see below). They all gave positive Ouchterlony tests with antibody against lac repressor. Mapping placed all 42 mutations in deletion groups 1 to 7 of the 39 deletion groups (Fig.1). The remaining 20 $i^{ql-op,s}$ mutants, which did not bind inducer in vitro were not transdominant. They gave no precipitation lines in a similar Ouchterlony test, indicating that repressor was either grossly deformed and/or destroyed by peptidases. Mapping placed 14 of these mutations between deletion groups 21-33, 2 mapped between deletion group 6-8, one between deletion group 9-10, 3 between deletion group 10-20 and one between deletion group 32-39 (14).

STRONG i^{q-d} MUTANTS HAVE DEFECTIVE OPERATOR BINDING SITES.

The above findings confirmed an analysis we had published earlier (8,9,10). We had isolated 247 transdominant lac constitutives (i^{q-d}) after NG mutagenesis. We had found that such

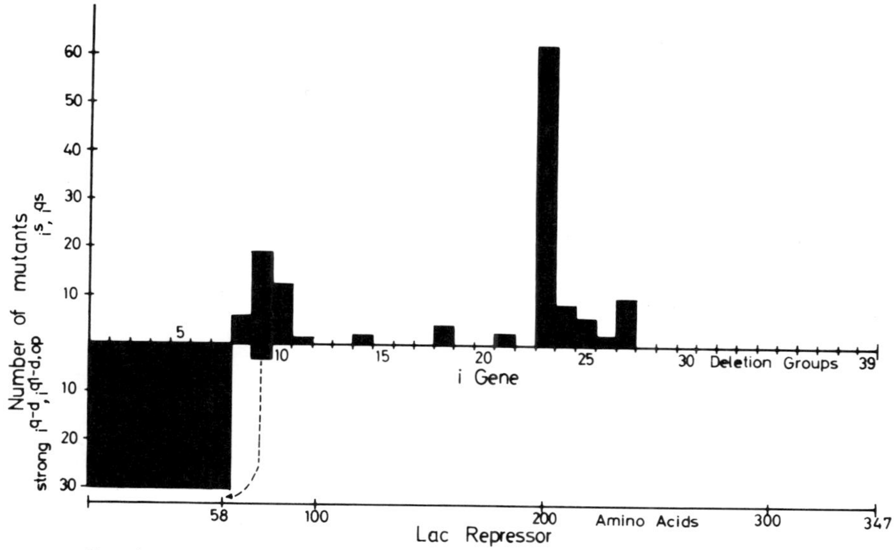

Figure 1. Map of the i gene and sequence of lac repressor. For details see figure 2.

transdominants did generally bind inducer, but not operator in vitro. Over 90% of our i^{q-d} mutants were strong constitutives and strong transdominants and mapped in the first nine deletion groups (Fig.1)! This has been recently confirmed by Sadler and his coworkers (12). With a selection system which included very weak constitutives they isolated and analysed 350 i^{q-d} mutants. They found all their strong i^{q-d} mutants to map in deletion group 1-7. They found, as we did (8,9,10), modestly weak i^{q-d} mutants to map in deletion groups 1-13. Even weaker ones they found to map all over the i gene.

There is a great deal of circumstantial evidence indicating that transdominance is the result of the formation of mixed oligomers between i^{+} and i^{-d} repressor. We have tested i^{-d} repressor for inhibition of operator binding of i^{+} repressor and found none, so subunit mixing must happen directly after synthesis. The fact that all 42 i^{ql-op} mutants were transdominant supports our contention (8) that strong transdominant constitutives are mutants with a defect in the operator binding site, but with the protein core and often the inducer binding site intact. $i^{q-d,op,s}$ mutants which have also a defective inducer binding site might be defective in the region where the effect inducer binding has upon operator binding is transmitted from one binding site to the other.

THE AMINO ACID EXCHANGES IN REPRESSOR FROM i^{q-d} MUTANTS.

The first i^{q-d} mutants have been sequenced by Weber and his colleagues (11,16,17,18). We have sequenced a few more. Fig. 2 shows all the exchanges which have so far been determined. We have also included exchanges which are known to occur by suppression of nonsense mutants (14,16,17,18). Some nonsense mutants can be effectively suppressed by various suppressors, indicating that the particular amino acid is not essential. We show these amino acids in brackets. We have underlined all amino acids which are potentially exchangable after NG mutagenesis (see below). To date, no mutants from deletion group six have been sequenced. It will be interesting to see whether they will occupy the entire region between residue 17 and 53. BG135 and BG78 are strong i^{q-d} mutants mapping closest to the C-terminus of lac repressor. Here we found both possible C/G to T/A exchanges. This makes us reasonably sure that we did not miss other strong i^{q-d} mutants to the right of amino acid 58. We conclude that the amino acids which make contact with the base pairs and the backbone -i.e. the active site for operator binding- is located somewhere between residue 1 and 58. BG124, BG185 and BG200, weak constitutives with similar amino acid exchanges, are probably examples for distortion of the binding site through moderate distance (see also below).

METHODOLOGICAL CRITIQUE: MUTAGEN SPECIFICITY, HOTSPOTS AND MAPPING ACCURACY.

Implicit in our analysis was the aim to analyse as many mutants as possible, i.e. all possible mutants of the particular type we were looking for. We used N-methyl-N'-nitro-N-nitroso-guanidine(NG) as mutagen since folklore had it that NG would do everything, transitions and transversions. Now we suspect that NG exchanges predominantly (we found in seven of

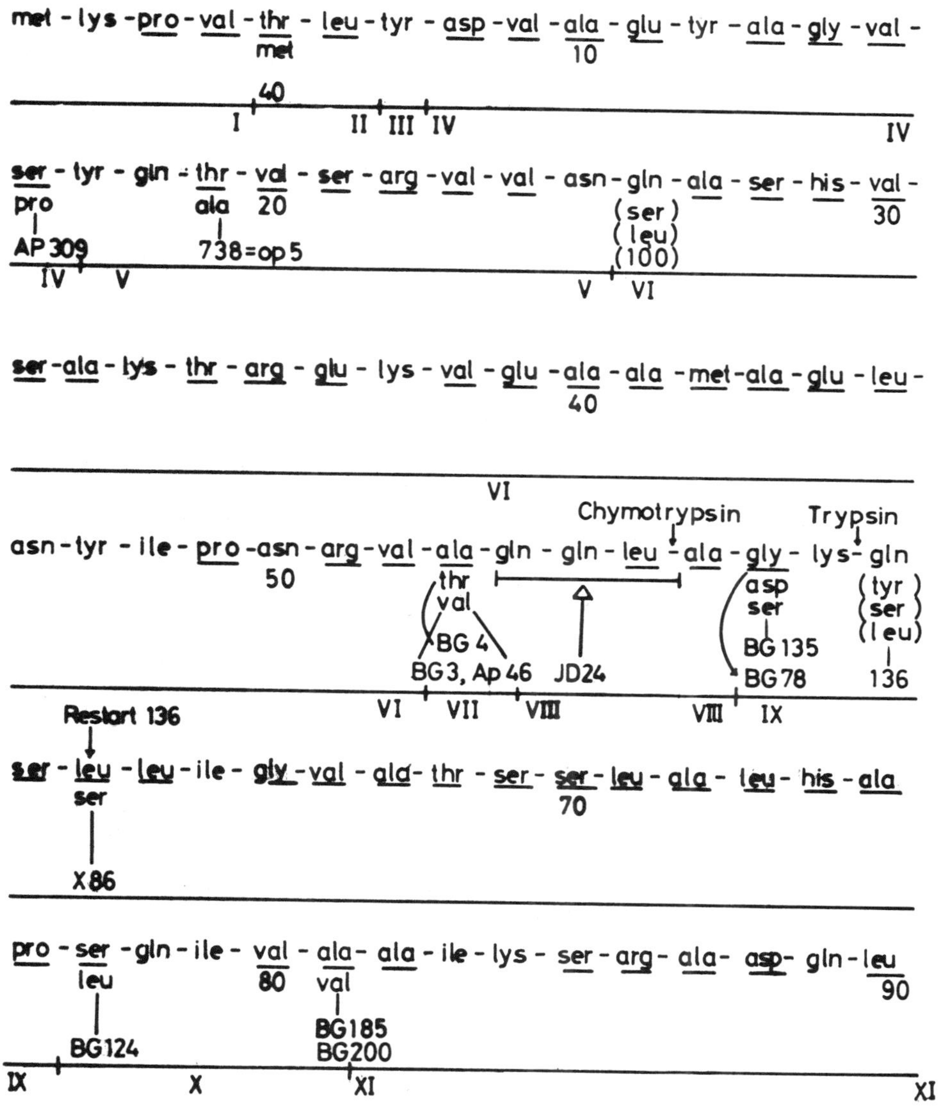

Fig.2: Sequence of lac repressor and map of the i gene. The amino acids underlined can be potentially exchanged after NG mutagenesis. Amino acids underneath the sequence indicate observed exchanges. Amino acids in brackets indicate exchanges which do not destroy operator binding. Roman numbers under the line indicate the deletion groups. Mutant names are given above the line. BG4; BG3; BG78; BG124; BG185; BG200 and JD24 (9,10) are i^{q-d} mutations whose lac repressor has been sequenced by us. The other mutations and exchanges are from the work of Miller, Weber and their colleagues (11,16,17,18).

seven cases) C/G pairs to T/A pairs. Therefore we expect the following amino acid exchanges: ala-> thr,val; arg-> cys,glu, his,lys,trp; asp-> asn; cys-> tyr; glu-> lys; gly-> arg,asp, glu,ser; his-> tyr; leu-> phe; met-> ile; pro-> leu,ser; ser-> asn,leu,phe; thr-> ile,met; val-> ile,met; we expect no exchange or nonsense codons with asn, gln, ile, lys, phe, tyr, trp and sometimes with leu. The exchanges are limited, but still quite a few amino acids which are inert to chemical modification may be exchanged for others. Since many amino acids can be exchanged we are rather certain that the location of strong i^{-d} or i^{s} mutants will not change quantitatively if we use other mutagens. We have probably not missed any major regions. Hotspots are also no real danger for this type of analysis of missense mutants. That we found an ala-> val exchange twice at residue 81 indicates that we have begun to saturate the map. The fact that this exchange was found in mutations BG200 and BG185 which mapped in the 10th or 11th deletion group shows that our mapping system was not perfect. Such wrong assignments are bound to happen if the resolution is not absolute. All deletion groups might be pushed a few base pairs to the left. We have been informed by J.Miller, who has developed a more refined mapping system, that this is true for deletion group 8.

SMALL AMOUNTS OF IODINE INACTIVATE THE OPERATOR BINDING SITE. We have analysed the inactivation of operator binding of lac repressor by iodine. We found that very low concentrations of iodine (2 bound iodine atoms per subunit) inactivate lac repressor binding to operator completely. Inducer binding remains intact under these conditions. Lac operator containing DNA gives a modest protection against inactivation. Digestion of native iodinated lac repressor with trypsin (as in ref. 20) located 90% of the radioactive iodine in the amino terminus. We found 39% of the label in tyr 7, 20% in tyr 12 and 31% in tyr 17 after automatic Edman degradation. Only 10% remained in the repressor fragment after 21 cycles of Edman degradation.

FAB FROM ANTIBODY AGAINST LAC REPRESSOR INACTIVATES OPERATOR BINDING.

We prepared Fab from rabbit antibody against lac repressor. Antibody or Fab inhibited operator binding of lac repressor. Inducer binding under the same conditions was increased 30%. We found that we could relieve half of the inhibition if we preincubated the antibody with cyanogen bromide peptides from lac repressor covering the sequences 2-42 and 43-98. It is not astonishing that each of the peptides had an effect since they were contaminated to about 10% with each other, but with no other ones from lac repressor (19).

DIGESTION OF NATIVE LAC REPRESSOR WITH PEPTIDASES.

Weber and his collaborators have shown that trypsin cleaves native repressor at position 59 (20). We found that there is another cut at position 312. We have also determined the effect of chymotrypsin on native repressor. It cleaves predominantly (52%) at position 56. But in minor fractions the core starts at positions 17 (26%) and 45 (22%). If we treat lac repressor in the presence of antiinducer (21) o-nitrophenyl-ß-D-fucoside (ONPF) with chymotrypsin, we find a pure core beginning with residue 57. IPTG has no such effect.

MOST i^s MUTANTS HAVE A DEFECT IN THE INDUCER BINDING SITE.

We isolated and mapped 135 i^s mutants which did not bind inducer, but still bound operator (10,14). Most of them were isolated after mutagenesis with NG and the use of a selection procedure which asks for total ($K_i > 10^{-1}$) inability to bind inducer (9,10). Fig.1 shows the map positions of these mutants. Most of the i^s mutants map in the middle of the i gene in deletion groups 21-27. They probably outline the cleft in which inducer binds. Others map in deletion groups 8-11. Below, we present arguments that these i^s mutations cause their effect not at the inducer binding site proper but at a region of lac repressor which is involved in allosteric transition.

THE AMINO TERMINUS AND ITS DISTORTIONS DURING INDUCTION AND REPRESSION.

There seem to be two subregions of the N-terminus which are involved in operator binding. One includes the first 25 amino acids, the other lies between amino acid 45 and 60. Induction probably occurs at the latter subregion. Let us recall here that leu56-ala57 are made more available to chymotrypsin by ONPF. We found several i^s mutations to map in deletion group 8 (10,14). If the mapping was correct, these mutants might have exchanges in leu56 or ala57. These i^s mutants make lac repressor which is frozen in the operator binding form. This is emphazied by the fact that restart 136 which begins with amino acid 63 binds IPTG normally (17). The mutation X86 produces repressor which binds operator only in the presence of IPTG (15); here leu 62 is exchanged by a serine.

Several of our i^{q-d} mutations produce repressor in which the inducer binding constant is decreased (BG7 in deletion group 7; BG21 in deletion group 9; BG56 in deletion group 10 ref.10.) The N-terminus might be so distorted in these mutant repressors that neither operator nor inducer bind properly. We should, however, not forget that some of these mutants might be double mutants since they were isolated after NG treatment. Finally, the two weak transdominant i^{q-d} BG200 and BG185 (mapping ref.10) having both ala81 exchanged by a valine produce repressor with a decreased binding constant for the anti-inducer ONPF. The decrease is modest, i.e. by a factor of 3 to 5, but we have not found a similar decrease in any of 20 other i^{q-d} repressors.

The experiments with the peptidases indicated that the N-terminus can be easily disconnected from the core and that the core has its own stable structure. We have asked ourselves whether the N-terminus is a separate, independent, ball-like structure or whether it is an arm that extends, in an ivy-like fashion, across the surface of the repressor core. If the former were true we should find temperature-sensitive i^{q-d} mutants. 150 i^{q-d} strains were checked between 25° and 37° and only one was found to be modestly temperature sensitive (22). In the absence of any comparable study this may be a hint for an extended, ivy-like structure.

THE CORE OF LAC REPRESSOR AND THE INDUCER BINDING SITE.
We know the genetic locations of the i^{s} mutations, but not much chemically about the inducer binding site. Besides i^{s} mutations there is another interesting mutant known to map in the core area: Novicks i^{TSS} maps in deletion group 23 (9). i^{TSS} mutants make lac repressor that is temperature sensitive only as it is made (23). Such mutants might define the sequences which are necessary for nucleation in the folding of the polypeptide chain.

THE C-TERMINUS.
The fact that we have not found strong i^{q-d} mutants to map at the C-terminus makes us doubt that it is directly involved in operator binding. It may, however, have some general structural function since i-z^{+} fusions (24), i-z^{-} fusions (14) and ICR derived i^{-}mutants (14) loose operator binding activity as soon as the defect enters the i gene beyond deletion group 39. Note, however, that operator binding is not harmed if ß-galactosidase is pinned to the tail of lac repressor. This happens both with active (24) and inactive (14) ß-galactosidase. We have analysed the product of one i^{+}-z^{+} fusion chemically and found the second from last tryptic peptide of lac repressor still present, whereas the last one had disappeared. This implies that no more than 5 amino acids have been cut from lac repressor in this fusion.

HOW DOES LAC REPRESSOR BIND TO LAC OPERATOR?
Two years ago we presented a model of lac repressor-lac operator interaction (8). Parts of the model stood the test of time, others were shown to be wrong.
We concluded at the time that the recognition between operator and repressor was made between amino acid 1 and 50 and nowhere

else. Today, we have to expand the binding site further to amino acid 58. We concluded that 50 amino acids would not make a traditional cleft-like recognition unit, but would act more like a protrusion. If we can call an ivy-like arm a protrusion we would like to stick to that. We assumed at the time that the DNA might be in the B conformation This may be wrong (25). We concluded that tyr 17, gln 18, ser 21, asp 25 and gln 26 were involved in binding. Gln 26 was shown to be inessential (11) and asp 25 was deamidated in our "prep" and is in fact an asn residue (11). The majority of the exchanges in the i^{q-d} repressors are found in region 4-17, so we still think that tyr 17, gln 18, ser 21 and asn 25 might be involved in binding - in addition to tyr 7 (tyr 12 has found to be inessential, personal communication from J.Miller). However, we missed the site around gln 54 and gln 55, which is not surprising since we did not know the sequence at the time.

The most serious mistake we made was to couple disbelief in Sadlers evidence for an operator with twofold symmetry (26) with a general dislike for such a symmetry. Now, after Gilbert and his colleagues have sequenced the operator and proven the twofold symmetry, we think that the N-terminus of one subunit winds itself "ivy-like" around one half of the operator. This would leave two of the subunits idle or almost idle, which is compatible with the model of Steitz (27). The possibility of helical protein regions interacting in a (modified?) deep grove of DNA seems still attractive to us, particularly since the helical parts of the lac repressor were shown to lie outside the core (Fasman, personal communication). Finally, there is yet no evidence for or against rules for DNA-protein recognition. An X-ray analysis of lac repressor obviously has to be made. Our mutant that makes 7 percent lac repressor might solve the logistic problem for those who want to crystallize.

We thank Deutsche Forschungsgemeinschaft for support through SFB 74.

REFERENCES

1) Jacob,F. Monod,J.: Genetic regulatory mechanisms in the synthesis of proteins. J.Mol.Biol. 3, 318-356 (1961)

2) Beyreuther,K. Adler,K., Geisler,N., Klemm,A.: The amino acid sequence of lac repressor. Proc.Nat.Acad.Sci.USA 70, 3576-3580 (1973)

3) Gilbert,W., Maxam,A.: The nucleotide sequence of lac operator. Proc.Nat.Acad.Sci.USA 70, 3581-3584 (1973)

4) Maizels,N.M.: The sequence of the lactose mRNA transcribed from the UV5 promoter mutant of E.coli. Proc.Nat. Acad.Sci.USA 70, 3585-3589 (1973)

5) Gilbert,W., Maizels,N., Maxam,A.: Sequences of controlling regions of the lactose operon. Cold Spring Harbor Symp. 38, 845-855 (1973)

6) Müller-Hill,B., Crapo,L., Gilbert,W.: Mutants that make more lac repressor. Proc.Nat.Acad.Sci.USA 59, 1259-1264 (1968)

7) Müller-Hill,B., Beyreuther,K., Gilbert,W.: Methods in Enzymology, eds. Grossman,L. and Moldave,K. (Academic Press, New York) Vol.21, Part D, pp 483-487

8) Adler,K., Beyreuther,K., Fanning,E., Geisler,N., Gronenborn,B., Klemm,A., Müller-Hill,B., Pfahl,M., Schmitz,A.: How lac repressor binds to DNA. Nature 237, 322-327 (1972)

9) Pfahl,M.: Genetic map of the lactose repressor gene (i) of Escherichia coli. Genetics 72, 393-410 (1972)

10) Pfahl,M., Stockter,C., Gronenborn,B.: Genetic analysis of the active sites of lac repressor. Genetics, in press

11) Weber,K., Platt,T., Ganem,D., Miller,J.H.: Translational restarts. AUG reinitiation of lac repressor fragment. Proc.Nat.Acad.Sci.USA 69, 897-901 (1972)

12) Miwa,J., Sadler,J.R., Smith,T.F.: A characterization of i^{-d} repressor mutations of the lactose operon. J.Mol. Biol., in press.

13) Davies,J., Jacob,F.: Genetic mapping of the regulator and operator genes of the lac operon. J.Mol.Biol. 36, 413-417 (1968)

14) Gho,D.: Dissertation, Köln (1974)

15) Chamnes,G.C., Willson,C.D.: An unusual lac repressor mutant. J.Mol.Biol. 53, 561-565 (1970)

16) Platt,T., Weber,K., Ganem,D., Miller,J.H.: Altered sequences changing the operator binding properties of the lac repressor: Colinearity of the repressor protein with

the i gene map. Proc.Nat.Acad.Sci.USA 69, 3624-3628 (1972)

17) Ganem,D., Miller,J.H., Files,J.G., Platt,T., Weber,K.: Reinitiation of lac repressor fragments at a codon other than AUG. Proc.Nat.Acad.Sci.USA 70, 3165-3169 (1973)

18) Files,J.G., Weber,K., Miller,J.H.: Translational re-initiation: Reinitiation of lac repressor fragments at three internal sites early in the lac i gene of Escherichia coli. Proc.Nat.Acad.Sci.USA 71, 667-670 (1974)

19) Schmitz,A.: Dissertation, Köln (1974)

20) Platt,T., Files,J.G., Weber,K.: Lac repressor: specific proteolytic destruction of the N-terminal region and loss of the deoxyribonucleic acid binding capacity. J.Biol. Chem. 248, 110-121 (1973)

21) Müller-Hill,B., Rickenberg,H.V., Wallenfels,K.: Specificity of the induction of the enzymes of the lac operon in Escherichia coli. J.Mol.Biol. 10, 303 (1964)

22) Meissner,H.: Diplomarbeit, Köln (1974)

23) Sadler,J.R., Novick,A.: The properties of repressor and the kinetics of its action. J.Mol.Biol. 12, 305-327 (1965)

24) Müller-Hill,B., Kania,J.: Lac repressor can be fused to ß-galactosidase. Nature 249, 561-563 (1974)

25) Chan,H.W., Wells,R.D.: Structural uniqueness of lactose operator. Nature, in press

26) Sadler,J.R., Smith,T.F.: Mapping of the lactose operator. J.Mol.Biol. 62, 139-169 (1971)

27) Steitz,T.A., Richmond,T.J., Wise,D., Engelman,D.: The lac repressor protein: molecular shape, subunit structure, and proposed model for operator interaction based on structural studies of microcrystals. Proc.Nat.Acad.Sci. USA 71, 593-597 (1974)

Received July 2, 1974.

Note added in proof: During this symposium we have been informed by J.Miller that mutant X 86 recombines with deletion 8632 (deletion group X, Fig. 2). This suggests that the functional amino acid exchange in X 86 occurs behind amino acid 70 and the exchange in position 62 is probably due to a strain variation.

DISCUSSION

Changeux: I was not clear about the quaternary structure of the fused repressor-ß-galactosidase molecule. Are there two repressor and two ß-galactosidase subunits per molecule or four of each?

Müller-Hill: We do not know whether the fusion is a dimer or a tetramer. Its sedimentation constant (determined on a sucrose gradient) is close to 16 S, the value we find also for galactosidase. We might expect a sedimentation constant of 17.5 S for the fusion, but this is almost within experimental error. We have not used the linking technique of Davis and Stark, but we will do so soon.

Sund: The ß-galactosidase-repressor fusion product is still associated. Therefore, one can imagine that either the ß-galactosidase part or the repressor part is associated as in the unmodified protein and that the other part does not - or only partly - interact with its other polypeptide chains:

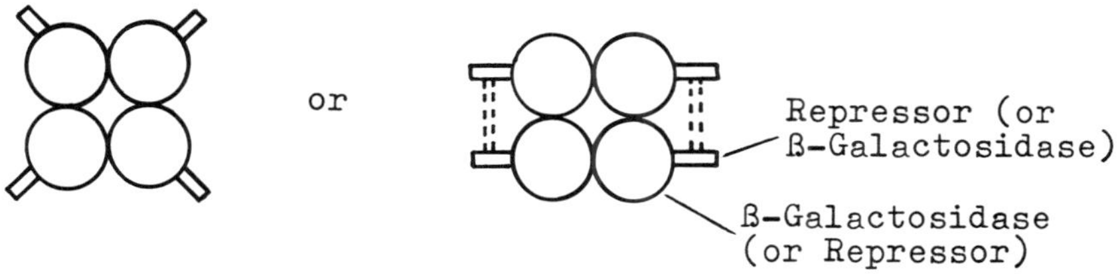

Müller-Hill: One can imagine various structures and indeed we find several peaks of fusion activity on phosphocellulose columns and also higher aggregates on sucrose gradients. If the 16 S material is tetrameric (which is not proven) it has to aggregate in a way that gives at least normal dimeric repressor or possibly even tetrameric repressor with the galactosidase ends sticking out. In any case, we would love to know its structure(s).

Riesner: Is also the unspecific binding to DNA destroyed in mutants, in which the binding to the operator is destroyed?

Müller-Hill: Nonspecific DNA binding can be measured in various ways: one can look for phosphocellulose binding, which is rather unspecific, or for DNA cellulose binding, which is already a bit better. We have found phosphocellulose binding comparable to wild-type repressor with repressors from the strong i^{q-d} strains BG1, BG3, BG4, BG78, BG135 and with the weak i^{q-d} strains BG56, BG126, BG200 and BG52. All these exchanges (with possible exception of BG1 which is not yet sequenced) happen to occur in amino acid residues later than 52. In fact, all late mapping i^{q-d} repressors bound normally to phosphocellulose. On the other hand, most of the mutants which mapped closer to the N terminus produced repressor which bound badly, if at all, to phosphocellulose. We have also measured the binding of some of these repressors to DNA cellulose and have found so far a close parallel to phosphocellulose binding. I might add here: the restart and core proteins which have been checked by Weber and us show no phosphocellulose binding. One gets the impression that the exchanges in amino acids 53 and 58 which do not change the unspecific DNA binding destroy one of the subregions where specific contacts are made, maybe the GlN54-GlN56 region.

Knippers: Do you know whether the quaternary structure of the mutant repressor molecules is changed?

Müller-Hill: All i^{-d} or i^{-op} lac-repressors we have looked at so far were tetrameric. The same holds true for the mutant repressors which Weber and his collaborators analysed.

Wagner: One could imagine that part of the affinity between repressor and operator is contributed by precise fitting of oppositely charged groups. Can you extract from your results information about the influence of the replacement of charged amino acids upon repressor-operator recognition?

Müller-Hill: Our data are not sufficient for such an analysis.

Kirschner: Has the (1-58) fragment been isolated? If so, does it retain any specific affinity for the operator? Is there any evidence that it has a folded structure (effects of temperature or urea on the spectrum)?

Müller-Hill: The BrCN fragments 2-42 and 43-98 bind both non-specifically to DNA in the filter assay (Schmitz, Thesis, Cologne 1974). So far, no physico-chemical measurements have been made with these or other fragments.

Bourgeois: You propose that the amino-terminal region of the repressor forms an "ivy-like arm" which would interact with the operator. How strong are the arguments for preferring such an "arm" to, say, a protrusion and how do you envisage this 60 amino-acid residue-long arm to be a stable structure in the absence of operator?

Müller-Hill: The first 62 amino acids are not necessary for folding of the core of lac repressor. We proposed first that they form an independent little structure - a protrusion, which does not need the core to find its own conformation. I have become a little doubtful about this recently. The reasons for my doubt are rather weak, its the negative evidence of not finding certain mutants one might expect: we did not find any temperature-sensitives among our large set of i^{q-d}'s, that is, no mutants which formed functioning lac repressor at low temperature (i.e. at 22°). One might expect an independent structure to heat-denature independently of the core. I tried also rather hard to obtain fusions between i and z which would be phenotypically i^s, that is putting the "protrusion" on a suitable galactosidase fragment. The failure to find such mutants may hint against a totally independent protrusion and for a structure which needs the presence of repressor core to function.

Lac Repressor

K. Weber, J. G. Files, T. Platt, D. Ganem and *J. H. Miller*

Lac repressor is a tetrameric molecule with 347 amino acids per polypeptide chain. Lac repressor has two different active sites, or to stay within the nomenclature of this symposium, the repressor has two different ligands: one site recognizes specifically the lac operator, the other the operon inducer, the small sugar derivative isopropyl-B-D-thiogalactoside. The amino-acid sequence of the lac repressor (1) and the DNA sequence of the operator (2) are known. The molecular interaction of the two macromolecules however, is only vaguely understood. We are summarizing here some properties of lac-repressor derivatives which are missing the amino terminal part of the molecule. These derivatives were either obtained *in vivo* as the product of translational re-start mutants (3,4,5) or *in vitro* by limited proteolytic digestion of lac repressor under native conditions (6,7). The properties of these molecules argue strongly, that the amino terminal sixty amino acids are not involved in IPTG binding or in the ability of the polypeptide to fold correctly and to form a tetramer, but that the amino-terminal part is necessary for DNA binding. Direct protein chemical mapping of several i^{-d} mutants supports this conclusion.

Translational Reinitiation Mutants

This class of mutants was originally detected in a collection of spontaneous nonsense mutants (3). Interestingly, some of these nonsense mutants mapped early in the i-gene but displayed negative complementation in diploid tests with the wild-type i-gene in a nonsuppressing (su-) background. Negatively complementing activity is thought to reflect an interaction of the mutant lac-repressor subunits with the wild-type lac-repressor subunits due to mixed oligomers. This suggests that early amber mutants which can negatively complement must give rise to translational reinitiation past the amber termination signal. In order to verify this model we isolated the negatively complementing gene product from a variety of these early amber mutants. Purification of the mutant lac repressor was achieved by immune precipitation of the mutant protein followed by SDS-polyacrylamide electrophoresis. Although this approach allows the isolation of only small amounts of proteins, it has the advantage of a rapid protein chemical characterization of the mutant protein by its polypeptide chain molecular weight. The mutant protein recovered from the gel can be used to obtain an amino-terminal sequence of the protein. The characterization of the mutant lac repressor by molecular weight and amino-terminal sequence together with the known wild-type amino acid sequence allows the identification of the translational reinitiation codon.

Our results (3,4,5) can be summarized as follows. Amber mutants corresponding to positions 7, 12 and 17 of the lac repressor polypeptide chain give rise to two translational restarts (5). One has been found to occur in the codon corresponding to valine residue 23 (Π_1), while the second site is the first internal in phase AUG codon corresponding to methionine residue 42 (Π_2). All three amber mutants activate both Π_1 and Π_2 simultaneously. If the amber codon is moved

1				5					10		12
met -	lys -	pro -	val -	thr -	leu -	tyr -	asp -	val -	ala -	glu -	tyr -
				(met)		amber					amber

13		15					20				24
- ala	- gly	- val	- ser	- tyr	- gln	- thr	- val	- ser	- arg	- val	- val
			(pro)	amber	amber	(ala)				Π_1 →	

25					30					35	36
- asn	- gln	- ala	- ser	- his	- val	- ser	- ala	- lys	- thr	- arg	- glu
	amber										

37			40					45			48
- lys	- val	- glu	- ala	- ala	- met	- ala	- glu	- leu	- asn	- tyr -	ile
					Π_2 →						

49	50					55					60
- pro	- asn	- arg	- val	- ala	- gln	- gln	- leu	- ala	- gly	- lys	- gln
				(val)							amber

61				65					70
- ser	- leu	- leu	- ile	- gly	- val	- ala	- thr	- ser	- ser -
amber	Π_3 →								

Fig. 1. Amino-terminal sequence of the Lac-repressor (1,3,4,5,6,8). Amino acids underneath the sequence (including "amber") indicate observed changes (3,4,5). Π_1, Π_2 and Π_3 indicate translational re-initiation codons (3,4,5). For a discussion of the patterns of suppression of the amber mutants see Miller, this symposium.

beyond the Π_1 boundary as in mutant 100, which maps at position 26, translation reinitiation occurs only at Π_2 and no other restart material is observed (3). Amber mutants mapping beyond the Π_2

boundary, like 136 (position 60), lose the $\overline{II}_2$ restart and open a new restart at the codon corresponding to leucine residue 62 ($\overline{II}_3$)(4).

The three restart lac-repressor mutant proteins $\overline{II}_1$, $\overline{II}_2$ and $\overline{II}_3$ are missing 23, 42 and 61 amino acid residues in comparison to the wild-type polypeptide chain. However, all three proteins show unimpaired binding of the inducer IPTG and form tetramers, i.e. they sediment with a sedimentation coefficient corresponding to a molecular weight, which is four times the value of their polypeptide chain molecular weight. Furthermore, at least the $\overline{II}_2$ protein can be reversibly denatured and renatured after exposure to 8M guanidine-HCl in the presence of 0.01 M ß-mercapto-ethanol.

Digestion of Native Repressor

When lac repressor is treated for a few minutes under native conditions with either trypsin or chymotrypsin a rather limited number of peptides are released (6,7). In an extensive study of the digestion products, we showed that trypsin cleaves the peptides at the amino terminus leaving a tryptic core starting with residue 60. A second set of three peptides spanning the carboxy terminal twenty residues is also released. Characterizations of the tryptic core shows that the core is still a tetramer and shows unimpaired IPTG-binding activity, but is unable to bind to DNA. Furthermore, the tryptic core can be reversibly denatured and renatured in 6M guanidine-HCl in the presence of 0.01 M ß-mercapto-ethanol.

Protein Chemical Mapping of Early Mutants

The limited action of trypsin on native repressor has provided a simple procedure to map point mutants in the first 59 residues of lac-repressor. We have used this system to map the early amber mutants of lac repressor after suppression with a suitable suppressor as well as a small number of i-^{d} point mutants. Following the isolation of the mutant repressor and "native tryptic digestion" the resulting peptides are separated by fingerprint techniques and the peptides are further characterized by amino-acid composition and sequence analysis. Usually, 2 mg of protein are sufficient to detect the point mutation. This approach to mapping of early mutants has been very effective and we have localized point mutations in positions 5, 7, 12, 16, 17, 18, 19, 26, 53, 60 and 61, so far. Mutants mapping early in the core can be found by automatic Edman degradation of the core (Files and Weber, unpublished).

Discussion

From the experiments reported above, we would like to draw two conclusions:

1. The three-dimensional structure of lac repressor has to be such that the amino terminal sixty residues form a loop, or globule, which is available to trypsin and chymotrypsin which is not necessary for folding, oligomerization or IPTG binding, but which is required for DNA binding (see discussion below).
2. Translational reinitiation past a nonsense codon appears to be a common event in E.coli since there are at least three such sites in the first 70 in-phase codons of the i-gene messenger RNA and all amber mutants found in this region induce translational restarts. Only one of these restarts involves the classic AUG codon; the other two involve an *in vivo* ambiguity of the genetic

code, in that the same codon can be translated into two different amino acids depending on whether it is recognized during initiation or elongation of protein biosynthesis. The two non-AUG codons are the codons corresponding to leucine residue 62 and valine residue 23. Thus it is likely that codons of the type XUG can act as translational initiation codons.

From our experiments on lac repressor we want to propose the following restriction on the possible three-dimensional structure of lac-repressor and its action in DNA binding and inducer binding.

1. The amino-terminal part of the first (at least) 60 residues forms a structure different from the rest of the molecule. This structure may be necessary for the correct conformation of the lac repressor to bind to DNA or it may be directly involved in DNA binding or both.
2. The amino-terminal part of the molecule is *not* necessary for binding of IPTG (see the properties of π proteins and of the tryptic core).
3. The amino-terminal part is *not* required for tetramer formation (see the properties of π proteins and of the tryptic core).
4. The amino-terminal part is *not* required for the correct folding of the polypeptide chain, i.e. repressor derivatives obtained *in vivo* (π proteins) and *in vitro* can be reversibly denatured and renatured.
5. i^{-d} mutants map in the first 60 residues of the lac-repressor polypeptide chain and repressor derivatives

missing the first 22, or 42, or 61 residues (τ proteins) and the tryptic core do not bind to operator. Although numerous DNA-binding mutants have been shown to occur in the first 60 residues, we cannot yet exclude the possibility that amino acid residues in the later part of the polypeptide chain can participate in DNA binding. Thus it seems likely that the original proposal that DNA binding resides in this part of the molecule is correct (8,3), but it is not excluded that other parts of the molecule may participate either directly or indirectly in DNA binding.

References

1) Beyreuther, K., Adler, K., Geisler, N. and Klemm, A.: The amino acid sequence of lac-repressor. Proc. Nat. Acad. Sci. USA 70, 3576-3580 (1973)

2) Gilbert, W., and Maxam, A.: The nucleotide sequence of lac-operator. Proc. Nat. Acad. Sci. USA 70, 3581-3584 (1973)

3) Platt, T., Weber, K., Ganem, D., and Miller, J.H.: Translational restarts. AUG reinitiation of a lac-repressor fragment. Proc. Nat. Acad. Sci. USA 69, 897-901 (1972)

4) Ganem, D., Miller, J. H., Files, J. G., Platt, T., and Weber, K.: Reinitiation of lac-repressor fragments at a codon other than AUG. Proc. Nat. Acad. Sci. USA 70, 3165-3169 (1973)

5) Files, J. G., Weber, K., and Miller, J. H.: Translational reinitiation: Reinitiation of lac-repressor fragments at three internal sites early in the lac i gene of Escherichia coli. Proc. Nat. Acad. Sci. USA 71, 667-670 (1974)

6) Platt, T., Files, J. G., and Weber, K.: Lac-repressor: specific proteolytic destruction of the amino-terminal region and loss of the deoxyribonucleic acid binding capacity. J. Biol. Chem. 248, 110-121 (1973)

7) Weber, K., Platt, T., Ganem, D., and Miller, J. H.: Altered sequences changing the operator binding properties of the lac-repressor: Colinearity of the repressor protein with the i gene map. Proc. Nat. Acad. Sci. USA 69, 3624-3628 (1972)

8) Adler, K., Beyreuther, K., Fanning, E., Geisler, N., Gronenborn, B., Klemm, A., Mueller-Hill, B., Pfahl, M., and Schmitz, A.: How lac-repressor binds to DNA. Nature 237, 322-327 (1972)

Received August 19, 1974.

DISCUSSION

Veeger: I wonder whether your elegant purification procedure does not have the danger to purify away the initial part of the peptide chain, e.g. from the N-terminus till the mutant position. One could visualize, especially when the mutation occurs around position 60, that this smaller peptide is in interaction, like in RNAse, with the rest of the protein. By isolating the protein by a method without SDS-treatment and heating, measurements of the molecular weight by sedimentation might give an answer.

K. Weber: I would like to give three arguments:
1. In the case of the amber mutant in tyrosine residue 7 the amino terminal amber peptide can only be a hexapeptide. Thus, in the Π_2 protein in this mutant residues 7 to 42 are missing.
2. In case of the native tryptic digest we know that gentle gelfiltration separates the core from the tryptic peptide spanning residues 1 to 22.
3. Denaturation followed by renaturation of the Π_2 protein as well as the tryptic core.
For more details, see references 3,4,5 and 6.

Bourgeois: Upon trypsinization of repressor, didn't you observe some trypsin core resulting from a cut after lysine residue 314, i.e. a core reduced by 33 $-COO^-$ terminal residues rather than by only 20 residues? If so, do you have any information about the properties of this smaller trypsin core?

K. Weber: We don't have such information at the moment.

Brahms: Do you have an idea about the isoelectric point of the fragment (of 60 amino acid residues)?

K. Weber: The amino terminal part is slightly more basic than the average part of the repressor polypeptide chain. Please note that the amino acid sequence shows that this part of the molecule is not like a histone.

Riesner: Can you protect the lac repressor against tryptic degradation by DNA?

K. Weber: We have not yet found the appropriate conditions.

Bourgeois: Did you measure the affinity of the trypsin core for an anti-inducer, such as ONPF?

K. Weber: No.

Woody: Is there any information about the rates of initiation with these abnormal initiation codons relative to the normal one?

K. Weber: If we use the total IPTG-binding activity found in the extract, we can put the level of internal initiation in these amber mutants at approximately 5 to 10 % of the corresponding wild-type level.

Wagner: Selecting your reinitiation codons you obviously made the assumption that the recognition of the initiator codon involves "wobble" in the first position of the codon. Is there more evidence for this?

K. Weber: Yes, but we do not have further evidence for the codon assignment. Final proof should come from a study of the appropriate frameshift mutants or from an extended study of missense mutants. The fact that mutant repressor X86 has a leu→ser exchange in position 62 (see Müller-Hill et al., this symposium) reduces the number of possible leu codons for Π3 from 6 to 2, one of which is UUG.

Heilmann: Is it possible to detect gene products in the wild type which are caused by starting translation at GUG or UUG codons?

K. Weber: We don't find such gene products in the wild type. This result can be interpreted to mean, that if such products do exist, then they must be made at a level lower than 2 % of the amount of the wild-type product, i.e. the full, normal polypeptide chain. In order to explain this finding, we assume that the internal translational initiation sites are not available to initiating ribosomes in the wild-type situation, since transcription and translation may be coupled, i.e. ribosomes bind directly to the normal initiation site and start translation and therefore involve the internal sites in elongation before they can act in initiation. If we introduce the amber mutation, we slow ribosomes down at the stop codon, which is localized prior to the internal start codon. We assume that now the II site is activated and can act in initiation since it is not involved in elongation. For a more detailed discussion, see references 3,4 and 5.

von Hippel: Ratios of sedimentation constants of core against "native" tetramer might tell you how loose and flexible the arms are.

K. Weber: Maybe such an answer could be obtained.

The Use of Suppressed Nonsense Mutations to Generate Altered Lac Repressor Molecules

J. H. Miller, C. Coulondre, U. Schmeissner, A. Schmitz and *P. Lu*

INTRODUCTION

How do specific changes in a protein's structure affect its function? This question takes on added interest when applied to proteins that interact specifically with DNA. One class of these proteins, repressors, controls the expression of genes by binding to specific regions of the DNA, operators, in response to intra-cellular signals. The lac repressor, the protein product of the lacI gene in E. coli, is particularly well suited for a study of how changes in structure result in changes in function, since the full amino acid sequence is known (1) and because genetic studies have provided a great deal of information concerning different types of mutations and their effects on the protein molecule (2,3,4,5,6,7).

A direct approach towards answering the above question would be to substitute different amino acids at many known positions in the repressor, and to examine the effects of these changes in vivo and in vitro. Although the sequencing of proteins altered by missense mutations, where single amino acids have been changed, can give some of this information (8,9,10), we have exploited a more flexible approach making extensive use of nonsense mutations.

The ultimate answer to the question of structure and function will require the eventual knowledge of the three-dimensional structure of repressor. However, the lesson of the studies on hemoglobin is that a number of variants or mutants of a protein and their effects are needed for us to understand the relation of structure to function (11). Fortunately, a parallel effort by a large number of laboratories over several decades provided such a variety of hemoglobins with altered structures and properties. We present here a method that is capable of collecting the same or perhaps greater variety of mutant variants of a protein with considerably less work. It is based on the suppression of nonsense mutations.

Nonsense mutations cause polypeptide chain termination resulting in the production of a protein fragment (12). Three nonsense codons (UAG, UAA, and UGA) exist in *E. coli*. Mutants have been isolated carrying suppressor mutations which reverse the effect of nonsense mutations. In these strains a tRNA molecule has been altered so that it now inserts a specific amino acid at a nonsense site (see review by Gorini, 13). By suppressing amber (UAG) mutations with characterized nonsense suppressors, as many as five different amino acids can be inserted at each site; serine, tyrosine, glutamine, leucine, and lysine (13). Ochre mutations (UAA) can be suppressed with suppressors inserting lysine, tyrosine, or glutamine, and can be converted to UGA mutations which would enable the insertion of tryptophan by a UGA suppressor (13). The lac repressor contains 347 amino acids, and we have now found over 70 different positions in the *lacI* gene which have been converted to amber or ochre codons. We have recently developed a system which enables us to assign many nonsense mutations to specific residues independent of and consistent with protein sequencing (14). As a result, any new amber and ochre mutations in the *lacI* gene can be correlated with specific amino acids in the lac repressor. In this paper we summarize the properties of some of the more striking mutants created by suppression of nonsense mutations. A full report analyzing all of the substitutions at each of the 70 sites will appear elsewhere (14).

Establishing the position of a series of nonsense mutations enables us to define in turn the absolute position of specific deletion groups. The resulting gene-protein map contains 65 deletion intervals corresponding to known intervals in the repressor protein (14,15). Because the average interval length is approximately 5 amino acids, additional mutations can be assigned to specific regions of the protein.

Although we can induce multiple substitutions by suppression at many amino acids, only those amino acids whose codons can be converted to nonsense (UAG, UAA, or UGA) can be examined. Therefore, to complement this study we have also isolated and characterized missense mutations. Several studies by different authors (10, 16) have focused on attempts to find fully constitutive mutations which result in repressors that still bind IPTG (the gratuitous inducer, isopropyl-thiogalactoside) and then to examine these for the ability to form aggregates. All of the mutations which fall into this category so far described have been found to map in the part of the gene coding for approximately the first 70 amino amino acids (7,10,16). Because only one mutagen was used in these studies (nitrosoguanidine), we decided to employ a variety of different mutagens and to map the mutations into one of 65 deletion groups prior to biochemical screening. By analyzing several mutations per deletion group derived by each respective mutagen, we attempt to maximize the number of different mutants being screened. (A complete description of the isolation, characterization, and mapping of these mutations will be published elsewhere) (15). We show in this paper that in addition to the early part of the protein, at least one other region (between amino acids 260 and 273) exists in which mutations can be found which are fully constitutive and result in repressors with partial or full IPTG-binding activity in crude extracts.

As we show here the combined use of suppressed nonsense mutations and missense mutations offers a wider range of substitutions than either technique alone.

RESULTS

A) Constitutive Mutations Created by Suppression

Each of the different nonsense mutations has been tested with 4 different amber suppressors and 3 different ochre suppressors. Because we have isolated the nonsense mutations on an F'lacpro factor carrying the i^Q mutation which overproduces the lac repressor 10-fold (17), we are able to compensate for the inefficiency of most of the suppressors. Therefore, we expect to overproduce the mutant repressor with respect to the wild-type i^+ promoter from 3-6 fold in amber suppressor strains when the episome carries an amber mutation in the lacI gene. Approximately wild-type amounts of repressor should be produced in strains carrying less efficient ochre suppressors when the episome carries an ochre mutation in lacI. Many sites in the repressor show various degrees of tolerance for alternate amino acids. In this preliminary report we mainly describe those mutations which fail to be well suppressed by all suppressors, except those inserting the original wild-type amino acid. If we consider amber sites, where suppression results in overproduction of the repressor and also confine ourselves to very large effects on the control of the lac enzymes, then we greatly lessen the chance that we are dealing with effects caused by the failure of a particular suppressor to work at full efficiency at a certain site. Demonstration of the presence of altered repressor molecules in these cases can of course eliminate this possibility.

Using a set of isogenic suppressor strains, the i^- phenotype was scored on Xgal indicator plates (18) and verified by direct assay of beta-galactosidase. Figure 1 depicts most of the positions in the lac repressor corresponding to points where nonsense mutations have been found. In cases where the mutations have been correlated with known sites in the protein, we have indicated the wild-type residue. Open circles represent unidentified nonsense mutations. Table 1 shows some of these results. A full report including assay results of all of the substitutions at each site will appear elsewhere (14).

SUBSTITUTION SITES IN THE LAC REPRESSOR

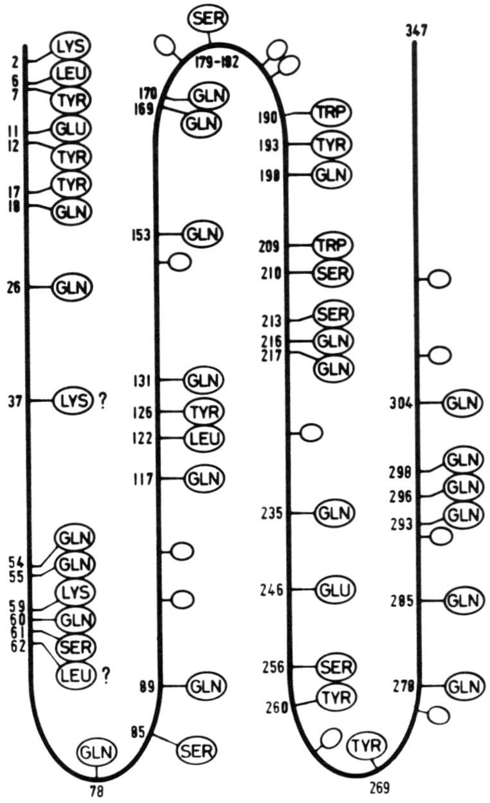

(The change at position 85 has been reassigned to position 77).

Figure 1.

Table 1

Wild-type amino acid-position	Suppressor- inserted amino acid						
	Su1 (Ser)	Su2 (Gln)	Su3 (Tyr)	Su6 (Leu)	SuB (Gln)	SuC (Tyr)	Su5 (Lys)
Tyrosine - 17	-	-	+	-			-
Glutamine - 18	-	+	-	-			-
Glutamine - 54					+	-	-
Glutamine - 55					+	-	-
Serine - 77	+	ts	ts	ts			-
Tryptophan - 190	-	-	ts	-			-
Glutamic Acid - 246	ts	ts	+/-	-			-
Serine - 256	+	ts	-	-			-
Tyrosine - 269	-	-	+	-			-

Table II

	Su1 (Ser)	Su2 (Gln)	Su3 (Tyr)	Su6 (Leu)	SuB (Gln)	SuC (Tyr)	Su5 (Lys)
Tyrosine - 7	+/-	+	+	+/-			-
Tyrosine - 12	+	+	+	+			-
Tyrosine - 17	-	-	+	-			-
Tyrosine - 126	+	+	+	+			-
Tyrosine - 193	+	+	+	+			+/-
Tyrosine - 260					i^r	+	-
Tyrosine - 269	-	-	+	-			-

It can be seen that specific sites (where only the original amino acid restores wild-type activity) are found at many positions throughout the protein. At this stage the data given here are not quantitative enough to permit many detailed conclusions regarding repressor structure. However, we can already ask which of the tyrosines are important. Table II groups the results of substitutions at each of 7 of the 8 tyrosines. It can be seen that tyrosine 12, 126, and 193 are dispensible, whereas tyrosines at positions 17, 260, and 269 are necessary for repression. The tyrosine at position 7 is replacable by some of the amino acids tested, but not by others. Both tryptophan residues have also been substituted. Position 190 shows a high degree of specificity with regard to constitutivity (Table 1), whereas the tryptophan at position 209 is essential for induction by IPTG (see next section, Table III).

B) Noninducible (i^s) Mutations

Several sites in the repressor yield i^s repressors (those which cannot be fully induced by IPTG) when substituted with certain amino acids. We have listed some of these in Table III. Here we score their properties on lactose-MacConkey indicator plates and on Xgal medium with and without IPTG. Beta-galactosidase assays have verified the i^s character of these mutations (14). It can be seen that some sites become i^s with all substitutions except the wild type, while others retain partial or full inducibility with some substitutions while becoming i^s with others. Several of these sites have not previously been found after missense mutagenesis. Possible explanations for this are considered in the discussion.

C) i^r Mutations

Mutations have been described which reverse the normal role of IPTG in induction, and have been termed i^r mutations (5). In the absence of IPTG, strains carrying these mutations are constitutive but certain concentrations of IPTG restore repression. One of these mutations, X86, results in repressor molecules which bind the lac operator more tightly (6). The ochre mutation which we have assigned to the codon normally specifying tyrosine at position 260 produces a repressor with an i^r phenotype when suppressed with suB, which inserts glutamine. At 30° in the absence of IPTG this mutant is constitutive, but increasing amounts of IPTG restore repression. At 37°, however, the mutant is constitutive even in the presence of IPTG.

Table III

Wild-type amino-acid position	Su1 (Ser)	Su2 (Gln)	Su3 (Tyr)	Su6 (Leu)	SuB (Gln)	SuC (Tyr)	Su5 (Lys)
	Suppressor - amino acid						
Serine - 61	+	ts	i^s	i^s			-
Glutamine - 78					+	+	i^s
Serine 179, 180 or 182	+	i^s	i^s	i^s			-
Tryptophan - 209	i^s	i^s	$+/i^s$	i^s			-
Glutamine - 235	i^s	+	i^s	i^s			i^s
Leucine 305, 306 310 or 317	$+/i^s$	$+/i^s$	i^s	+			-

D) Constitutive Missense Mutations

Previous genetic studies on the active sites of repressor have focused on mutants which have lost one activity of the repressor (such as operator binding) but still retain other functions (IPTG binding, tetramer formation). Until now, the only point mutations that have been described which are fully constitutive and which still bind IPTG map in the beginning of the protein (7,9,10,16). Most of these have been selected by screening for transdominant i^- mutations after NG mutagenesis, or by screening simply for i^- mutations which still bind IPTG. We have also screened missense mutations for IPTG-binding activity but taking note of the limited specificity of nitrosoguanidine (see Discussion) and of the tendency for mutagens to preferentially induce mutations at certain sites ("hotspots"), we have decided to examine a cross section of missense mutations in lacI induced by a variety of mutagens; 2-aminopurine (2AP), ultraviolet light (UV), mutT. To permit the maximum number of different sites to be screened, we first eliminated nonsense mutations and then mapped the remaining mutations into one of 65 deletion intervals, as well as testing for different levels of negative complementation. Mutations in one deletion group could often be divided into different subgroups based on their levels of negative complementation. A total of 700 independently induced mutant (at least 100 mutants for each mutagen) were analyzed in this manner. Subsequently, one mutation per deletion group or per subgroup derived from each mutagen was tested for IPTG-binding activity. In this manner we have screened over 80 mutations mapping throughout the lacI gene for IPTG-binding activity. The results for the 2AP, UV, and mutT-derived mutants are shown in Figure 2. It can be seen that there is a region late in the protein in which mutations can be found that produce repressors unable to repress but still able to bind IPTG. This region extends from amino acids 260-273 based on a correlation of the relevant deletion groups with this region of the protein (14). Because tyrosine 269 appears to be an essential amino acid (Tables I,II), we might expect that some of these mutations actually effect substitutions at this position.

DISCUSSION

In this paper we have shown that the use of suppressed nonsense mutations is a powerful tool for creating altered proteins. By inserting different amino acids at specific sites with

MISSENSE SUBSTITUTIONS IN THE LAC REPRESSOR

(△ = UV ● = mutT ○ = 2AP)

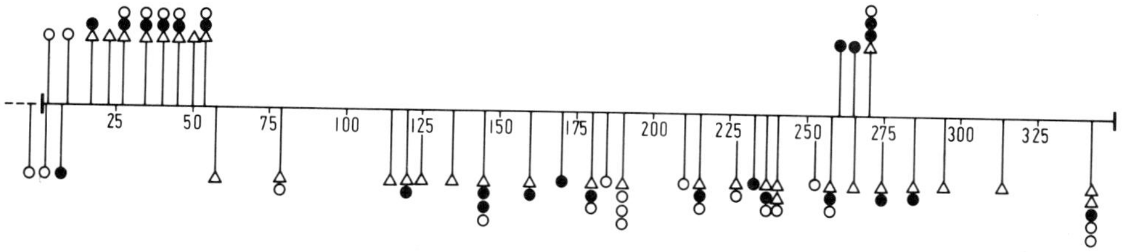

Figure 2: Each vertical line represents a set of mutants resulting from mutations in a specific deletion group or subgroup. Circles and triangles depict mutants isolated after treatment with different mutagens that have been screened biochemically for IPTG-binding activity. All mutants above the line have partial or full IPTG-binding activity. Mutants below the line have no detectable activity (less than 8% of wild type). The distances are given in terms of amino acids of the lac repressor protein and have been determined to a degree of accuracy of approximately 5 amino acids (14,21).

nonsense suppressors a wide variety of interesting mutants can be obtained. In fact, with this technique we have been able to find mutants corresponding to most of the previously described lac repressor mutants. Careful examination of the properties of these altered molecules will be important for the study of which amino acids of the repressor are involved in each of its active sites, since we can deduce the position of each altered residue in the protein and also the exact nature of the amino acid change (14). The information obtained from all of these suppressed derivatives should greatly enhance the understanding of different aspects of the eventual three-dimensional structure of the repressor determined by X-ray crystallography.

The use of suppressed nonsense mutations as a means of obtaining altered proteins has several advantages when compared with conventional missense mutagenesis, particularly when used on an extensive scale and when coupled with a system enabling the exact location of the nonsense sites. Some advantages:
1. Multiple substitutions are possible at each site. In fact, 5 different residues can be inserted at each nonsense site. The isolation of new suppressor strains would allow additional amino acids to be inserted.

2. Certain codons are difficult to convert to specific missense mutations with many mutagens (see discussion below) but can easily be converted to nonsense. The UGG triplet, which specifies the amino acid tryptophan, is a prime example.

3. Some mutations may be difficult to select for, since they differ only slightly from wild type. Examination of suppressed nonsense mutations allows the detection of this set of mutants.

4. Suppressing nonsense mutations allows the determination of which substitutions do not produce detectable changes in the activity of the protein.

5. Some substitutions made by suppression of nonsense mutations could not occur by a single base change, and thus these substitutions would rarely, if ever, be found by conventional mutagenesis. For instance, glutamine and tyrosine can be exchanged for one another through suppression, but not by missense mutagenesis.

There are several disadvantages to the sole use of suppressed nonsense mutations, however:

1. Only those sites which can go to nonsense can be examined.

2. Missense mutations can effect certain substitutions which nonsense suppression cannot.

Because these two approaches complement one another, their combined use is clearly the preferred strategy for those attempting to explore as many mutational alterations in a protein as possible. We have employed this strategy and some of the properties of the missense mutations we have isolated are included here.

There are several important considerations concerning mutagenesis and the isolation of mutations which are often overlooked by investigators. Most mutagens, and particularly nitrosoguanidine, have a very limited mutagenic spectrum. In fact, nitrosoguanidine has been shown to induce primarily GC → AT transitions (10,20,21). This would make it impossible, for example, to induce substitutions at any of the tryptophan, tyrosine, lysine or phenylalanine residues. This situation is complicated by a second and less well understood phenomenon termed "hotspots", the tendency of a mutagen to induce changes at certain sites more frequently than other sites despite the fact that the same transition or transversion is involved (22). In the lacI gene we can see these two phenomena at work by examining and comparing amber mutations (changes to UAG) induced by two different mutagens, 2AP and mutT (14). Because each of these two mutagens has a different molecular specificity [2AP induces changes to UAG from CAG and UGG codons (14,23); mutT from UAU codons (14,24)] none of the 650 ambers isolated after 2AP mutagenesis are at the same site as any of the 150 ambers induced by mutT (14). However, among the 2AP ambers (all of which arise from the same transition) there is a marked preference for certain sites which can vary by as much as 100 fold (14).

Considering the above points, it can be seen that studies attempting to determine the active sites of repressor which depend on the sole use of NG (7,10,16) or the sole use of missense mutations are quite limited. In fact, in some cases the authors are analyzing NG hotspots rather than particularly important sites in the protein. Therefore, our finding of new

i^s sites by the use of nonsense suppression or of new regions late in the protein where alterations cause full loss of repression, but not of IPTG-binding activity, is not surprising. Continued efforts in this direction should detect additional interesting mutational sites.

The results reported here summarize the initial findings of a project aimed at determining the structure-function relationship of the lac repressor. In order to achieve this, a system enabling extensive substitution of repressor amino acids has been developed. Certain residues important for the various repressor configurations have already been identified. Future work should provide a better understanding of how these residues are involved in the active sites of the repressor.

This work was supported by a grant from the Swiss National Fund (F.N. 3.800.72). P. Lu was aided by a travel Fellowship from EMBO.

REFERENCES

1. Beyreuther, K., Adler, K., Geisler, N., Klemm, A.: The amino acid sequence of lac repressor. Proc. Nat. Acad. Sci. USA 70, 3576-3580 (1973).

2. Jacob, F., Monod, J.: Genetic regulatory mechanisms in the synthesis of proteins. J. Mol. Biol. 3, 318-356 (1961).

3. Wilson, C., Perrin, D., Cohn, M., Jacob, F., Monod, J.: Non-inducible mutants of the regulator gene in the lactose system of E. coli. J. Mol. Biol. 8, 582-592 (1964).

4. Gilbert, W., Müller-Hill, B.: Isolation of the lac repressor. Proc. Nat. Acad. Sci. (Wash.) 56, 1891-1898 (1966).

5. Chamnes, G.C., Willson, C.D.· An unusual lac repressor mutant. J. Mol. Biol. 53, 561-565 (1970).

6. Jobe, A., and Bourgeois, S.: The lac repressor-operator interaction. VII. A repressor with unique binding properties: the X86 repressor. J. Mol. Biol. 72, 139-152 (1972).

7. Pfahl, M., Stockter, C., Gronerborn, B.: Genetic analysis of the active sites of lac repressor. Genetics, in press.

8. Weber, K., Platt, T., Ganem, D., and Miller, J.H.: Altered sequences changing the operator binding properties of the lac-repressor: Colinearity of the repressor protein with the i gene map. Proc. Nat. Acad. Sci. USA 69, 3624-3628 (1972).

9. Weber, K., Files, J.G., Platt, T., Ganem, D., and Miller, J.H.: Lac repressor. Symposium on Protein-Lingand Interactions. Walter de Gryter, Berlin (1974).

10. Müller-Hill, B., Fanning, T., Geisler, N., Gho, D., Kania, J., Kathmann, P., Meissner, H., Shlotmann, M., Schmitz, A., Triesch, I., and Beyreuther, K.: The active sites of lac repressor. Symposium on Protein-Ligand Interactions. Walter de Gruyter, Berlin (1974).

11. Perutz, M.F.: Stereochemistry of Cooperative Effects in hemoglobin. Nature, 228, 726-734 (1970).

12. Sarabhai, A.S., Stretton, A.O.W., Brenner, S., Bolle, A.: Colinearity of the gene with the polypeptide chain. Nature 201, 13-17 (1964).

13. Gorini, L.: Informational Suppression. Ann. Rev. Genet. 4, 107-134 (1970).

14. Miller, J.H., Ganem, D., Lu, P., Schmitz, A.: Construction of a gene-protein map for the lac repressor. In preparation.

15. Schmeissner, U., Ganem, D., and Miller, J.H.: Fine structure map of the lacI gene. In preparation.

16. Adler,K., Beyreuther, K., Fanning, E., Geisler, N., Gronenborn, B., Klemm, A., Müller-Hill, B., Pfahl, M., and Schmitz, A.: How lac repressor binds to DNA. Nature 237, 322-327 (1972).

17. Müller-Hill, B., Crapo, O., and Gilbert, W.: Mutants that make more lac repressor. Proc. Nat. Acad. Sci. USA 59, 1259-1264 (1968).

18. Miller, J.H.: Experiments in molecular genetics. Cold Spring Harbor Laboratory (1972).

19. Myers, G.L., and Sadler, J.R.: Mutational inversion of control of the lactose operon of Escherichia coli. J. Mol. Biol. 58, 1-28 (1971).

20. Prakash, L., and Sherman, F.: J. Mol. Biol. 79, 65-82 (1973).

21. Coulondre, C., and Miller, J.H.: Mutagenic Specificity and the isolation of new nonsense mutations in the lacI gene. In preparation.

22. Benzer, S.: On the topography of the genetic fine structure. Proc. Nat. Acad. Sci. (Wash.) 47, 403-416 (1961).

23. Yanofsky, C., Ito., J., and Horn, V.: Amino acid replacements and the genetic code. Cold Spring Harbor Symp. Quant. Biol. 31, 151-162 (1966).

24. Cox, E.C., and Yanofsky, C.: Altered Base ratios in the DNA of an E. coli mutator strain. Proc. Nat. Acad. Sci. (Wash.) 58, 1895-1902 (1967).

Received September 13, 1974

DISCUSSION

Veeger: Do you agree with my conclusion derived from your slides that mutations are only allowed at positions not involving hydrophobic amino acid residues?

Miller: No. The method I used enabled me to isolate mutants more easily at positions coding for hydrophilic residues.

Woody: What are the prospects for obtaining enough of these mutant proteins to study the effect of specific amino acid substitutions on polypeptide chain folding?

Miller: The prospects are very good.

von Hippel: How will you follow the purification of the inactive (and partially unfolded) polypeptide in some of the replacements which don't fold properly?

Miller: Either by immunological techniques or by the use of conditional mutants.

Effect of Alterations in Lac Operator DNA on Repressor Binding

Suzanne Bourgeois, Mary D. Barkley, Alan Jobe, John R. Sadler and *James C. Wang*

The lac repressor-operator interaction has been characterized by binding studies in which the effect of a variety of parameters on the equilibrium (1) and the kinetics (2) of the reaction can be determined. In particular, the effects of variables such as temperature, pH and ionic strength, and the effect of small-molecular weight effectors, inducers and anti-inducers (3), have been measured quantitatively. Binding studies can also be used to examine the results of specific alterations in the reactants, the repressor and the operator, upon their interaction, and the binding properties of a number of lac repressors altered by mutations (4) have been described. We present and discuss here our recent studies on the effect of two types of alterations in the lac operator DNA upon its binding by repressor: mutational alterations in the operator region and variations in the degree of operator DNA superhelicity.

AFFINITY OF THE REPRESSOR FOR OPERATORS CONTAINING 0^c MUTATIONS

Mutational events which introduce single-base changes in the operator, can result in decreased affinities of the repressor for the altered operator regions and, thus, in constitutive expression of the lac operon (synthesis of the lac operon products in the absence of inducer). Many 0^c (operator constitutive) mutants have been isolated and characterized *in vivo* (5), and the 0^c mutations have been mapped within the operator region (6). 0^c mutations were found not only to increase the constitutive level of β-galactosidase synthesis but also to affect the level of enzyme synthesized in the presence of saturating amounts of the inducer isopropyl-β-D-thiogalactoside (IPTG). The fully induced enzyme level is either higher or lower than in the wild-type (O^+) strain, depending on the O^c mutant being examined. Therefore, the ratio of the constitutive to the fully induced enzyme level, or P value, was taken as the best measure of the effect of each O^c mutation on repressor affinity. On the basis of P values virtually all base-substitution O^c mutants could be arranged into six discrete classes, numbered II to VII. Each class comprises two members, called a and b, characterized by the same P value but having different constitutive and fully induced β-galactosidase levels (5). The enzyme activities and P values of the different classes of O^c mutants are described in the first three columns of Table 1.

O^c class	Z/B	$(Z/B)_{max}$	P $(Z/B)/(Z/B)_{max}$	$t_{1/2}$(min) i^q R I=0.05M	$t_{1/2}$(min) X_{86} R I=0.2M
O^+	0.000537	1.28	0.00042	75	480
II_a	0.101	2.22	0.0455	0.95	30
II_b	0.0766	1.59	0.0482	0.95	16
III_a	0.0308	1.15	0.0268	1.00	19
III_b	0.0234	0.93	0.0252	1.10	24
IV_a	0.0443	2.23	0.0199	1.00	60
IV_b	0.0170	0.98	0.0173	1.16	75
V_a	0.0148	2.01	0.00736	2.42	63
V_b	0.0103	1.49	0.00691	2.33	63
VI_a	0.00734	1.76	0.00417	2.50	92
VI_b	0.00552	1.27	0.00435	2.60	95
VII_a	0.00333	1.41	0.00236	3.80	180
VII_b	0.00285	1.25	0.00228	4.80	-

Table 1. Characteristics of the O^c mutants in vivo and in vitro. Z/B and $(Z/B)_{max}$ represent the specific activities of β-galactosidase in vivo in the absence of inducer and in the presence of 10^{-3}M IPTG respectively. The growth and assay conditions and the enzyme units of activity have been described elsewhere (7). The half-lives ($t_{1/2}$) of the different repressor-operator complexes formed by binding either i^q repressor of X_{86} repressor to the O^+ or O^c operator DNA's were measured as shown in Figure 1.

The two members of each 0^c class were found to map at different sites of the operator. Because of ambiguities in the relative location of some of the 0^c's, four different 0^c map orders were found to be consistent with the mapping data (6). Of these four possibilities, one map order was such that the two members of each 0^c class were disposed about a plane of reflection. These observations were taken by Sadler and Smith to indicate a two-fold symmetry in the operator. This interpretation rested on the assumption that the two members of each 0^c class, characterized by the same P value, were equivalent as far as repressor binding was concerned,in spite of having different constitutive and maximal levels of lac operon expression.

We have tested this hypothesis by measuring directly in vitro the affinity of the lac repressor for each class of 0^c operator DNA. Each 0^c mutation was introduced into λφ80 dlac transducing phages which were used as the source of 0^+ or 0^c operator DNA's for binding experiments. Kinetic measurements (2) were used to characterize the interaction between lac repressor and the altered 0^c operators. Since the presence of an 0^c mutation was shown not to affect the rate of association of repressor and operator (7), the rate of dissociation of the repressor - 0^c operator complexes can be taken as a direct measurement of the affinity of repressor for the different 0^c operators. Measurements of kinetics of dissociation of such complexes are illustrated in Fig. 1 and the results are listed in the last two columns of Table 1. The lac repressor purified from a repressor overproducing strain (i^q) does not distinguish between the operators from the 0^c mutants of classes II, III and IV, with which it forms complexes having half-lives of about 1 minute, (corresponding to an affinity decreased about 75-fold as compared to its affinity for a normal 0^+ operator) or between the operators of the 0^c classes V and VI, with which it forms complexes characterized by half-lives of about 2.5 minutes. To separate these different 0^c classes in vitro a repressor altered by an i gene mutation, called X_{86}, which results in a 40-fold increase in affinity for the 0^+ operator, was used (8). The X_{86} repressor forms a complex with the 0^+ operator characterized by a half-life of 480 minutes (Table 1) at the ionic strength of 0.2M. This high ionic strength was chosen in the case of this repressor in order to accelerate the very slow dissociation at

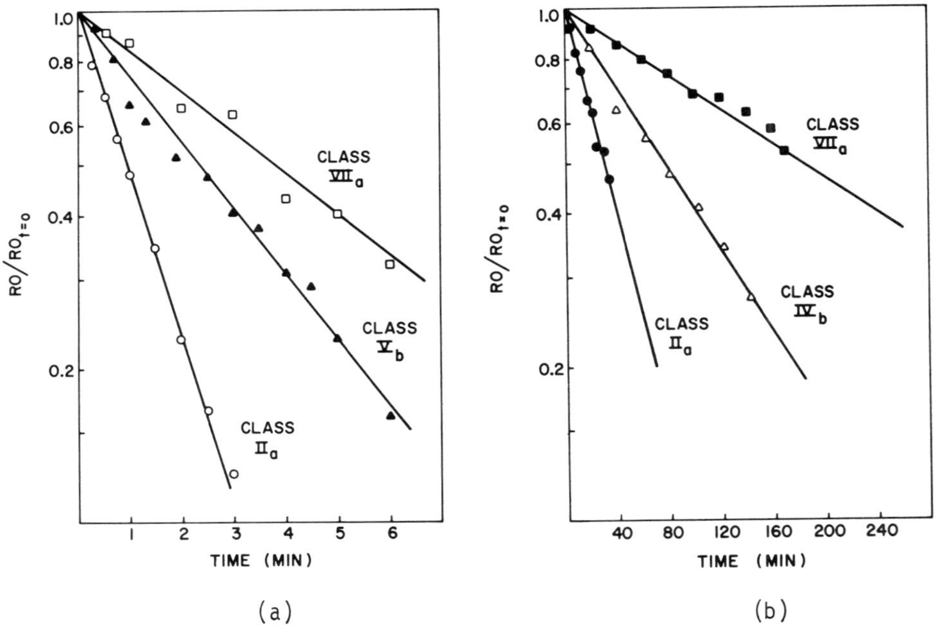

Fig. 1. Half lives of different repressor-O^c operator complexes. The membrane filter technique used to measure repressor-^{32}P-labeled operator complexes and the conditions of the binding assay have been described elsewhere (2,7). The data are expressed as the fraction of the amount of repressor-operator complex (RO) present at time zero. (a) binding of purified i^q repressor. (b) binding of purified repressor altered by an i gene mutation, called X_{86}, and described elsewhere (8).

lower salt concentration and make these measurements more convenient and accurate. By taking advantage of these two repressors, all the different classes of O^c operators could be distinguished from one another (Table 1). Within experimental error, the two members, a and b, of each O^c class are equivalent in their affinity for repressor. The only clear-cut exception found to that rule is the case of the O^c class II of which the two members, II_a and II_b, form complexes with half-lives of 30 and 16 minutes, respectively, with X_{86} repressor. However, substantial mutagenic evidence was obtained (5) that these two O^c sites represent different types of substitution mutations which are distinguished by the X_{86} repressor.

AFFINITY OF THE REPRESSOR FOR OPERATOR DNA'S WITH VARYING DEGREES OF SUPER-HELICITY

The study of the second type of alteration introduced in the operator DNA is aimed at answering the question whether or not repressor binding involves unwinding of the DNA strands in the operator region. If unwinding were observed, the measurement of the angular alteration would help choose between different models for the recognition of the lac operator by repressor. A model which could involve considerable unwinding of the operator DNA strands is the one first proposed by Gierer (9) and according to which a repressor would interact specifically with a branched operator structure rather than with the normal double-helical structure. This model is of special relevance in the case of the lac system since the sequence of the lac operator region (10) manifests the symmetry required for the formation of Gierer-type structures, as shown in Fig. 2.

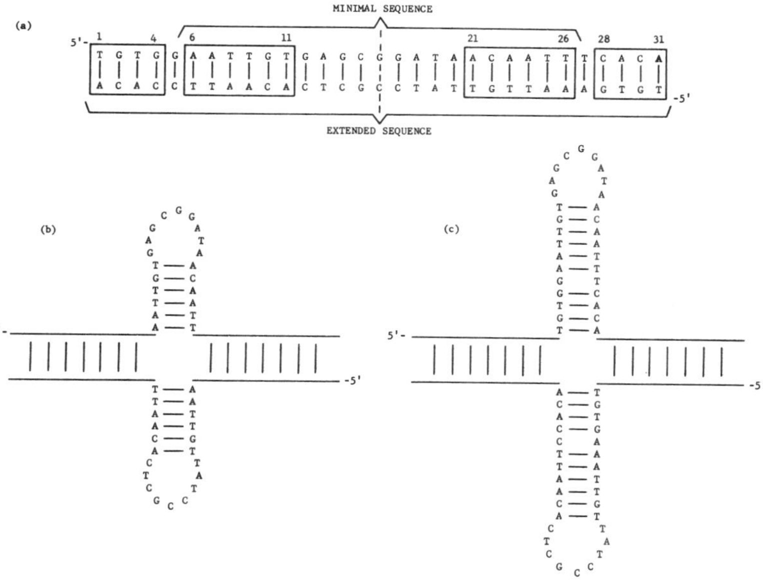

Fig. 2. Lac operator à la Gierer. (a) Fully paired structure of the lac operator. The minimal sequence is taken from Gilbert and Maxam (10); the extended sequence includes some data from Abelson and coworkers (J. Abelson, personal communication). The regions boxed are the symmetric sequences forming the base pairs of the stems of the structures shown in (b) and (c); (b) and (c) represent Gierer-type structures obtained from the minimal and extended lac operator sequence, respectively.

The conversion of the fully-paired minimal sequence of the lac operator, shown in Fig. 2 (a), to the branched structure shown in Fig. 2 (b) involves the unpairing of twenty one base pairs, reducing the number of helical turns between the complementary strands by about two full turns. Recent results from Abelson and coworkers (personal communication) indicate that this minimal sequence of the lac operator is flanked by sequences of five base pairs, of which four pairs are symmetrically disposed, and which might be part of the operator region. This extended operator sequence, shown in Fig. 2(a), can be converted to the Gierer-type structure shown in Fig. 2(c), having stems eleven base-pairs long (with one mis-match) and resulting from the unwinding of the double-helical structure by three full turns. Since the Gierer-type structures (b) and (c) include loops of nine unpaired bases, these would be expected to be less stable than the fully-paired structure (a). The melting temperature of 67° in 0.15M NaCl-0.015M Na citrate, determined for the operator fragment protected by the repressor from digestion with pancreatic DNase (10), indicates that the operator is in the double-helical form in the absence of repressor. Binding of repressor, then, would stabilize the Gierer-type structure of the operator through interactions with unpaired bases of the loops, and result in the unwinding of the DNA strands by two or three turns.

Our experimental approach to the measurement of the angular alteration of the lac operator upon binding of the repressor is based on the fact, which is well documented in the case of intercalative dyes (11,12), that the removal of DNA superhelical turns is favored thermodynamically. Thus, unwinding is accompanied by a negative free-energy change which raises the net chemical free energy associated with the binding of a ligand and results in an increase in the binding affinity. Any ligand capable of unwinding of the DNA is, therefore, expected to have a higher affinity for a negative superhelical DNA than for the same DNA without superhelical turns. For two DNA's with superhelical densities σ and zero (the superhelical density is defined as the number of superhelical turns per ten base pairs) it can be shown that the ratio of the association constants K_a^{σ} to K_a^{0} for the binding of ligand to the two DNA's, respectively, is

(1) $$K_a^{\sigma}/K_a^{0} = \exp\left(-\frac{1}{RT}\frac{dG_\tau}{d\tau}\Delta\tau\right)$$

where $dG_\tau/d\tau$ is the free-energy change per unit change in the number of superhelical turns, and $\Delta\tau$ is the number of turns unwound by the ligand(13). The function $dG_\tau/d\tau$ is evaluated from the binding curves of the intercalative dye ethidium to superhelical and non-superhelical DNA's. Bauer and Vinograd (12) give $\frac{1}{RT}(dG_\tau/d\tau)=844\underline{\sigma} - 2640\,\underline{\sigma}^2$. The more recent unpublished results of Hsieh and Wang gave $\frac{1}{RT}(dG_\tau/d\tau)=505\underline{\sigma}$. Substituting the results of the latter authors into Eq. (1) yields

(2) $K_a^{\sigma}/K_a^{o} = \exp(-505\underline{\sigma}\Delta\tau)$

A measurement of K_a^{σ}/K_a^{o} may be used to determine whether binding of a ligand does in fact cause unwinding, and to evaluate the magnitude of the angular alteration $\Delta\tau$. For example, if the lac repressor unwinds the operator by two turns ($\Delta\tau=2$), one can calculate from Eq. (2) that the ratio of its affinity for operator DNA with $\underline{\sigma} = -0.01$ to that for relaxed operator DNA, $K_a^{\sigma}/K_a^{o} = 2.4\times10^4$. In other words, a very large increase in the association constant is expected for the binding of lac repressor to a negative superhelical operator DNA, if the operator structure is changed from the double-helical form to the Gierer-type structure shown in Fig.2(b).

Kinetic measurements were used to determine the affinity of lac repressor for five covalently closed circular λplac DNA samples with superhelical densities -0.001, -0.017, -0.028, -0.035 and -0.049 (or 5, 80, 130, 160, and 230 negative superhelical turns, respectively) under the binding conditions. The experimental data for the rates of dissociation of the different repressor-operator complexes formed with these operator DNA's are shown in Fig. 3. The presence of negative superhelical turns in the DNA clearly stabilizes the repressor-operator complex, which has a half-life of 22 minutes with the essentially untwisted DNA ($\underline{\sigma}$= -0.001) and of 105 minutes in the case of the DNA with -160 superhelical turns ($\underline{\sigma}$= -0.035). The values obtained for the rate constants of dissociation k_b of the different complexes are listed in Table 2. The data for the rate constants of association k_a are shown in Fig. 4. These measurements are somewhat less accurate than those for the half-lives. However, a small increase in k_a is seen with increasing negative superhelicity. Table 2 presents all the values obtained for the rate constants, k_a and k_b, and for the equilibrium constants of association, K_a.

One remark that should be made is that no difference in affinity, beyond experimental errors, was observed for the binding of repressor to linear

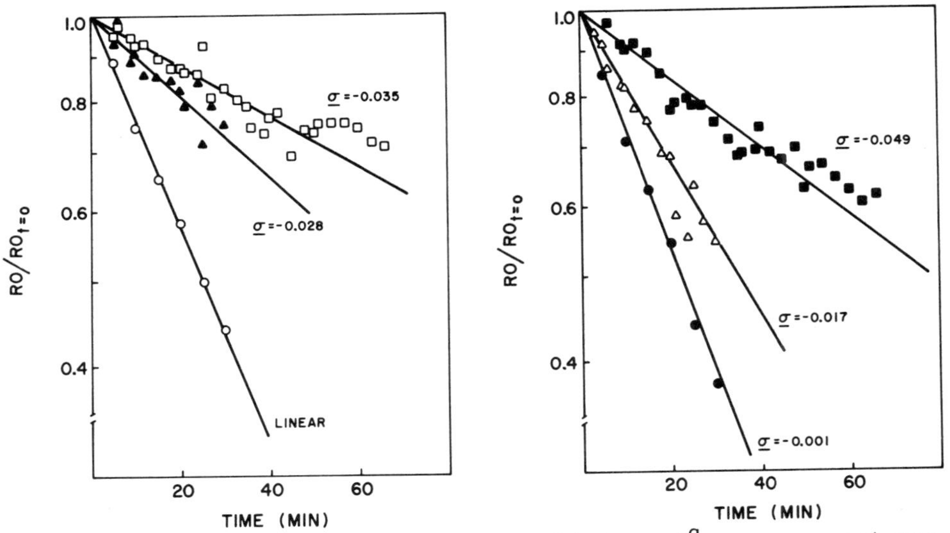

Fig. 3. Half-lives of the complexes formed between i^q repressor and operator DNA's with different superhelical densities (σ). The cyclization and the covalent closure of λplac 5 DNA in the presence of varying amounts of ethidium bromide were done according to published procedures(14). The value of the superhelical density, σ, in the conditions of the repressor binding assay was calculated from the binding curve of ethidium determined by sedimentation for each DNA(14). The conditions of the repressor binding assay have been described elsewhere(2,15). The data shown, expressed as the fraction of repressor-operator complex(RO) present at time zero, represent averaged values obtained in several independent experiments for each DNA.

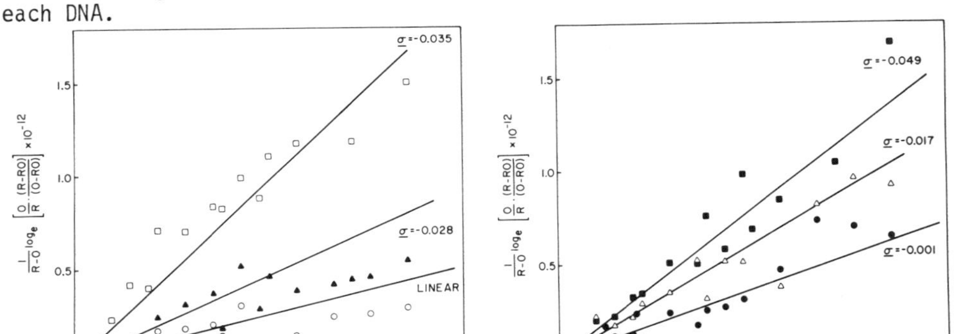

Fig. 4. Rates of association of i^q repressor to operator DNA's with different degree of superhelicity (σ). The conditions of the binding assay have been described elsewhere(15). The data are plotted according to the rate equation for a bimolecular reaction (2), with R and O denoting the total repressor and operator concentrations, respectively. The data shown represent averaged values obtained in several independent experiments.

λplac DNA as compared to the untwisted covalently closed DNA (σ = -0.001). This is an important control since so far all repressor binding studies have used linear operator DNA. Table 2 also shows that the affinity of the repressor for operator increases with increasing DNA negative superhelicity up to a maximum of about 14-fold for the DNA with σ = -0.035. Further unwinding of the operator DNA to a value of σ = -0.049 decreases somewhat its affinity for the repressor. One can only speculate on the reasons for that decrease. A similar phenomenon has, however, been observed previously (16) for highly twisted PM2 DNA's which manifest a decreased activity as templates for transcription by E.coli RNA polymerase when their negative superhelical density is greater than ~ 0.045. For such highly twisted DNA's the tight interwining of the strands possibly influences the binding of macromolecules not only through the free energy of superhelix formation (Eq.2), but in other ways as well, such as steric hindrance, change of hydration of the DNA or structural distortion of the helix(17). From the values for K_a, the number of helical turns unwound by a bound repressor, $\Delta\tau$, can be calculated directly from Eq.2. The unwinding angle θ in degrees, which is equal to $360\Delta\tau$, is tabulated in the last column of Table 2. The results indicate that the binding of lac repressor unwinds the operator by about 40°. However, in obtaining Eq.2 it has been implicitly assumed that the intercalative dye ethidium unwinds the DNA helix by 12°. Recent results of one of us (J.C.W.) indicate that it is considerably larger, with 26° being the best estimate at present. This would make the unwinding angle of the lac operator by repressor about 90° instead of 40°, or approximately one fourth of a superhelical turn. That the value of θ cannot be much larger than 90° is confirmed by attempts to measure the unwinding of the lac operator by the method of Saucier and Wang (18). No unwinding could be clearly detected in those conditions, which is not unexpected since our 90° estimate for θ is approximately the limit of sensitivity of that technique.

DISCUSSION AND CONCLUSIONS

Our results on the binding of lac repressor to operators altered by 0^c mutations have been discussed in detail elsewhere (7). However, some points deserve comment, in particular, our puzzling observation that the i^q

σ	$t_{1/2}$ min	$k_b x10^4$ sec^{-1}	average $k_b x10^4$ sec^{-1}	$k_a x10^{-9}$ $M^{-1}sec^{-1}$	average $k_a x10^{-9}$ $M^{-1}sec^{-1}$	$K_a x10^{-13}$ M^{-1}	K_a^{σ}/K_a^o	θ degrees
Linear	23	5.0 ± 0.6	4.7 ± 0.3	3.1 ± 0.7	2.4 ± 0.5	0.5		
	29	4.0 ± 0.4		1.3 ± 0.3				
	27	4.3 ± 0.3		3.5 ± 0.8				
	22	5.3 ± 0.4		1.7 ± 0.3				
-0.001	20	5.8 ± 0.8	5.4 ± 0.2	2.6 ± 0.6	3.4 ± 0.6	0.6		
	23	5.0 ± 0.4		2.6 ± 0.6				
	23	5.1 ± 0.7		2.9 ± 0.5				
	20	5.7 ± 0.6		5.5 ± 1.0				
-0.017	40	2.9 ± 0.4	3.3 ± 0.2	6.4 ± 0.7	5.7 ± 0.5	1.7	2.8	43
	34	3.4 ± 0.5		5.1 ± 0.9				
	37	3.2 ± 0.3						
	31	3.8 ± 0.2						
-0.028	42	2.8 ± 0.6	1.8 ± 0.3	8.0 ± 1.0	4.4 ± 1.1	2.5	3.9	35
	73	1.6 ± 0.6		2.8 ± 0.3				
	90	1.3 ± 0.2		3.4 ± 0.4				
	78	1.5 ± 0.3		3.5 ± 0.7				
-0.035	130	0.9 ± 0.3	1.1 ± 0.3	7.6 ± 0.8	9.2 ± 1.0	8.5	13.6	53
	106	1.1 ± 0.2		11.6 ± 9.1				
	78	1.5 ± 0.2		8.4 ± 1.2				
	146	0.8 ± 0.1						
	106	1.1 ± 0.1						
-0.049	73	1.6 ± 0.3	1.5 ± 0.1	9.4 ± 0.5	7.5 ± 1.0	5.1	8.7	31
	73	1.6 ± 0.3		5.1 ± 0.9				
	78	1.5 ± 0.1		8.1 ± 0.8				
	90	1.3 ± 0.1						

Table 2. Affinity of i^q repressor for operator DNA's with different degrees of superhelicity (σ). The rate constants of dissociation, k_b, and of association, k_a, were calculated from a least squares fit to a straight line of data such as those shown in Figs. 3 and 4. The errors shown on each measurement of k_b and k_a represent the 70% confidence limit. The values obtained from N independent experiments for each DNA were averaged and the errors on the mean values of k_b and k_a were estimated as $1/N \sqrt{(X-\bar{X})^2}$ were $X - \bar{X}$ is the deviation from the mean for each experiment. The half-life ($t_{1/2}$ min.) of each repressor-operator complex was calculated from the k_b as $t_{1/2}=1.17x10^{-2}/k_b$. The equilibrium constant of association, $K_a=k_a/k_b$, was calculated for each DNA and the ratios(K_a^{σ}/K_a^o) of the association constants for the binding to a DNA of superhelicity σ and zero, respectively, were used to calculate θ, the unwinding angle of the operator DNA due to the binding of lac repressor, according to Eq.(2) given in the text.

repressor exhibits nearly the same affinity in vitro for all the operators derived from the highly constitutive 0^c mutants of classes II, III and IV. This led us to consider and to test the effect of some of the variables in the system which could be different from the in vivo situation, and to discuss several relevant binding mechanisms (7). However, none of our models could account quantitatively for this result; the major part of that effect could be due to the existence of a second repressor-binding site within the lac z gene of λφ80 dlac DNA, analogous to the site independently discovered by Reznikoff and by Gilbert (personal communications). This operator-like region would be non-functional in blocking lac transcription in vivo but, inasmuch as the repressor has a stronger affinity for that region than for the high 0^c operators, the binding to that site would be predominant in vitro. Recently Pfahl (unpublished result) found another pseudo-operator site at the operator proximal end of the i gene. This second pseudo-operator is made functional as the result of an i gene mutation. This observation raises the interesting question of the existence of several operator-like sequences in the lac region.

Since our results substantiate the assumption made by Sadler and Smith about the equivalence of the two members of each 0^c class, they also imply that the effect of 0^c mutations on the fully induced level of lac operon products is not directly related to the binding of repressor to operator. This effect has been recently investigated by Carlson and Smith(19), and it appears to be due largely to changes in lac mRNA lifetimes resulting from 0^c mutations. This indicates that the operator region must be transcribed in vivo into RNA and that lac operon expression is affected by a translational phenomenon characteristic of each particular 0^c mRNA species, the precise mechanism of that phenomenon being still unclear. Our in vitro study of 0^c operators confirms that base changes at distinct sites in the operator can result in equivalent changes in repressor affinity and quantifies the affinity of the repressor for each 0^c operator. The maximum decrease in affinity observed for X_{86} repressor binding to an 0^c (class II_b) as compared to the 0^+ operator is about 30-fold, representing a decrease in binding energy of about 2kcal./mole. The sequencing of these 0^c operators will reveal the exact location and the type of base changes involved.

Our study of the binding of repressor to superhelical operator DNA's provides information on the possible role of the operator's symmetry in its recognition by repressor. The various mechanisms proposed to date for the recognition by a protein of a unique sequence of bases in a double-stranded DNA can be divided into two categories: (I) The binding involves the disruption of base pairing between the DNA strands. Either bases exposed in the single-stranded region or Gierer-type structures formed subsequently are recognized by the protein. (II) The binding has little effect on the base pairing of the DNA strands. The recognition of a unique base sequence is achieved by processes such as (a) interactions between the protein molecule and groups in the grooves of the DNA helix, (b) intercalation of certain groups of the protein into the DNA helix at certain nearest neighbor sequences, and (c) a transition of the DNA helix into a different helical form, which then provides unique structural features for specific interactions with the protein.

The 90° unwinding angle observed in this case indicates that mechanism (I) is not likely, since the disruption of base pairing of a segment large enough for recognition would result in a much greater unwinding angle. The same conclusion can be drawn for the λ repressor operator-interaction. Maniatis and Ptashne (20) have shown, by measuring the amounts of λ repressor bound at equilibrium to linear and superhelical λ DNA's, that virtually no unwinding of the λ operator occurs upon binding of the λ repressor. The equilibrium method they used is, however, less sensitive than the kinetic measurements done here for the _lac_ system. It cannot be ruled out at this time that the λ repressor-operator interaction also involves a small unwinding of the DNA helix. The source of superhelical DNA used in the study of the λ repressor binding was the natural λ superhelical DNA which, like most naturally occurring superhelical DNA's, has a superhelical density of approximately -0.03 (21). The range of superhelical densities achieved by the ethidium bromide procedure and used in our experiments with the _lac_ operator covers the degree of superhelicity of circular DNA's _in vivo_.

We take our results to indicate, in particular, that the symmetry of the _lac_ operator may allow the binding of two or four repressor subunits, to amplify the affinity and the specificity of the interaction, but does not serve the purpose of forming a Gierer-type structure upon binding of the repressor. One should, however, introduce a word of caution about this interpretation since the branched structures could conceivably be stabilized by modification of unpaired bases of the loops. In such case, the binding of lac repressor to a preexisting Gierer-type structure could take place without considerable unwinding of the DNA strands. As mentioned earlier, the existence of a Gierer-type structure, stable in the absence of repressor, is made very unlikely by the fact that the operator has the melting temperature (10) expected for double-helical DNA.

It is of interest to compare the mechanism of operator binding by repressors with that of promoter recognition by RNA polymerase. For the E.coli RNA polymerase, the measured unwinding angle is about ten times that of ethidium, or 260° (18). Due to the uncertainty of the determination of the number of RNA polymerase molecules bound, the uncertainty in the unwinding angle measured is estimated to be a factor of two. With this fairly large uncertainty, it cannot be ruled out that a segment as large as ten base pairs is disrupted. The result for RNA polymerase is more complex and difficult to interpret than the results obtained with the λ and lac repressors because of the necessity to differentiate binding from chain initiation. However, we do not believe either that the recognition of the promoter sequence is by mechanism (I), on the basis of two experiments. First, it has been shown recently (22) that the formation of the RNA polymerase-DNA complex, with significant unwinding of the DNA helix, has a steep temperature dependence: no unwinding of the helix is detected below 18°C. Moreover, Mangel and Chamberlin (23) found that at low temperature (or high ionic strength) a complex between RNA polymerase is formed which cannot initiate RNA synthesis efficiently. A large activation energy is required for the transition of that complex to a form which is capable of rapid initiation of RNA synthesis. These results indicate that this transition involves the disruption of several base pairs and it, therefore, appears that the unwinding is necessary primarily for the reading of the genetic code, rather than a direct consequence of promoter recognition. Secondly, the recognition of the promoter depends on the presence of the sigma factor. It has been deduced, however, that RNA synthesis by core polymerase (which does not recognize the promoter) is also associated with a substantial unwinding (16). This again indicates that at least the bulk of unwinding of the helix due to RNA polymerase is not directly related to promoter recognition.

Thus, in the case of the RNA polymerase as well as in the cases of the lac and λ repressors, recognition of a specific DNA sequence by a mechanism of type (II), which does not require the disruption of base pairing, is probable. Since the lac operator is only 20 or 30 base pairs long, the 90° unwinding angle measured in that case corresponds to a significant structural change in the operator resulting from the binding of repressor. However, additional information is necessary to specify the contributions of the three factors listed in the mechanism of type (II).

This work was supported by research grants from the National Institutes of Health and by Career Development Awards to S.B. and J.R.S.

1. Riggs, A.D., Suzuki, H., Bourgeois, S.: J.Mol.Biol. 48,67-83(1970).
2. Riggs, A.D., Bourgeois, S., Cohn, M.: J.Mol.Biol. 53,401-417(1970).
3. Riggs, A.D., Newby, R.F., Bourgeois, S.: J.Mol.Biol. 51,303-314(1970).
4. Jobe, A., Riggs, A.D., Bourgeois, S.: J.Mol.Biol. 64,181-199(1972).
5. Smith, T.F., Sadler, J.R.: J.Mol.Biol. 59,273-305(1971).
6. Sadler,J.R., Smith,T.F.: J.Mol.Biol. 62,139-169(1971).
7. Jobe, A., Sadler, J.R., Bourgeois, S.: J.Mol.Biol. 85,231-248 (1974).
8. Jobe, A., Bourgeois, S.: J.Mol.Biol. 72,139-152(1972).
9. Gierer, A.: Nature 212,1480-1481(1966).
10. Gilbert, W., Maxam, A.: Proc.Nat.Acad.Sci.U.S.A. 70,3581-3584(1973).
11. Bauer, W., Vinograd, J.: J.Mol.Biol. 33,141-171(1968).
12. Bauer, W., Vinograd, J.: J.Mol.Biol. 47,419-435(1970).
13. Davidson, N.: J.Mol.Biol. 66,307-309(1972).
14. Wang, J.C.: Procedures in Nucleic Acid Research(Cantoni, G.L. and Davies, D.R. editors). 2,407-416 (Harper and Row, New York, 1971).
15. Wang, J.C., Barkley, M.D., Bourgeois, S.: Nature in press (1974).
16. Wang, J.C.: J. Mol.Biol. in press(1974).
17. Maestre, M., Wang, J.C.: Biopolymers 10,1021-1030(1971).
18. Saucier, J.M., Wang, J.C.:Nature New Biology 239,167-170(1972).
19. Carlson, H., Smith, T.S.: J.Mol.Biol. submitted (1974).
20. Maniatis, T., Ptashne, M.: Proc.Natl.Acad.Sci.U.S.A. 70,1531-1535(1973).

21. Wang, J.C.: J.Mol.Biol. 43,263-272(1969).

22. Jacobsen, J., Saucier, J.M., Wang, J.C.: Fed.Proc. 33, Abstract No.750 (1974).

23. Mangel, W., Chamberlin, M.: J.Biol.Chem. in press (1974).

Received June 24, 1974.

DISCUSSION

Wagner: According to your results the binding of the repressor involves a slight unwinding of the operator region. This unwinding energy should grossly depend upon the ratio of GC to AT base pairs present. Replacement of an AT pair by a GC pair in a mutated operator should (besides its very specific contribution to recognition) increase the unwinding energy and thus decrease the binding affinity. Does a comparison of the properties of the known mutated operators reveal such an interrelation? On the average, an AT ⟶ GC replacement should reduce the binding affinity more than a GC ⟶ AT replacement does.

Bourgeois: In principle, your prediction is correct. In fact, the two AT ⟶ GC replacements which have been analyzed so far do not reduce repressor-binding affinity more than GC ⟶ AT replacements. This must mean that the specific contributions of the base pairs to recognition overrides their contribution to the unwinding energy.

G. Weber: Since the high rate of association of repressor reflects a very rapid "unspecific" addition to λDNA, the ratio K_-/K_+ does not directly give a measure of the free energy of binding to the specific site. Could this be responsible for some of the differences, of the order of 1 kcal, observed in the binding of repressor to mutated operator DNA?

Bourgeois: It is true that the non-specific DNA binding step, preceding the operator binding, contributes some free energy to the value we give as an estimate of the free energy of binding to the operator. Unfortunately, as long as we do not know the mechanism involved in accelerating the repressor-operator association we cannot correct the value of the binding energy to take into account the non-specific binding step. Whatever the non-specific binding mechanism which accelerates the repressor-operator interaction might be, however, it must take place both on the DNA carrying a mutated or the wild-type operator. Therefore, this "accelerating" step cannot be responsible for the differences in binding energy observed for the binding to mutated, as compared to the wild-type operator.

Maass: The rate constants of association k_a between repressor and operator DNA are by about one order of magnitude faster than one would expect for a diffusion-controlled reaction between molecules of this size. What is the explanation for this accelerated reaction relative to a diffusion-controlled one?

Bourgeois: The association of repressor and operator must involve at least two steps, the specific binding to the operator region being preceded by a weaker binding to the non-operator DNA, mostly by electrostatic interactions. Several mechanisms have been proposed to explain how this non-specific binding would accelerate the finding by repressor of its target, the operator. These mechanisms could involve "sliding" or "hopping" of repressor along the DNA and are discussed in the following lecture by Dr. von Hippel.

Kirschner: How can you distinguish between DNA packaging (supercoils) versus unwinding as being the physical basis of increasing the affinity of repressor for operator?

Bourgeois: In the range of low superhelical densities in which we are working, I believe DNA packaging is negligible.

Veeger: Which value of molecular weight did you use in the calculation of your k_a-values? I am surprised about the extreme high rate of association, being almost as fast as H^+ addition to a base. Can this value become higher?

Bourgeois: Since there is only one operator region per λ genome, the molecular weight used to calculate operator molarity is, simply, the molecular weight of λ DNA i.e. 30×10^6 daltons. This value of molecular weight has been precisely determined and cannot be responsible for the very high rate of association calculated.

Veeger: I get the impression from Fig. 4 of your preprint, that the reaction pattern deviates considerably from a pure second-order reaction and is in fact biphasic. Could you comment on this?

Bourgeois: The deviation from a straight line of our results is within experimental error of our measurements and cannot be taken to indicate a biphasic kinetics.

Knippers: Repressor binds weakly to AT sequences and there is a good chance that AT sequences appear in λ DNA. Do you observe a reduction in superhelicity of λ DNA without the lac operon?

Bourgeois: Unwinding of non-operator sequences, such as poly-dAT, by repressor cannot easily be measured because of the uncertainty in estimating the total number of repressor molecules bound unspecifically per λ genome.

<u>Mannherz</u>: Can you give an estimate of how much the rate of dissociation of repressor from the operon is increased by inducer?

<u>Bourgeois</u>: At saturating concentration of inducer, the rate of dissociation of the repressor-operator complex is increased about one thousand-fold.

Interaction of Lac Repressor with Non-Specific DNA Binding Sites*

Peter H. von Hippel, Arnold Revzin, Carol A. Gross and *Amy C. Wang*

I. INTRODUCTION

The strong binding of the lac repressor to the sequence of DNA base-pairs which make up the lac operator region of the E. coli genome, and the change in this binding affinity on interaction of the repressor with small-molecule inducers produced as an initial consequence of lactose metabolism, comprise the central feature of the in vivo control of the function of the lac operon. A variety of molecular and genetic approaches to an understanding of the details of these interactions are considered in other contributions to this volume. (For a recent physico-chemically oriented review, see von Hippel and McGhee, 1972.)

Our purpose here will be to point out that the above description is incomplete. In addition to operator, repressor and inducer, the system also contains a great deal of other DNA to which (as first shown by Lin and Riggs, 1972), repressor and the repressor-inducer complex also bind, though much more weakly than to operator. In this paper we will describe briefly the measurement, by straightforward physico-chemical techniques, of the affinity of repressor (and repressor-inducer complexes) for non-specific DNA sites. We will then outline some aspects of the possible thermodynamic, kinetic and molecular consequences of this binding for the control of the function of the lac operon, and for our understanding of the mechanisms of the repressor-operator interaction.

II. MEASUREMENT OF BINDING OF REPRESSOR AND REPRESSOR-INDUCER COMPLEXES TO NON-SPECIFIC DNA

Association constants, K, and binding site size, n, for lac repressor (R)

*These studies were supported in part by USPHS research grants GM-15792 and GM-15423, a John Simon Guggenheim Memorial Fellowship (to PHvH), and postdoctoral fellowships GM-55928 (to AR) and GM-43987 (to CAG). We are grateful to Ms. Pamela O'Conner for expert and tireless preparation of lac repressor, and other technical assistance.

binding non-specifically to DNA were directly measured by glycerol gradient sedimentation techniques. Gradients were prepared in which labeled (^{32}P) λDNA and R (from i^{sq} strain CSH 46) were uniformly distributed throughout the centrifuge cell at the start of the run. Fractions were assayed for repressor by measuring intrinsic protein fluorescence, and for DNA, by radioactive counts. Results from a typical run (in 0.11 M NaCl, 0.012 M Tris, pH 7.8) are shown in Fig. 1. In this experiment the DNA, to

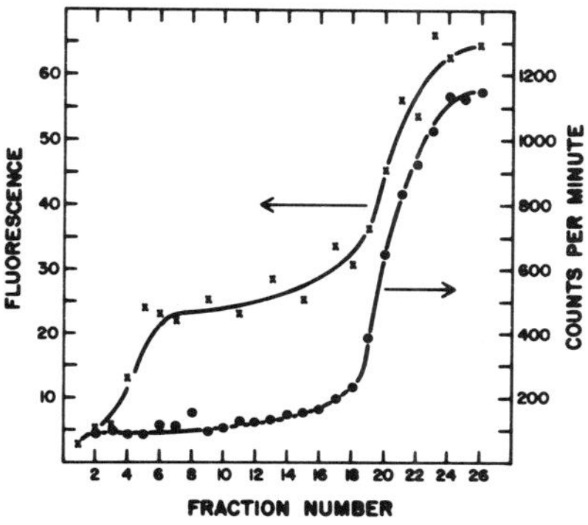

Figure 1

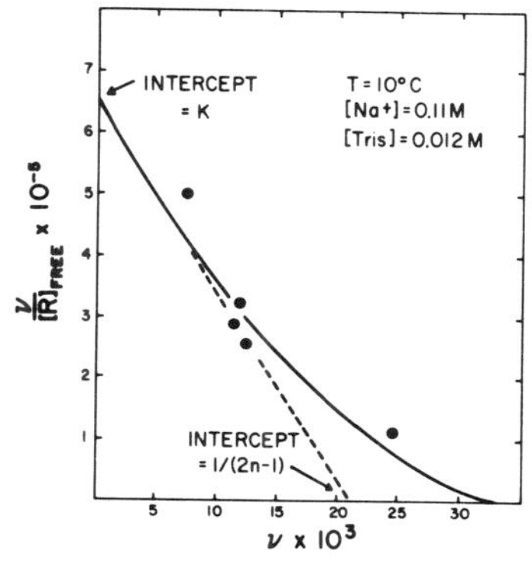

Figure 2

which a fraction of the R is bound, has moved about 3/4 of the way down the tube. (We emphasize that several R molecules may be bound non-specifically to the same DNA molecule.) The level of fluorescence at the right of the graph measures the total concentration of R (free and bound) while the fluorescence plateau value at the left reflects the (slowly sedimenting) unbound repressor.

From the data in Fig. 1 one can easily extract values for the concentration of free R ([R]), and for ν, the binding density in units of moles R tetramer/mole DNA base pairs. These parameters have been measured using gradients at several total R and DNA concentrations, and Scatchard binding plots constructed (an example is shown in Fig. 2). Since each DNA base pair represents the start of a potential binding site, and since bound R covers several base pairs, this plot will be non-linear even in the absence of cooperative interactions between the bound ligands. It can be shown theoretically for such systems (McGhee and von Hippel, 1974), that the Scatchard curve intersects the vertical axis at K, and that a linear extension of the initial slope intersects the horizontal axis at 1/(2n-1). While it is technically difficult to obtain experimental points at either very low or very high values of ν, it is straightforward to fit the available data to the theoretical curve, using a non-linear least-squares procedure with K and n as parameters. For the ionic conditions shown in Fig. 2, we obtain $K \simeq 6.5 \times 10^5\ M^{-1}$ and $n \approx 25$ base pairs. Performing the fit using data adjusted to correspond to our most extreme error estimates changes the calculated value of K by less than a factor of 2, and that of n by less than 5 base pairs. The association constant is extremely sensitive to ionic strength, increasing markedly at lower salt concentrations.

Quantitative support for these measurements of site size can be derived from circular dichroism data. The shape of the CD spectrum of λDNA is little altered by the presence of non-specifically bound R molecules, but the peak value of the ellipticity (at λ=275nm) is increased. Fig. 3 shows the ellipticity (in arbitrary units) at λ = 275 nm as a function of the DNA/R ratio, under ionic conditions at which there is virtually stoichiometric binding. (All data correspond to the same R concentration.) The slope of the DNA+R curve at low values of the DNA/R ratio reflects the change in DNA ellipticity due to R binding. At high values of DNA/R (an excess of DNA over R), the DNA+R curve is essentially parallel to that for DNA alone. The position of the break in the DNA+R data yields an estimated site size, $n \simeq 30$ base pairs/R tetramer, in good agreement with the glycerol gradient results. Full analysis of the observed CD spectral changes accompanying non-specific R binding should reveal details of possible changes in DNA (and/or repressor) conformation as a consequence of this binding.

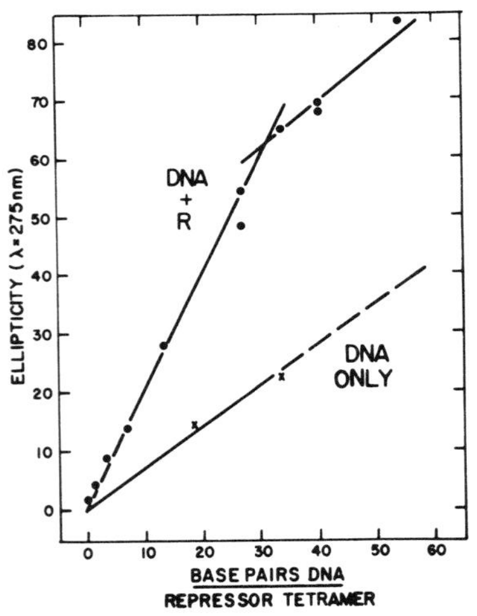

Figure 3

Binding parameters may also be measured by using data obtained from DNA melting profiles in the presence of various concentrations of R. This approach is based on the observation that the non-specific binding of R to the alternating copolymer, poly d(A·T), stabilizes the polynucleotide double helix [and, of course, the R molecules are stabilized when bound to poly d(A·T)]. Thus at $[Na^+]$ = 0.002 M (Fig. 4), poly d(A·T) alone melts at 25°, while R denatures at 44°. In a mixture of poly d(A·T) plus R, the R denatures at 55° and the poly d(A·T) melts broadly over a temperature range considerably above its melting temperature when alone. In these solutions essentially all R molecules are bound to poly d(A·T). [The d(A·T)/R ratio is expressed as base pairs per R tetramer.]

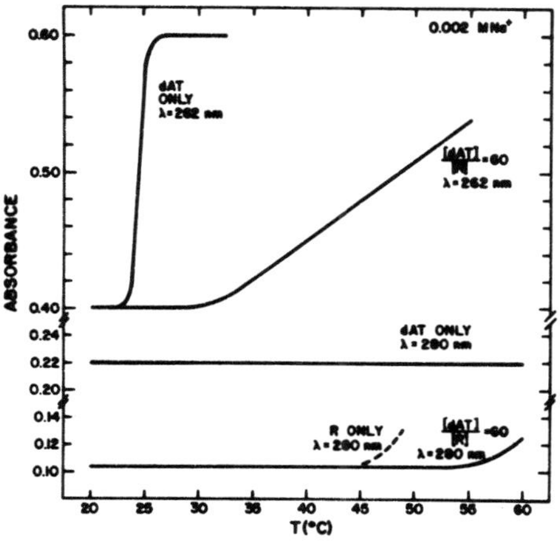

Figure 4

A series of equilibrium melting curves has been obtained for poly d(A·T) in the presence of R at various d(A·T)/R ratios. The shapes of these curves can be analyzed in terms of a theory of helix-coil transitions for DNA in the presence of large ligands (J.D. McGhee, in preparation). This analysis, now in progress, provides us with yet another means for deriving values of K and n. The data are again consistent with a binding site size of about 30 base pairs.

We have shown that R can transfer between different poly d(A·T) molecules in solution. This fact permits us to evaluate <u>relative</u> non-specific binding affinities of R for different DNAs by the following procedure. We add a competing DNA (e.g., λDNA) to a solution containing a particular d(A·T)/R ratio. Even at the low ionic strengths used, the competing DNA is stable to T > 55°, so that the melting observed at T < 55° reflects only that of poly d(A·T). The melting curve followed in the presence of the competing DNA will not be that characteristic of the input d(A·T)/R ratio, but rather will reflect a higher d(A·T)/R ratio since some R will have transferred to the competing DNA. Analysis of such data shows that DNAs of different base composition have very similar affinities for R under the conditions of these experiments. This approach provides a sensitive method for comparing the binding of R to various DNAs, and thus complements the direct, but more tedious, glycerol gradient method.

Finally, we have found that the changes in R fluorescence upon IPTG binding (Laiken, Gross and von Hippel, 1972) are the same whether R is free in solution or is bound non-specifically to DNA. Titrations with IPTG, monitored by the fluorescence change, show that non-specifically bound R binds IPTG in a non-cooperative manner (Hill coefficient of unity) and with the same binding constant as does free R. Furthermore, the glycerol gradient results are unaffected by the presence of excess IPTG in the solution, demonstrating directly that saturating levels of bound inducer have no effect on the R-DNA non-specific binding constant.

III. COUPLED EQUILIBRIA IN THE IN VIVO CONTROL OF THE LAC OPERON

In this section we describe the results of a computer modeling study of equilibrium aspects of the *in vivo* binding of *lac* repressor and repressor-inducer complexes to the *lac* operator and to non-specific DNA, as well as the concomitant binding of RNA polymerase to the *lac* promoter, and to other DNA sites. It will be shown (see also von Hippel *et al*., 1974) that only by considering the consequences of non-specific binding can one account quantitatively for the level of repression of *lac* enzyme synthesis, the effectiveness of inducer-modulated derepression, and other functional aspects of the *in vivo* system.

Any proposed model must be consistent with the following "facts" which summarize our present understanding of the control of the *lac* system. Limitations of space prevent us from providing detailed justification or the necessary qualification of these points here; for this we refer the reader to other papers in this volume, as well as to the relevant "*lac* literature"; e.g., Beckwith and Zipser (1970) and the papers of Bourgeois, Gilbert, Novick, Riggs, Sadler and their co-workers.

(i) Repression of the lac operon of E. coli is a consequence of the binding of repressor (R) to operator (O), and induction (derepression) is due to inducer (I) binding to R, with the formation of an RI_n complex characterized by decreased affinity (relative to R) for O. (ii) The constitutive rate of lac enzyme production in wild-type (w.t.) E. coli cells is approximately the same in the presence of saturating inducer and in i^- (defective R) mutant cells; i.e., full derepression is possible with attainable inducer concentrations. (iii) The constitutive rate of lac enzyme synthesis is $\sim 10^3$ times greater than the fully repressed (basal) rate in w.t. cells. (iv) An increase in intracellular R levels decreases the basal rate of the lac enzyme synthesis in direct proportion. (v) Inducer concentrations necessary to achieve fully constitutive rates range from $\sim 10^{-1}$ to $\sim 10^{-3}$ M for inducers with repressor association constants (K_{RI}) ranging from 10^4 to 10^6 M^{-1}; R mutants exhibiting decreased values of K_{RI} require increased concentrations of I to achieve full derepression. (vi) Cells carrying operator single- (or double-) base-pair mutations (O^c) show an increase of 10- to 500-fold in basal rate; the isolated O^c DNAs show a parallel 10- to 500-fold decrease in w.t. repressor binding constant. (vii) The affinity of the (I saturated) RI_n complex for O is $\sim 10^3$ less than that of R for O. (viii) R and RI_n bind to non-specific DNA with approximately equal affinity.

A. The Model

The following parameters relevant to the wild-type E. coli cell have been used to establish a "base-set" of constants and constraints to describe the coupled equilibria which apply in vivo. The internal volume of the cell is taken as 10^{-15} liters. We assume a total operator concentration $[O_T]$ of $\sim 2 \times 10^{-9}$ M ($\sim$1/cell); a total repressor concentration $[R_T]$ of $\sim 2 \times 10^{-8}$ M ($\sim$10/cell); a total non-specific DNA site concentration $[D_T]$ of $\sim 2 \times 10^{-2}$ M ($\sim 10^7$/cell; the E. coli chromosome contains $\sim 10^7$ base pairs, and in principle every base pair represents the beginning of a separate non-specific DNA binding site); and a fully-derepressing total inducer concentration $[I_T]$ of $\sim 10^{-3}$ M (this applies to the gratuitous inducer, isopropylthiogalactoside, IPTG). The definitions of the necessary binding constants, together with the "base-set" values used, are:[1]

$$
\begin{aligned}
&K_{RO} &=& [RO]/[R]\cdot[O] &\simeq& 10^{14}\ M^{-1}\\
&K_{RD} &=& [RD]/[R]\cdot[D] &\simeq& 10^{5}\ M^{-1}\\
(1)\quad &K_{RI} &=& [RI]/[R]\cdot[I] &\simeq& 10^{6}\ M^{-1}\ \text{(for IPTG)}\\
&K_{RIO} &=& [RIO]/[RI]\cdot[O] &\simeq& 10^{-3}\ K_{RO}\\
&K_{RID} &=& [RID]/[RI]\cdot[D] &\simeq& K_{RD}
\end{aligned}
$$

with conservation constraints:

$$
\begin{aligned}
&[O_T] &=& [O] + [RO] + [RIO]\\
&[R_T] &=& [R] + [RO] + [RD] + [RI] + [RIO] + [RID]\\
(2)\quad &[D_T] &=& [D] + [RD] + [RID]\\
&[I_T] &=& [I] + [RI] + [RIO] + [RID]
\end{aligned}
$$

Figure 5 presents a model of the system, showing the relevant equilibrium constants and molecular species.[2] We have calculated the concentrations

---

[1]The exact internal ionic milieu of the _E. coli_ cell is unknown. It is usually assumed that the cell is ∿0.2 M in KCl and several millimolar in Mg^{++}, but since most of these ions are doubtless bound to the various macromolecular constituents of the cell, we have used estimates for _in vivo_ binding constants derived from _in vitro_ measurements at somewhat lower ionic strengths.

[2]Note that we consider only a single binding site for I to R in these equilibria. Of course, free _lac_ repressor actually contains four sites to which inducer binds independently. It is very likely that more than one inducer must be bound per repressor to bring about complete derepression, but since we define K_{RIO} (and K_{RID}) as the observed binding constants for the repressor-inducer complex at _saturating_ inducer concentrations, for present purposes we can represent the situation in terms of a single inducer binding site. By this definition of K_{RIO}, the concentration range of I over which the system is derepressed will be independent of the actual value of n involved in the derepressing RI_n complex. On the other hand, the _shape_ of the induction curve as a function of I concentration _will_ depend on n. We also list only a single mass-action equation (and constant) for the RD interaction, since direct repressor-binding experiments (see above) show that the predominant class of sites can be characterized by a single binding constant of about the order of magnitude indicated, and which is essentially independent of DNA nucleotide composition and sequence. However, in principle, we can subdivide the D sites

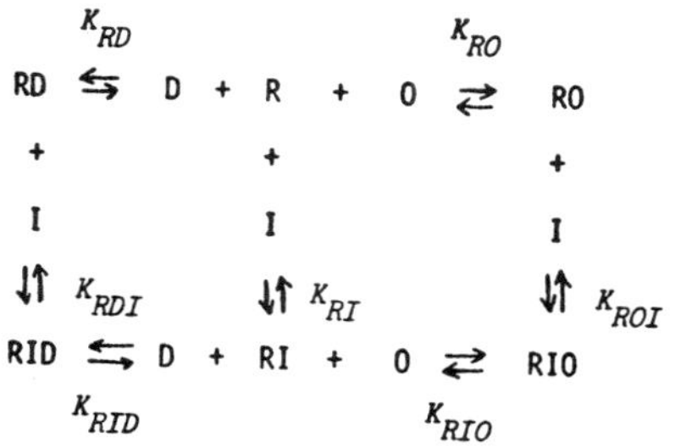

Figure 5

of all species in the system by means of a computer program for solving the simultaneous equations (1) with constraints given by equations (2). The results are generally expressed as fraction of free operator present in the cell ($[O_{free}]/[O_T]$), which is assumed (e.g., see Sadler and Novick, 1965) to be directly proportional to the parameter actually measured in vivo, namely the ratio of the repressed to constitutive intracellular β-galactosidase activity.

B. Results

The basal level. The role played by non-specific DNA binding in controlling the magnitude of the observed repression of the lac operon is illustrated in Fig. 6, in which is plotted the fraction of free operator

---

2(con't) into i different types, each characterized by a mass action relation: $K_{RD_i} = [RD_i]/[R]\cdot[D_i]$. The effects of sub-classes of tighter binding sites on the repression system are considered in Fig. 7.

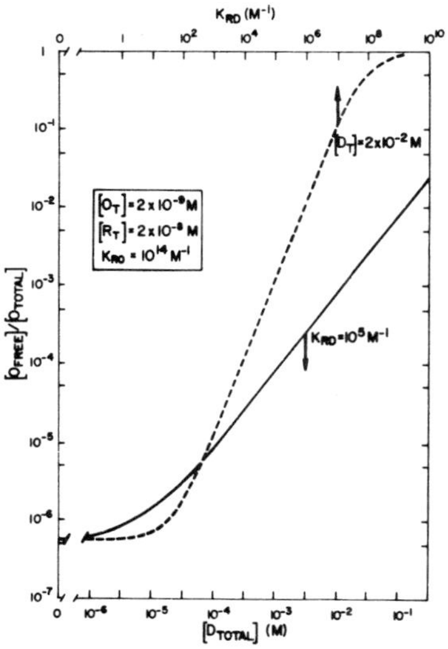

Figure 6

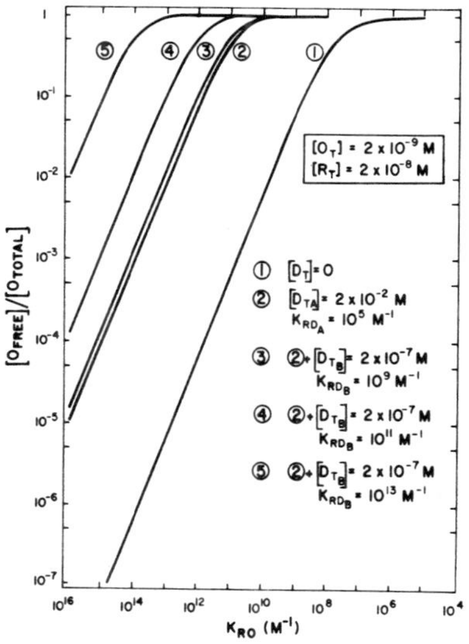

Figure 7

sites as a function of either non-specific repressor-binding constant (K_{RD}) for a total concentration $[D_T]$ of 2×10^{-2} M non-specific sites (upper abscissa and dashed curve), or $[D_T]$ at a K_{RD} of 10^5 M^{-1} (lower abscissa and solid curve). No inducer is present, and there is assumed to be only one class of non-specific binding site. At low K_{RD} (or low $[D_T]$), there is no non-specific binding and the system is repressed to a calculated basal level $([O]/[O_T]) \simeq 5 \times 10^{-7}$ of the constitutive level. As we increase K_{RD} (or $[D_T]$) the extent of repression decreases, approaching asymptotically the totally unrepressed state (i.e., constitutive levels of lac enzyme production) at $K_{RD} > 10^9$ M^{-1}, or $[D_T] > \sim 10$ M, under the conditions of the calculation. Note that the value of $[O]/[O_T]$ predicted for the levels of $[D_T]$ and K_{RD} which presumably apply in the *E. coli* cell is $\simeq 10^{-3}$, which is close to the basal level actually measured *in vivo* [fact (iii) above]. Thus Fig. 6 shows that in the absence of R binding to non-specific DNA sites, the basal rate of lac enzyme synthesis should be 3 to 4 orders of magnitude smaller than is actually observed, and that the observed basal level is primarily established by non-specific binding of R to D sites, the latter acting as "sinks" for R, in competition with O.

Effects of increasing $[R_T]$. Calculations (data not shown) designed to determine the basal level of lac enzyme synthesis as a function of $[R_T]$ show that in mutants over-producing repressor by 10- to 200-fold, the basal level is expected to decrease linearly with increasing $[R_T]$, from $\sim 10^{-3}$ for wild-type cells to $\sim 3 \times 10^{-5}$ of the constitutive level. This is in excellent accord with physiological measurements [fact (iv), above; Sadler and Novick, 1965; Smith and Sadler, 1971].

Effects of additional strong non-specific sites. Fig. 7 shows the calculated effects on $[O]/[O_T]$ of adding 100 stronger binding sites to the cell, in addition to the 10^7 sites/cell having $K_{RD} = 10^5$ M^{-1}. From Fig. 7 it can be seen that to have a significant effect on the basal level, such sites must exhibit repressor affinities in excess of 10^{11} M^{-1} (compare curves 2-5, for $K_{RO} = 10^{14}$ M^{-1}). Even a few (< 10) sites having K_{RD} as large as 10^{13} M^{-1}, would have little effect on $[O]/[O_T]$.

Operator-constitutive (O^c) mutations. Fig. 7 also shows that decreases of one to two orders of magnitude in K_{RO} (O^c-mutations) would still not bring $[O]/[O_T]$ to the observed basal level in the absence of competitive R binding to non-specific sites (curve 1), but would indeed show just the effects observed on the basal level with O^c mutants [fact (vi), above; Smith and Sadler, 1971; Jobe, Sadler and Bourgeois, 1974] when non-specific binding is included (curve 2).

Inducer-binding effects. In Fig. 8 are plotted the calculated effects of inducer binding on the fraction of lac operator which is free. The lower (solid) curve shows that, in the absence of non-specific binding of R and RI complex, derepression of the lac operon, even at supersaturating concentrations of I, cannot take place. This follows because the RI complex binds to O with only ∿ three orders of magnitude less affinity than does R itself (Jobe and Bourgeois, 1972); thus in the absence of the non-specific sites as "sink" for the RI complex, the latter itself binds to O and maintains repression at calculated values below the basal level. It is only if one makes the thermodynamically impossible assumption that I binding can reduce the affinity of the RI complex for O to values close to zero (Fig. 8, dashed curve), that total derepression in the absence of non-specific binding can be achieved, and even then only at total inducer levels $\sim 10^3$ times greater than actually required in vivo.

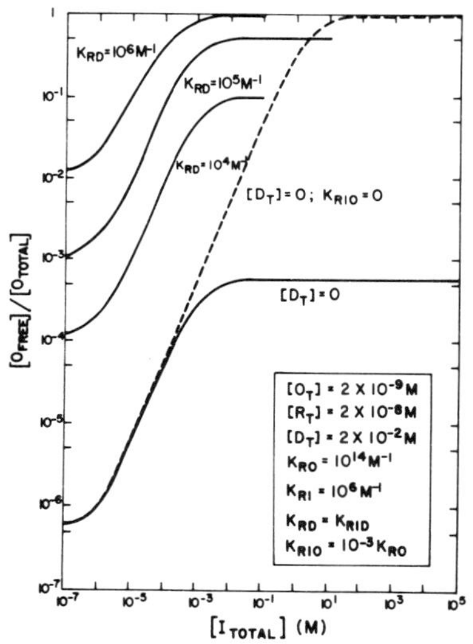

Figure 8

The upper curves in Fig. 8 show the calculated effects of added I on the basal level at three different non-specific binding affinities. (In constructing these curves we assume that $K_{RIO} = 10^{-3} K_{RO}$ [fact (vii), above] and that $K_{RID} = K_{RD}$ [fact (viii), above].) We note that only the curve for $K_{RD} = 10^5 M^{-1}$ starts at the observed basal level at low $[I_T]$ and reaches approximately the constitutive level (within a factor of two) at saturating concentrations of $[I_T]$. This observation provides an independent confirmation that the value of K_{RD} chosen for the "base-set" of parameters is indeed a reasonable representation of the *in vivo* situation or, more accurately, that the *ratio* of K_{RO} to K_{RD} is approximately correct.[3] Fig. 8 also shows that derepression of the *lac* operon occurs at the I concentrations observed *in vivo* [fact (v), above] when non-specific binding of R and RI is taken into account, but not otherwise.

Finally, we have carried out a series of calculations (not shown) in which we vary the ratio of K_{RIO}/K_{RO} to K_{RID}/K_{RD}. It can easily be shown by such procedures that if $K_{RIO}/K_{RO} > K_{RID}/K_{RD}$, the effect of increasing I concentrations is to further *repress* the *lac* operon. Thus a mutation-induced change in repressor structure which results in a greater decrease in K_{RID} than in K_{RIO} could at least partially explain the behavior of the i^r mutant phenotype (Myers and Sadler, 1971; Jobe and Bourgeois, 1972).

C. Repressor-Polymerase Interactions

We have extended these calculations to include the coupled equilibria involving the binding of RNA polymerase to the *lac* promoter. Here we consider that polymerase binds to the *lac* promoter (in equilibrium with other promoters in the cell), that R binds to O, and that *both* polymerase and repressor bind to non-specific DNA. Making reasonable assumptions for the concentrations of RNA (core) polymerase, sigma factor and promoter sites in the *E. coli* cell, we may use the relevant binding constants measured by Hinkle and Chamberlin (1972a) for polymerase to promoters and non-specific sites, together with the comparable parameters for repressor-operator-non-specific DNA interactions, to calculate the expected basal rate of *lac* enzyme synthesis for two simple models.

If we assume that polymerase and repressor bind *competitively*, then the ratio of the basal to the constitutive rate of *lac* enzyme synthesis can be

---

[3] We note that for the X86 mutant repressor (Jobe and Bourgeois, 1972) in which K_{RO} is approximately an order of magnitude *greater* than for w.t. repressor, but K_{RIO}/K_{RO} and K_{RD} are unchanged from the w.t. values, the fully induced rate is about an order of magnitude *smaller* than the w.t. constitutive rate of *lac* enzyme synthesis, as predicted from our model.

described by a modified form of the equations governing the competitive inhibition of enzymes by reversibly-bound inhibitors. If the binding of R to 0 and of polymerase to promoter is totally non-competitive, then this ratio simply equals $[0_{free}]/[0_{total}]$. Calculations made with currently available (or estimated) binding constants and component concentrations show that the *in vivo* basal rate is compatible with either model (for details see von Hippel *et al*., 1974), but when more complete data are obtained (especially for the polymerase system), these (and related) models should be distinguishable since they show quite different dependencies on the relevant system parameters. We note that for any competitive binding model, the present data are already adequate to show that repression of the *lac* operon is impossible in the absence of non-specific binding of *both* functional protein components.

IV. KINETICS OF THE *LAC* OPERATOR-REPRESSOR INTERACTION

Binding of *lac* repressor to non-specific DNA can also play a major role in the kinetics of the interaction of R with 0. We will consider only one aspect of this problem here; namely a possible mechanism for the transfer of R to 0 in which R binding to non-specific DNA is invoked to account for the anomalously fast apparent forward rate constant for formation of the RO complex. Other aspects of the kinetics of the R-0 interaction have been considered, and will be reported elsewhere.

The apparent R-0 forward rate constant (k_f) has been measured by means of a filter-binding technique (Riggs *et al*., 1970a). This assay uses very dilute solutions, thereby slowing the association rate sufficiently to make measurements possible. The experimental k_f can be compared with theoretical values of k_f generated by assuming that the rate-limiting step is diffusion-controlled, and then using the Debye-Smoluchowski equation:

$$(3) \qquad k_f = 4\pi\kappa a_f\,(D_R + D_0)\,N_0/1000 \qquad [M^{-1}\ sec^{-1}]$$

where N_0 is Avogadro's number, D_R and D_0 are the free-volume diffusion constants for R and 0, respectively (in cm^2/sec), κ is a unitless steric interaction factor, and a_f is an interaction radius (in cm). If the value used for a_f corresponds to the estimated size of the operator region, the predicted k_f is more than an order of magnitude smaller than that observed. This has led to the promulgation of a variety of models proposing that the "landing site" for R is considerably larger than the O region itself, and that the diffusion of R to O is then "facilitated" by various special mechanisms (see summary in von Hippel and McGhee, 1972; also, P. Richter and M. Eigen, personal communication). Using the measured k_f [e.g., 7×10^9 M^{-1} sec^{-1} at an ionic strength of 0.05 M, pH 7.4, 24° (Riggs et al., 1970a)] together with reasonable estimates of D_O ($< 10^{-8}$ cm^2/sec) and D_R ($\sim 2 \times 10^{-7}$ cm^2/sec), we calculate from eq. (3) that the value of the κa_f product is $\sim$500 Å. Fig. 9, which depicts schematically the situation that actually prevails in the DNA-repressor solution under the conditions of the filter-binding measurement, suggests an interpretation of this large apparent interaction radius in molecular terms. Since the operator-containing DNA is essentially a stiff random coil, its segments are not uniformly distributed throughout the solution (at the dilutions used in the assay). Rather, one can show by considering polymer chain statistics that at a certain level of dilution the DNA molecules will separate into domains with a characteristic radius (radius of gyration, r_g) of $\sim$5000 Å for λDNA under the conditions of the filter assay. The intra-domain concentration of non-specific DNA binding sites will be quite high ($\sim 2 \times 10^{-4}$ M). This suggests that a repressor molecule striking anywhere within a DNA domain will have a considerable probability of remaining bound within that domain. We visualize that the transfer of R within the domain proceeds rapidly by a direct DNA to DNA mechanism involving "ring-closure" events with the R molecule transiently bound between two DNA sites. If both sites are of the non-specific type, there is a 50% probability of inter-site transfer. A series of such events will eventually result in transfer of R to O; this process may be very fast because the proposed

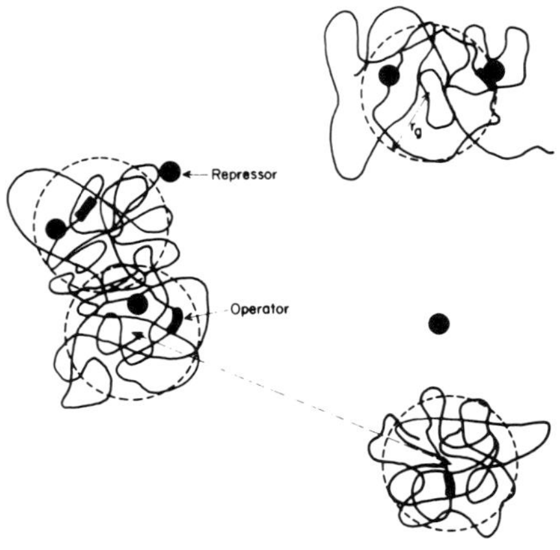

Figure 9

mechanism circumvents the need to overcome the activation barrier for dissociation into solution of non-specifically bound R. Since the RO affinity is much greater than that of RD, the probability of subsequent "ring-closure" events removing the R from O is small.

These considerations provide a rationalization for the large value observed for κa_f. We identify a_f with r_g (the average radius of the polymer domain; see Fig. 9) and κ with the probability that an initial collision is fruitful (i.e., leads to an RO complex). Under the usual filter assay conditions we obtain $\kappa = 0.1$, which implies that there is an ∿90% probability of an R molecule being lost back into solution following collision with a DNA domain, and a 10% chance that an R molecule will successfully find the O region (via intra-domain transfers).

We are presently attempting to "model" the intra-domain transfer process, and are pursuing experiments designed to test this interpretation. This mechanism may also play a role in the binding of other regulatory proteins to DNA (cf. RNA polymerase to promoters, Hinkle and Chamberlin, 1972b; and λ repressor to its operator sites, Chadwick et al., 1971).

V. MOLECULAR ASPECTS

Study of the non-specific binding mechanisms of R to D sites may also yield molecular information related to the specific binding of R to O. The experiments described above (Section II) show that the size of the non-specific binding site (25 to 30 base pairs) is comparable to that of the operator region (Gilbert, this volume). Furthermore, the melting data reveal that R binds preferentially to double-helical non-operator DNA (i.e., R is a helix-stabilizing protein) as seems also to be the case for R binding to O (Riggs et al., 1970b). In addition, Platt et al. (1971) have shown that limited tryptic digestion of R leaves a "core" fragment which retains binding activity for I, but no longer binds to O; we have shown that this core fragment also has lost its affinity for non-specific sites. Thus, both specific and non-specific binding seem to share common elements of the binding region of the protein.

On the other hand, we have shown directly that I binds to non-specifically bound R without affecting the affinity of R for D sites. Furthermore, the number of I binding sites and the I binding affinity are approximately the same as for unbound repressor; and I binds non-cooperatively to non-specifically bound R. Work in our laboratory, and in others, on IPTG-dependent dissociation of the RO complex (using filter-binding techniques) shows that the Hill coefficient for this process is surely greater than unity (Bourgeois and Jobe, 1970; A.D. Riggs, personal communication; C. Gundlach and J.R. Sadler, personal communication). Determination of the exact extent of the cooperativity may shed light on the mode of R subunit binding to O and, by inference, on the mechanisms by which R may be transferred to the O region from non-specific binding sites.

REFERENCES

Beckwith, J.R. and Zipser, D. (eds): The Lactose Operon. Cold Spring Harbor Laboratory, Cold Spring Harbor, New York (1970).

Bourgeois, S. and Jobe, A.: Superrepressors of the lac operon. In The Lactose Operon, J.R. Beckwith and D. Zipser, eds. Cold Spring Harbor Laboratory, Cold Spring Harbor, New York, pp. 325-341 (1970).

Chadwick, P., Pirrotta, V., Steinberg, R., Hopkins, N. and Ptashne, M.: The λ and 434 phage repressors. Cold Spring Harbor Symp. Quant. Biol. 35, 283-294 (1970).

Hinkle, D.C. and Chamberlin, M.J.: Studies of the binding of Escherichia coli RNA polymerase to DNA. I. The role of sigma subunit in site selection. J. Mol. Biol. 70, 157-185 (1972a).

Hinkle, D.C. and Chamberlin, M.J.: Studies of the binding of Escherichia coli RNA polymerase to DNA. II. The kinetics of the binding reaction. J. Mol. Biol. 70, 187-195 (1972b).

Jobe, A. and Bourgeois, S.: The lac repressor-operator interaction. VII. A repressor with unique binding properties: the X86 repressor. J. Mol. Biol. 72, 139-152 (1972).

Jobe, A., Sadler, J.R. and Bourgeois, S.: lac repressor-operator interaction. IX. The binding of lac repressor to operators containing O^c mutations. J. Mol. Biol. 85, 231-248 (1974).

Laiken, S.L., Gross, C.A. and von Hippel, P.H.: Equilibrium and kinetic studies of E. coli lac repressor-inducer interactions. J. Mol. Biol. 66, 143-155 (1972).

Lin, S.-Y. and Riggs, A.D.: lac repressor binding to non-operator DNA: Detailed studies and a comparison of equilibrium and rate competition methods. J. Mol. Biol. 72, 671-690 (1972).

McGhee, J.D. and von Hippel, P.H.: Theoretical aspects of DNA-protein interactions: Cooperative and noncooperative binding of large ligands to a one-dimensional homogeneous lattice. J. Mol. Biol. 86, 469-489 (1974).

Myers, G.L. and Sadler, J.R.: Mutational inversion of control of the lactose operon of Escherichia coli. J. Mol. Biol. 58, 1-28 (1971).

Platt, T., Files, J.G. and Weber, K.: Lac repressor. Specific proteolytic destruction of the NH_2-terminal region and loss of the deoxyribonucleic acid-binding activity. J. Biol. Chem. 248, 110-121 (1973).

Riggs, A.D., Bourgeois, S. and Cohn, M.: The lac repressor-operator interaction. III. Kinetic studies. J. Mol. Biol. 53, 401-417 (1970a).

Riggs, A.D., Suzuki, H. and Bourgeois, S.: lac repressor-operator interaction. I. Equilibrium studies. J. Mol. Biol. 48, 67-83 (1970b).

Sadler, J.R. and Novick, A.: The properties of repressor and the kinetics of its action. J. Mol. Biol. 12, 305-327 (1965).

Smith, T.F. and Sadler, J.R.: The nature of lactose operator constitutive mutations. J. Mol. Biol. 59, 273-305 (1971).

von Hippel, P.H. and McGhee, J.D.: DNA-protein interactions. Ann. Rev. Biochem. 41, 231-300 (1972).

von Hippel, P.H., Revzin, A., Gross, C.A. and Wang, A.C.: Non-specific DNA binding of genome regulating proteins as a biological control mechanism. I. The lac operon: Equilibrium aspects. Proc. Natl. Acad. Sci., U.S.A. (1974) in press.

Received July 13, 1974

DISCUSSION

Changeux: It was not clear to me what evidence you have for the homogeneity of the non-specific binding sites. In addition, did you try the interaction of the repressor with synthetic polynucleotides?

von Hippel: The shape of the binding isotherm suggests that the distribution of non-specific binding constants is not too wide. Also, at low ionic strengths (under the conditions of the competitive binding experiments) DNAs of varying base composition bind about equally. We find, at both low and intermediate ionic strengths, that the synthetic (double-stranded) polynucleotides we have tested bind with comparable affinities to natural native DNA, though in qualitative agreement with the competitive filter-binding results of Riggs and co-workers, we find that poly-d(A·T) binds somewhat (factors 3 to 5) better than natural DNAs. Single-stranded DNAs (natural or synthetic) bind much more weakly.

Riesner: In the interchain-transfer model you assume at least two binding sites on the repressor for DNA. Is this a reasonable assumption? Can you check it experimentally by short dAT pieces?

von Hippel: Yes, one possible model for repressor is a tetramer, with two DNA binding sites on opposite sides involving two R subunits each. This would lead to a nice transfer system, but is not required by the model in general. If the protein has at least two sites on opposite sides of the protein with some DNA affinity, then transfer could take place by the proposed means, though, if the sites were of unequal affinity, transfer would be more efficient in the direction of the "stickier" site. In principle, one could show that the ratio of very short pieces of poly-(dA·T) or other short DNA stick to repressor extrapolates to 2:1 under limiting conditions, if there are indeed two sites for DNA binding per R tetramer. People (Sadler and coworkers and others) are trying such experiments, but they are difficult.

G. Weber: According to your model of trapping the repressor inside a 'DNA cage', the binding constant will represent the ratio of the rates of entering into and escaping from the cage and not that of association and dissociation of repressor and operator. The dissociation constant measured by the chemical potential of the repressor necessary for half-saturation of the operator could then turn out to differ from the one measured by rates, even by orders of magnitude.

von Hippel: It may be that what is measured by ratios of rates could in some way turn out to differ from the directly measured equilibrium constant. It is hard to say, since the details of the filter binding experiment are hard, to analyze in molecular terms. On the other hand, ultimately only those DNA molecules which are bound to repressor via operator at the moment of filtration will be measured, in determinations of either forward or backward rate constants, so the other effects probably cancel out. It is interesting to conjecture the fact that, at the moment of filtration, the repressor will be distributed between non-specific and operator sites on the basis of the ratio of site concentrations times binding constants, and, that only those bound to operator will be measured, may account for the fact that 100 % operator binding is never attained in these experiments.

Sund: You analyzed the binding of the repressor to dA·T by circular dichroism titration. The graphical extrapolation, which yields the base pairs DNA/repressor ratio, gives correct values only if at the beginning of the titration the ligand is completely bound, that means the dissociation constant has to be lower than µM. Is this the case?

von Hippel: Yes, the titration curves shown were done under ionic conditions of complete ligand binding. At higher concentrations of salt, where binding under these conditions is not complete, one gets the expected rounded titration curve shape.

Veeger: You showed that the repressor binds to 30 base pairs. I could visualize a model in which the repressor recognizes any base pair on the operator. This would lower the observed second-order rate constant because of statistical reasons by about 1/30. Furthermore, in assuming that the four subsites have equal chance of recognizing a base pair another decline in rate constant by 1/4 has to be taken into account, giving an actual intrinsic rate constant of $\sim 10^7 sec^{-1}M^{-1}$, a much more reasonable value.

This would not interfere with the high value of the equilibrium constant of $10^{14}M^{-1}$ because the isomerisation leading to interaction with all bases of the operator is responsible for this:

$$R + O \underset{k_{-1}}{\overset{k_1 = 10^7}{\rightleftharpoons}} RO \underset{k_{-2}}{\overset{k_2}{\rightleftharpoons}} R^*O$$

The state of dissociation will be determined by the slowest step in the backward reaction according to:

$$K = \frac{[R^*O]}{[R]\cdot[O]} = \frac{k_1k_2}{k_{-1}k_{-2}}$$

von Hippel: I believe that your proposal is mathematically sound, but it strikes me as molecularly unlikely. Thus the repressor would have to discriminate base pairs associated with operator from others, and isomerize appropriately. Loosely stated, your model is a modified "sliding" model with a "landing site size" of ~88 bases (~ 30 for operator and 29 on each side, in the extreme). As such, I think, it could be tested.

Maass: I don't see yet quite the advantage of your model relative to the sliding model, because in your model you have to break at least one interaction between repressor and DNA in order to be able to transfer the repressor, involving the activation energy of dissociation. In the sliding model, whatever it looked like, the activation barriers may be smaller.

von Hippel: Perhaps so - but it seems to us that the segmental moving apart of the DNA segments bound to the R tetramer on opposite sides may help to provide this activation energy, since a repressor "cross-bridge" would essentially be competing against entropic forces to hold these two segments together. In addition, as you and Dr. Riesner suggested privately, there could be an anti-cooperative aspect to the binding, which would also lower the activation barrier. The molecular details of a sliding model (taking into account the details of specific and non-specific binding sites and DNA topology) would have to be worked out before one could assess the magnitude of the activation energy to sliding.

Steinhardt: As I understand it the diffusion constants are used essentially only to calculate collision rates between R and O. However, (speaking only from memory) Flory showed long ago that when two macromolecules react, very much higher collision rates result than the diffusion constants would predict. The reason is that numerous collisions occur in each encounter, due to rapid local conformational fluctuations. If this were not so, condensation polymerization would be exceedingly slow at high molecular weights.

von Hippel: Yes, but if every collision (as with this sticky domain model) were fruitful in, at least initially, sticking R to the non-specific DNA domain, then the second-order part of the reaction would be over and possible subsequent collision would not occur or be necessary.

Riesner: The speeding-up effect is not only related to theoretical considerations but to comparison of other polymer-polymer or small molecule-polymer association which have been determined experimentally to be slower.

Wagner: Did you compare the ionic strength dependence of the unspecific and the specific repressor-DNA interaction, and what are the results?

von Hippel: Riggs and coworkers have done such experiments, using the filter-binding operator-competition method, in more detail than we have up to now. I believe they find the specific and non-specific binding constants vary approximately in parallel with changes in ionic strength.

Sund: Is the melting of the repressor reversible?

von Hippel: Not totally under usual conditions, but it can be partially reversed, and perhaps with considerable care (avoiding aggregation, etc.) conditions might be found to permit total reversal. But this has not yet been accomplished.

Section IV. Receptors

Chairman: *W. Gilbert*

Binding and Functional States of the Cholinergic Receptor Protein from Torpedo Marmorata

H. Sugiyama, J. L. Popot, J. B. Cohen, M. Weber and *J. P. Changeux*

The cholinergic (nicotinic) receptor protein has now been characterized and purified in milligram quantities both from Electrophorus (1-7) and Torpedo (8-11) electric organs. Since the most characteristic chemical and physical properties of the purified protein in detergent solution are known (6, 8, 11), we shall be primarily concerned, here, by the receptor protein in its membrane environment and by the mechanism of its control of cation translocation through the excitable membrane.

1) THE PHYSIOLOGICAL RESPONSE TO ACETYLCHOLINE

The response of Electrophorus (12-17) and Torpedo (18) electroplaque to acetylcholine and its congeners can be monitored by electrophysiological methods following two different approaches : either through the neurally evoked release of acetylcholine or upon application in bath of known concentrations of cholinergic ligands. In both cases the effect is an increase of membrane conductance for Na^+, K^+ and possibly Ca^{++} (14, 17).

To account for the selectivity of recognition for both cholinergic ligands and alkali cations, it was proposed that the most elementary unit, or protomer, involved in the electrogenic action of acetylcholine comprises at least two elements of structure : a receptor moiety which binds cholinergic ligands and an ionophore involved in the selective translocation of ions (19). Both elements might as well be carried by different polypeptide chains or constitute different parts of the same polypeptide chain. The coupling between ionophore and receptor would then be mediated by a change of structure analogous but not necessarily identical to the well documented "allosteric" transition of regulatory enzymes. The protomer would undergo a transition between at least two states : a resting state, R, favored in the membrane at rest, stabilized by the antagonists and impermeable to cations and an active state, A, exhibiting a preferential affinity for agonists and with the ionophore in an open conformation.

The time course of the response differs markedly when the agonist is released by the nerve terminals and when it is applied in bath. In the first case, the conductance reaches its peak in less than a millisecond, remains at this high value for a few milliseconds and then returns to its resting value again for times of the order of the millisecond. Little if any refractory period exist after the endplate potential.

When cholinergic agonists are applied in bath the membrane potential decreases until, after a few minutes, it reaches a plateau. The amplitude of the steady state depolarisation can be taken as a measure of the amplitude of the response to the agonist. Antagonists block this response in a competitive manner, local anesthetics such as prilocaine or tetracaine in a non-competitive one.

When the membrane conductance is measured instead of the potential, then the observed response is transient, even though the concentration of agonist remains constant. After a few minutes of contact the conductance starts to decrease with a rate constant in the order of the minute (20-21). This is the phenomenon called receptor "desensitization" which has been studied extensively in the case of the neuromuscular junction and found also with *Electrophorus* (22) and *Torpedo* (18) electroplaque. In the case of the neuromuscular junction, local anesthetics and divalent cations such as Ca^{++} enhance the rate of desensitization. To account for this phenomenon, Katz and Thesleff (20) have postulated that the receptor protomer might exist under a third "desensitized" state D, which would bind the agonists with a high affinity but would not transport cations.
The overall scheme would then be :

Resting state (R)	⇌	Active state (A)
binds antagonists ionophore shut		binds agonists ionophore open

⇅ Desensitized state (D) ⇅

binds agonists (high affinity)
local anesthetics, Ca^{++} ions
ionophore shut

2) THE SUBCELLULAR FRACTIONATION OF TORPEDO ELECTRIC ORGAN AND THE RECEPTOR-RICH MEMBRANE FRAGMENTS.

A useful step in the analysis of this mechanism at the molecular level was the development of a preparation simpler than the electroplaque and with which the binding of cholinergic ligands and the corresponding permeability response could be followed in parallel. It was first provided by a subcellular membrane fraction from Electrophorus electric tissue (23). However, the electric organ from Torpedo, which is about 10 times richer in nerve terminals than that of Electrophorus, appeared to be a much more convenient starting material. The procedure we have developped in this laboratory (24) and which has been subsequently used with success by Raftery and coworkers (this symposium), consists in the homogenization of the electric tissue in distilled water followed by ultracentrifugation in sucrose gradient. It yields membrane fragments which band at 39 % (w/v) sucrose and contain about 30 % of all the tissue toxin binding sites. Their specific activity is 2-3,000 nmoles of α-toxin sites per g protein and 100 times less in acetylcholinesterase catalytic site (24). As much as 20-30 % of their proteins consists of the receptor protein.

The gel electrophoresis in SDS of these membrane fragments (34) reveals a rather simple pattern of proteins with a dominant band of apparent molecular weight 40,000 which is precisely the value reported for the subunit molecular weight of the receptor protein purified from Torpedo (8) and Electrophorus (3-6, 25). This band is the only one labelled by Karlin's reagent MBTA (3).

Both freeze-etching (26) and negative staining (26, 27) reveal, in these receptor-rich membrane fragments, a high density of particles of dimension and shape similar to those of Electrophorus purified receptor protein (26). They are ring-like particles 8-9 nm in diameter consisting of 5-6 subunits surrounding an electron dense pit. These particles make hexagonal arrays with a surface density of 10,000-15,000 particles per μm^2. In the experiments done with purified membrane fragments, the particles appear only after deep etching and therefore are exposed to the membrane surface (26). In the intact electric tissue a lattice of particles with similar center-to-center distances is seen in the subsynaptic membrane, but in its fracture plane (28). The implication of this finding is that homogenization and fractionation of the electric tissue leads to a reorganization of the structure of the subsynaptic membrane.

After orientation by centrifugation the receptor-rich membrane fragments give X-ray diffraction patterns (29) with both equatorial and meridional reflections. The equatorial reflections confirm the presence of lattice structure in the plane of the membrane with a 8-9 nm center-to-center distance of the diffracting units. The lattice organisation appears rather labile and collapses below 35 % relative water content. The meridional reflections indicate a thickness of 90-150 Å which is rather unusual for a cytoplasmic membrane.

These membrane fragments constitute a particularly convenient preparation to study the functional properties of the receptor protein in its membrane environment.

3) THE PERMEABILITY RESPONSE OF RECEPTOR-RICH MICROSACS

If homogenized in the presence of Ca^{++}, the Torpedo receptor-rich membrane fragments reseal into closed vesicles or microsacs, and $^{22}Na^{+}$ efflux from these microsacs can be followed by the method of Kasai and Changeux (23). This passive efflux is accelerated in the presence of carbamylcholine and other agonists, and this effect is blocked by d-tubocurarine and α-toxin (30). Torpedo microsacs are therefore excitable in vitro. The concentration-effect curve for carbamylcholine reaches a plateau around 10^{-3}M, the half-maximal response taking place between 10^{-5}M and 10^{-4}M.

Prolonged exposure (minutes) of the excitable microsacs to agonists (carbamylcholine or PTA (31)) causes a subsequent decrease of the amplitude of the response to carbamylcholine. The magnitude of the effect increases with the concentration of agonist and the length of the preincubation period, and varies rather largely from one preparation to another. In many cases, a 10 minutes exposure to 10^{-4}M carbamylcholine abolishes completely the response. The preparation of Fig. 1 was even more sensitive. The

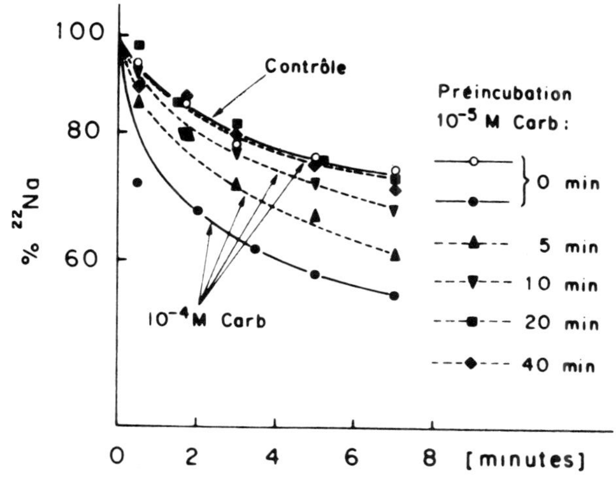

Fig. 1 - Evidence for long lasting "desensitization" of the response of <u>Torpedo</u> membrane fragments to carbamylcholine. From J.L. Popot <u>et al.</u> (31).

effect is blocked by d-tubocurarine and dilution after an exposure to carbamylcholine leads to a restoration of the response. The effect is thus reversible and presents strong analogies with the phenomenon of <u>desensitization</u> (31).

4) THE BINDING OF CHOLINERGIC LIGANDS TO RECEPTOR-RICH MICROSACS

Since with the receptor-rich membrane fragments, micromolar concentrations of receptor site are accessible in the test tube, binding of radioactive cholinergic ligands to the membrane-bound receptor protein can be measured with great accuracy. When acetylcholine and decamethonium are used as ligands, preincubation of the membrane fragments with α-toxin completely blocks the binding which, therefore, takes place almost exclusively at the level of the cholinergic receptor site (32). In the case of decamethonium, the binding curve is an hyperbola and half saturation takes place around 8×10^{-7}M. The interaction with antagonists such as d-tubocurarine or flaxedil is, in first approximation, competitive. The binding curve of acetylcholine in the presence of an acetylcholinesterase inhibitor, Tetram, exhibits with some preparations of membranes a slightly sigmoid shape (Hill coefficient : 1.3-1.4) and half-saturation occurs around 10^{-8}M. For carbamylcholine, the dissociation constant is 0.5-1 x 10^{-6}M. It is at least one order of magnitude smaller than the concentration which gives half of the maximal response following $^{22}Na^+$ efflux. Equilibrium dissociation constants for a set of about 10 ligands have been determined. The specificity of the binding is that expected for the nicotinic receptor (32).

Binding of cholinergic ligands to the receptor site present in the receptor rich membrane fragments can also be followed by fluorescence spectroscopy using an environment sensitive fluorophore designed by Weber et al. (33) : DNS-chol. This compound presents structural analogies with both cholinergic agonists and local anesthetics. It exhibits indeed both pharmacological activities on Electrophorus electroplaque (34). The interaction of DNS-chol with the membrane fragments can be followed by energy transfer from the membrane proteins (λ ex = 287 nm) in a differential spectrofluorimeter (35). At micromolar concentrations of DNS-chol and receptor site, the fluorescence intensity emitted at 540 nm (maximum emission) decreases by about 70 % in the presence of high levels of cholinergic effector or snake α-toxin. No such effect was observed with membrane fragments which did not contain the cholinergic receptor site. This decrease was interpreted as due to the displacement of DNS-chol from the receptor site. As expected, dissociation constants estimated by this method agreed with those measured directly with radioactive ligands (35). Binding of cholinergic ligands to the membrane-bound receptor can, therefore, be monitored by fluorescence spectroscopy.

One of the most interesting observations made with DNS-chol, concerns the spectral characteristics of the residual fluorescence emitted in the presence of an excess of cholinergic ligand or α-toxin. The maximum emission of this residual fluorescence takes place at 537 $\pm$ 3 nm when the cholinergic ligand occupying the receptor site is an antagonist or the α-toxin, but it is shifted to the blue by 15 nm when the ligand is an agonist on Electrophorus electroplaque. Antagonists reverse the shift caused by the agonists and conversely (35). It was concluded that DNS-chol binds in addition to the cholinergic receptor site, to "secondary" sites located on or in the vicinity of the receptor protein. There, the fluorescence of DNS-chol would be sensitive to the pharmacological activity of the ligand interacting with the receptor site and would, therefore, monitor a structural transition of the receptor associated in some manner with the physiological response of the membrane.

Since DNS-chol is active as a local anesthetic on Electrophorus electroplaque, the effect of local anesthetics such as prilocaine, lidocaine or dimethisoquin on the binding of cholinergic ligands to the receptor site was studied. At concentrations where they block non competitively the response of Electrophorus electroplaque to bath applied carbamylcholine, they do not bind to the cholinergic receptor site (32) but reverse the 15 nm shift observed in the presence of agonists. In other words, the "secondary" sites for DNS-chol are closely related if not identical to the local anesthetic binding site.

It was further shown that local anesthetics control the binding properties of the receptor site itself (34, 36). In the presence of 3 mM prilocaine the affinity of the receptor site for acetylcholine (and other agonists or antagonists) increase by a factor of 3. Moreover, when the membrane preparation is such that the binding of acetylcholine appears cooperative, then, the same concentration of prilocaine causes a change of shape of the acetylcholine binding curve from a sigmoid (n_H = 1.4) into an hyperbola (n_H = 1.0). This conversion of shape, which accompanies the affinity increase, is reminiscent of a characteristic property of regulatory proteins (37).

Calcium ions are known to affect the pharmacological action of local anesthetics. Like them, calcium ions cause an increase of affinity of the membrane-bound receptor protein for cholinergic ligands but, at variance with them, they do not modify the sigmoid shape of the binding curve (34, 36).

5) CONSEQUENCES OF SOLUBILIZATION BY DETERGENTS ON THE BINDING PROPERTIES OF THE RECEPTOR PROTEIN.

Franklin and Potter (38) have shown that detergent solubilisation of Torpedo membrane fragments causes a decrease of affinity for a cholinergic agonist. We have confirmed this finding and extended it. As already men-

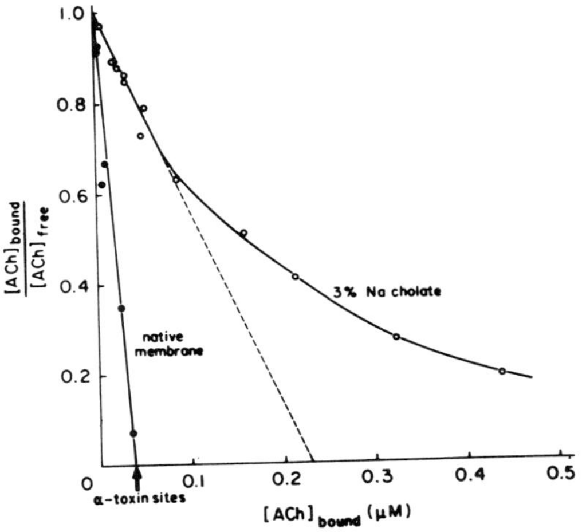

Fig. 2 - Consequences of solubilisation by the cholate or the binding properties of Torpedo receptor-rich membrane fragments. From H. Sugiyama and J.P. Changeux (39).

tioned, the binding curves for decamethonium or acetylcholine to the receptor-rich membrane fragments can be interpreted in terms of an homogeneous (or slightly interacting) population of binding sites. Dissolution by Na Cholate causes a decrease of affinity of the receptor site which is accompanied by a marked heterogeneity of the binding constants (34, 36). The binding data can be fitted simply on the basis of two populations of independent sites with dissociation constants for acetylcholine (in the presence of Tetram) close to 10^{-7}M (medium affinity) and larger than 10^{-6} M (low affinity)(Fig. 2).

When the concentration of detergent is varied, the data are better interpreted in terms of a change in the ratio of the low and medium affinity sites than by a continuous change of binding constants (39). Binding of acetylcholine to these low and medium affinity sites is no longer sensitive to local anesthetics and calcium ions. The binding curve of acetylcholine or decamethonium to the receptor protein purified from detergent extracts does not exhibit any sign of cooperativity (Fig. 3).

Interestingly, when the detergent concentration is decreased by dilution below a critical concentration close to 0.5 % (w/v), the receptor protein

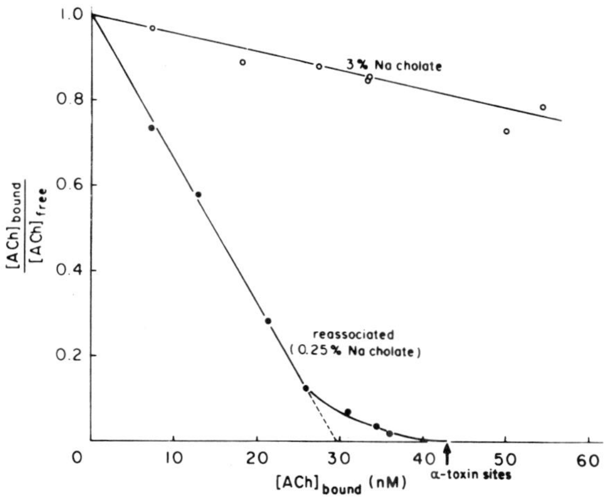

Fig. 3 - Recovery of high affinity sites upon reassociation of cholate solubilized receptor-rich membrane fragments. From H. Sugiyama and J.P. Changeux (39).

reaggregates from the crude membrane extract and high affinity sites are recovered in significant amounts. The sensitivity of these high affinity sites to local anesthetics or Ca ions is recovered as well.

The binding properties of the cholinergic receptor site and their control by local anesthetics and calcium ions seem, therefore, highly constrained by the membrane environment.

6) RECONSTITUTION OF AN EXCITABLE MEMBRANE FROM ITS DISSOCIATED COMPONENTS.

Using a rather sophisticated method to eliminate the detergent after cholate solubilization of receptor-rich membrane fragments, it becomes possible to reconstitute microsacs which trap $^{22}Na^{+}$ (30). The release of $^{22}Na^{+}$ from these microsacs is enhanced by carbamylcholine and the α-toxin blocks this effect. The reconstitution of an excitable membrane from its dissociated components can therefore be achieved. Such a method should lead to the identification of the ionophore and of the factors which critically control the binding properties of the receptor site and its coupling with the ionophore (Fig. 4).

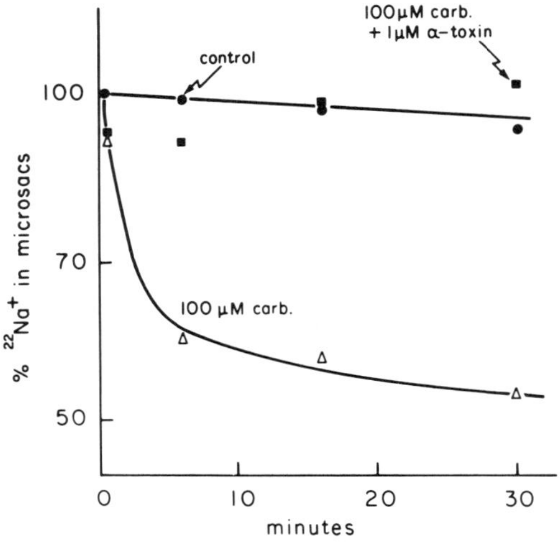

Fig. 4 - Reconstitution of excitable microsacs from detergent solubilized receptor-rich membrane fragments from Torpedo. From G.L. Hazelbauer and J.P. Changeux (30).

CONCLUSION

The cholinergic receptor protein presents several properties typical of globular regulatory proteins. In addition, it behaves like an "integral" membrane protein and its binding properties are under the control of its membrane environment lipidic and/or proteic. It is a membrane-bound regulatory protein.

A rather puzzling observation is the occurrence of several states of affinity of the receptor protein for cholinergic agonists. Another one is the fact that, following $^{22}Na^+$ efflux, half of the maximal response takes place around 10^{-5}M carbamylcholine while the equilibrium dissociation constant of carbamylcholine for the same membrane fragments is 5×10^{-7}M. Such a discrepancy might be accounted for at least two models :

1) Several independent receptor units carrying one receptor site share a common ionophore and the ionophore opens only when at least 90 % of the sites are occupied by the agonist.
2) The high affinity sites are not functional but sites with a lower affinity, present in small amounts in the membrane fragments, account

for the permeability response. Since Ca^{++} ions and local anesthetics, are known to enhance the rate of desensitization of the receptor, it might be proposed that the "high affinity" sites are carried by a "desensitized" state of the receptor protein. The functional state of the receptor protein involved in the increase of permeability of the membrane would rather be "medium" and/or "low affinity" states.

A naive remark renders the second model the most plausible. Let assume that the active, permeable state of the cholinergic protomer exhibits a dissociation constant close to 10^{-8}M for the physiological transmitter, acetylcholine. Then, if we further assume that the "on" rate constant of association is diffusion controlled, the "off" rate constant of dissociation of acetylcholine from its complex with the receptor site would be in the order of 0.1 sec. The synapse would be blocked for a time much longer than the duration of the end-plate potential. As mentioned earlier (34) this is not seen in vivo. Experiments are presently designed which should lead to the distinction between the two models considered.

This work was supported by grants from the National Institutes of Health, United States Public Health Service, the Centre National de la Recherche Scientifique, the Délégation Générale à la Recherche Scientifique et Technique, the Fondation pour la Recherche Médicale Française, the Collège de France, and the Commissariat à l'Energie Atomique.

REFERENCES

1 - Olsen, R., Meunier, J.C. & Changeux, J.P. : Progress in purification of the cholinergic receptor protein from Electrophorus electricus by affinity chromatography. FEBS Letters 28, 96-100 (1972).

2 - Klett, R.P., Fulpius, B.W., Cooper, D., Smith, M., Reich, E. & Possani L.D. : The acetylcholine receptor. I. Purification and characterization of a macromolecule isolated from Electrophorus electricus. J. Biol. Chem. 248, 6841-6853 (1973).

3 - Karlin, A. & Cowburn, D. : The affinity-labeling of partially purified acetylcholine receptor from electric tissue of Electrophorus. Proc. Nat. Acad. Sci. USA 70, 3636-3640 (1973).

4 - Lindstrom, J. & Patrick, J. : Purification of the acetylcholine receptor by affinity chromatography. In Synaptic transmission and nerve interaction, 191-216. Raven Press, New-York.

5 - Biesecker, G. : Molecular properties of the cholinergic receptor purified from Electrophorus electricus. Biochemistry 12, 4403-4409 (1973).

6 - Meunier, J.C., Sealock, R., Olsen, R. & Changeux, J.P. : Purification and properties of the cholinergic receptor from Electrophorus electricus electric tissue. Europ. J. Biochem. 45, 371-394 (1974).

7 - Chang, H.W. : Purification and characterization of acetylcholine receptor-I from Electrophorus electricus. Proc. Nat. Acad. Sci. USA 71, 2113-2117 (1974).

8 - Eldefrawi, M.E. & Eldefrawi, A.T. : Purification and molecular properties of the acetylcholine receptor from Torpedo electroplax. Arch. Biochem. Biophys. 159, 362-373 (1973).

9 - Karlsson, E., Heilbronn, E. & Widlund, L. : Isolation of the nicotinic acetylcholine receptor by biospecific chromatography on insolubilized Naja naja neurotoxin. FEBS Letters 28, 107-111 (1972).

10 - Schmidt, J. & Raftery, M.A. : Purification of acetylcholine receptors from Torpedo californica electroplax by affinity chromatography. Biochemistry 12, 852-856 (1973).

11 - Potter, L. : Acetylcholine receptors in vertebrate skeletal muscles and electric tissue. In Drug Receptors, H.P. Rang ed., 295-312 (1973) Mac Millan, London.

12 - Keynes, R.D. & Martins-Ferreira : Membrane potentials in the electroplates of the electric eel. J. Physiol. 119, 315-351 (1953).

13 - Higman, H., Podleski, T.R. & Bartels, E. : Apparent dissociation constants between carbamylcholine, d-tubocurarine and the receptor. Biochim. Biophys. Acta 75, 187-193 (1963).

14 - Higman, H., Podleski, T.R. & Bartels, E. : Correlation of membrane potential and potassium flux in the electroplax of Electrophorus. B.B.A. 79, 138-150 (1964).

15 - Karlin, A. : Permeability and internal concentration of ions during depolarization of the electroplax. Proc. Nat. Acad. Sci. USA 58, 1162 (1967).

16 - Changeux, J.P. & Podleski, T.R. : On the excitability and cooperativity of the electroplax membrane. Proc. Nat. Acad. Sci. USA 59, 944-950 (1968).

17 - Ruiz-Manresa, F. & Grunfest, H. : Synaptic electrogenesis in eel electroplaques. J. Gen. Physiol. 57, 71-92 (1971).

18 - Bennett, M.V., Wurtzel, M. & Grundfest, H. : The electrophysiology of electric organs of marine electric fishes. I. Properties of electroplaques of Torpedo nobiliana. J. Gen. Physiol. 44, 757-804, (1961).

19 - Changeux, J.P., Podleski, T.R. & Meunier, J.C. : On some structural analogies between acetylcholinesterase and the macromolecular receptor of acetylcholine. J. Gen. Physiol. 54, 225-244 S (1969).

20 - Katz, B. & Thesleff, S. : A study of the "desensitization" produced by acetylcholine at the motor endplate. J. Physiol. London 138, 63-80 (1957).

21 - Magazanik, L. & Vyskočil, F. : Desensitization at the motor endplate. In Drug Receptors, H.P. Rang ed., 105-120, Mac Millan London (1973).

22 - Changeux, J.P. & Lester, H. : Conductance increases produced by bath application of cholinergic agonists to *Electrophorus* electroplaques. Manuscript in preparation.

23 - Kasai, M. & Changeux, J.P. : In vitro excitation of purified membrane fragments by cholinergic agonists. I, II, III. J. Membrane Biol. 6, 1-80 (1971).

24 - Cohen, J.B., Weber, M., Huchet, M. & Changeux, J.P. : Purification from *Torpedo marmorata* electric tissue of membrane fragments particularly rich in cholinergic receptor. FEBS Letters 26, 43-47 (1972).

25 - Hucho, F. & Changeux, J.P. : Molecular weight and quaternary structure of the cholinergic receptor protein extracted by detergents from *E. electricus* electric tissue. FEBS Letters 38, 11-15 (1973).

26 - Cartaud, J., Benedetti, L., Cohen, J.B., Meunier, J.C. & Changeux, J.P. : Presence of a lattice structure in membrane fragments rich in nicotinic receptor protein from the electric organ of *Torpedo marmorata*. FEBS Letters 33, 109-113 (1973).

27 - Nickel, E. & Potter, L.T. : Ultrastructure of isolated membranes of *Torpedo* electric tissue. Brain Research 57, 508-517 (1973).

28 - Orci, L., Perrelet, A. & Dunant, Y. : A peculiar substructure in the postsynaptic membrane of *Torpedo* electrophorax. Proc. Nat. Acad. Sci. USA 71, 307-310 (1974).

29 - Dupont, Y., Cohen, J. & Changeux, J.P. : X-ray diffraction study of membrane fragments rich in acetylcholine receptor protein prepared from the electric organ of *Torpedo marmorata*. FEBS Letters 40, 130-133 (1974).

30 - Hazelbauer, G.L. & Changeux, J.P. : Reconstitution of a chemically excitable membrane. Proc. Nat. Acad. Sci. USA 71, 1479-1483 (1974)

31 - Popot, J.L., Sugiyama, H. & Changeux, J.P. : Démonstration de la désensibilisation pharmacologique du récepteur de l'acétylcholine *in vitro* avec des fragments de membrane excitable de Torpille. To be published in C.R. Acad. Sci. Paris, Série D. (1974)

32 - Weber, M. & Changeux, J.P. : Binding of *Naja nigricollis* ^3H-α-toxin to membrane fragments from *Electrophorus* and *Torpedo* electric organs. I, II, III. Mol. Pharmacol. 10, 1-40 (1974).

33 - Weber, G., Borris, D., de Robertis, E., Barrantes, F., La Torre, J. & de Carlin, M. : The use of a cholinergic fluorescent probe for

the study of the receptor proteolipid. Mol. Pharmacol. 7, 530-537 (1971).

34 - Cohen, J., Weber, M. & Changeux, J.P. : Effects of local anesthetics and calcium on the interaction of cholinergic ligands with the nicotinic receptor protein from Torpedo marmorata. Mol. Pharmacol. (in press).

35 - Cohen, J. & Changeux, J.P. : Interaction of a fluorescent ligand with membrane bound cholinergic receptor from Torpedo marmorata. Biochemistry 12, 4855-4864 (1973).

36 - Cohen, J., Weber, M. & Changeux, J.P. : Effet des anesthésiques locaux sur les propriétés de liaison du récepteur cholinergique de Torpille. C.R. Acad. Sci. Paris 278, Série D, 1269-1272 (1974).

37 - Changeux, J.P. : Allosteric interactions on biosynthetic L-threonine deaminase from E. Coli K 12. Cold Spring Harbor Symp. Quant. Biol. 28, 497-504 (1963).

38 - Franklin, G.I. & Potter, L.T. : Studies of the binding of α-bungarotoxin to membrane bound and detergent dispersed acetylcholine receptors from Torpedo electric tissue. FEBS Letters 28, 101-106 (1972).

39 - Sugiyama, H. & Changeux, J.P. Manuscript in preparation. (1974).

Received September 6, 1974

DISCUSSION

Antonini: Do you have any information on the kinetics of the system?

Changeux: Not yet.

Veeger: Your final conclusion that the low-affinity sites rather than the high-affinity sites are finally responsible for the physiological effects on the receptor is appealing. Nevertheless I have difficulties to visualize the role of acetylcholinesterase in the intact system in relation to receptor-acetylcholine affinities of about 10^6 M^{-1}. The affinity of acetylcholine for the esterase is around 10^4 M^{-1}, so I do not see how it can compete successfully with the receptor. Could the enzyme be bound as a third constitutive component of a protein-complex in order to facilitate the attack on acetylcholine?

Changeux: This question raises an interesting problem, but we do not know the answer yet. What should be critical in the relative binding between receptor site and esterase catalytic site are the on rates constants of binding of acetylcholine to these two classes of sites. Possibly this rate is faster for the receptor site than for the esterase.

Veeger: There is accumulating evidence that in closed biomembrane fragments the proton-movement through the membrane has an opposite direction with respect to the intact membrane. Is this also true for your fragments, furthermore are they energisable by ATP?

Changeux: The answer to this question is no.

Helmreich: Do you attribute functional significance to the lattice structures? Because it is tempting to speculate that acetylcholine receptor might fire in clusters which might be arranged and concentrated in patches. Would you care to comment on that point?

Changeux: We do not have, yet, any evidence for a physiological function of the lattice structure, for instance for a role in the observed cooperative binding of acetylcholine.

Helmreich: I would like to add that the phenomenon which you described namely that the receptor which is a membrane protein recalls its encounter with a ligand even after the ligand is removed, that this hysteresis seems to be rather universal and might be of fundamental importance. In this connection, I would like to inquire whether the high- or low-affinity binding sites are responsible for this effect. Incidentally, this rather peculiar response to a ligand quite likely, might be a rather general consequence of slow association-dissociation phenomena in a hydrophobic environment.

Changeux: One possibility is, as mentioned in the presentation, that the high-affinity state is the long lasting "desensitized" state.

Helmreich: I would like to ask you whether you considered the possibility of covalent modification of the acetylcholine receptor as a possible mechanism governing association-dissociation or more generally modifying structure and function of the receptor. For example, one could speculate that Ca^{2+} is an indirect modifier which exerts its action as an activator of a protein kinase which in turn phosphorylates the receptor or an other membrane protein.

Changeux: We have no evidence yet for any covalent modification of the receptor protein but obviously such modifications might very well take place.

Sund: In some of your experiments with membrane fragments you found cooperativity corresponding to a Hill coefficient of 1.47. Local anaesthetics as prilocaine abolishes the cooperativity and increases the affinity. What is the effect of prilocaine on systems which do not show cooperativity?

Changeux: They cause an increase of affinity as well.

Buehner: You have shown that you are able to reconstitute functional membrane fragments. Can you use, or have you tried to use this ability to handle and control your system so well, to obtain three-dimensional crystals? Certainly, centrifuging membrane fragments into oriented gels is not a very promising approach to a structure analysis.

Changeux: First, I think that limited but useful information can be obtained after orientation of membrane fragments by centrifugation; many laboratories are using this kind of technique, of course, these studies should be complemented by other ones done with the pure protein and well defined lipids.

Hinz: Are there any explanations for the unusual thickness of the membranes as derived for X-ray analysis?

Changeux: The very high content in receptor protein and the manner the receptor is integrated to the hydrophobic phase of lipid bilayer are most likely responsible for the unusually large distance between polar head groups.

Raftery: Have you been able to determine any fraction of the total acetylcholine binding sites in purified membrane fragments as having a K_d 10^{-4}-10^{-5} M and if so are they cooperative binding sites?

Changeux: The Eldefrawi's have reported several classes of acetylcholine binding sites, including low-affinity sites, in crude membrane preparations from *Torpedo*. With the purified, receptor-rich membrane fragments, we have mainly studied the high-affinity sites. We are presently looking for conditions under which the low-affinity sites could be studied more easily.

Kempfle: Have you any suggestion about the concentration ratio between the esterase and the receptor in the synapse?

Changeux: Barnard and coworkers have shown that the ratio between esterase and receptor is close to one in the subsynaptic membrane of the vertebrate neuromuscular junction. The fact that there exists much less acetylcholinesterase than receptor in the purified membrane fragments from *Torpedo* electric organ most likely means that the esterase has been separated from the receptor in the course of the homogeneization of the electric organ.

Ramirez: I would like you to compare the characteristics of the extrasynaptic receptors which appear after denervation of skeletal muscles with the characteristics presented by the receptor isolated from electric organs of *Torpedo*.

Changeux: Pharmacological and electrophysiological differences have been reported between extrasynaptic and subsynaptic cholinergic receptors, see for instance the work by Beranek, Feltz and Maillart etc. However, *in vitro*, no differences have been reported at least as far as the binding properties and the immunological reactivity are concerned.

Sund: Differ your electron micrographs of the purified receptor protein if you use a preparation from frozen instead from fresh material?

Changeux: Most of the studies done with purified *Electrophorus* receptor were carried out from material originating from frozen electric tissue.

Subunit Structure and Binding Sites of the Acetylcholine Receptor

F. Hucho, A. Gordon and *H. Sund*

INTRODUCTION

Enzymes, as well as other proteins such as hemoglobins, immunoglobulins and repressors are molecules which exhibit, in many cases, quaternary structure [1-3]. These molecules can be formed by either identical or different polypeptide chains. Very often, the number of polypeptide chains equals the number of active sites. However, there are proteins such as the allosteric enzyme aspartate transcarbamylase [4] in which only half of the polypeptide chains contain a catalytically active site. The other polypeptide chains have a regulatory function. A special case are the enzymes showing half-site reactivity [5]. Possibly as an extreme consequence of negative cooperativity, only half of the otherwise identical polypeptide chains of these enzymes become catalytically active. Another interesting situation is represented by the aminoacyl-tRNA synthetases. Usually, the number of their substrate binding sites is the same as the number of polypeptide chains [6]. However, the E. coli phenylalanyl-tRNA synthetase [7], composed of four identical polypeptide chains, binds only one molecule of each substrate, phenylalanine, ATP and $tRNA^{Phe}$. The tetrameric E. coli methionyl-tRNA synthetase [8] has four ATP- but only two methionine-binding sites. The reason for this functional asymmetry is unknown.

The function-subunit structure relationship of acetylcholine receptor proteins is, as of many regulatory proteins, not simple. Receptor proteins have been isolated from Electrophorus electricus, Torpedo californica and Torpedo marmorata [9-13]. The molecular weight of the pure receptor from Electrophorus electricus was found to be 230 000 to 260 000 [10,13].

This protein molecule appears to consist of at least five polypeptide chains [10,12,13]. Sodium dodecylsulfate-polyacrylamide gel electrophoresis shows two bands with molecular weights of 45 000 and 54 000, indicating the existence of two different kinds of subunits in the receptor protein molecule [12,13].

The maximum number of substrate molecules bound to the receptor protein (including the snake venom neurotoxins α-bungarotoxin and those from _Naja naja_ or _Naja nigricollis_) is always lower than the number of subunits. The receptor from _Electrophorus electricus_ binds one molecule of the neurotoxin from _Naja nigricollis_ per 140 000-150 000 Daltons [13]; values of 90 000 [11] and 260 000 [10] Daltons are also reported for the neurotoxin from _Naja naja_. The _Torpedo californica_ receptor protein also shows an equivalent weight of 160 000 Daltons for the α-bungarotoxin binding site [9]. In all cases, the binding capacity for acetylcholine was half of that for snake toxins [13-15]. Treatment of the receptor, e.g. from _Electrophorus californica_, with α-bungarotoxin prior to the equilibrium dialysis experiments abolishes the binding of the small ligands (e.g. acetylcholine), suggesting that the small ligands bind to the same site as the snake neurotoxins.

Most of the results indicate that the receptor-ligand binding does not include positive or negative cooperative effects and that the receptor protein has only one kind of binding site [11,13,14], although results obtained with the receptor from _Torpedo marmorata_ [16] do appear to indicate cooperativity. It would appear unlikely that species differences or the presence of inactive membrane protein can account for this discrepancy, but rather that purification procedures, especially with respect to the use of organophosphate compounds which may affect the protein-ligand binding studies [14] could account for these observed differences. Experiments with membrane fragments from _Electrophorus_ and _Torpedo_ electric organs are also consistent with the hypothesis that the neurotoxin binds to a homogeneous class of binding sites [17].. However, experiments performed with a total homogenate of Torpedo californica electric tissue give evidence for heterogeneity of α-bungarotoxin binding sites [18]. This discrepancy may be explained by the assumption that the homogenate contains, in addition to specific binding sites, nonspecific ones.

To clarify some of the questions arising from the problems discussed above, we studied the covalent binding of radioactive-labelled Naja naja siamensis neurotoxin to the receptor protein from Torpedo marmorata electric tissue. The results obtained show that only one of the two different kinds of subunits of this receptor protein binds the neurotoxin.

RESULTS AND DISCUSSION

Disc electrophoresis of the purified Torpedo marmorata receptor protein * under non-denaturing conditions shows a single band (Fig. 1), indicating homogeneity of the protein. Sodium dodecylsulfate-polyacrylamide gel electrophoresis yields, in contrast, several bands (Fig. 2). The relative amounts of the higher molecular weight components depends on the age of the preparation i.e., they increase with time. The molecular weights of these components, as calculated from the calibration curve in Fig. 4, are 37 000, 49 000, 74 000, 93 000, and 148 000. The predominating bands at molecular weight 37 000 and 49 000 correspond to the two subunits of the receptor protein isolated from Electrophorus electricus [12,13].

A [^{125}I]-Naja-naja toxin-receptor complex was cross-linked with suberimidate [12,21]. This complex was then electrophoresed in the presence of sodium dodecylsulfate and subsequent-

* The acetylcholine receptor was purified from the electric organ of Torpedo marmorata by a modification of the method published in Ref. [19] using an affinity gel which was prepared with the principal neurotoxin of Naja naja siamensis (purified according to [20]).

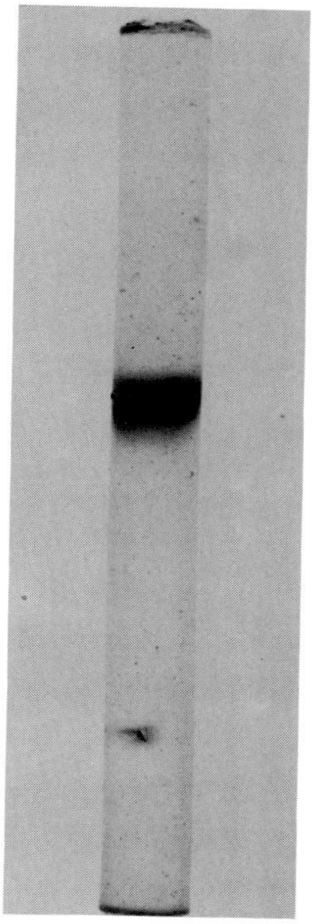

Fig. 1. Disc gel electrophoresis of purified receptor protein from *Torpedo marmorata* electric tissue. 5 % polyacrylamide gel, 50 µg receptor protein.

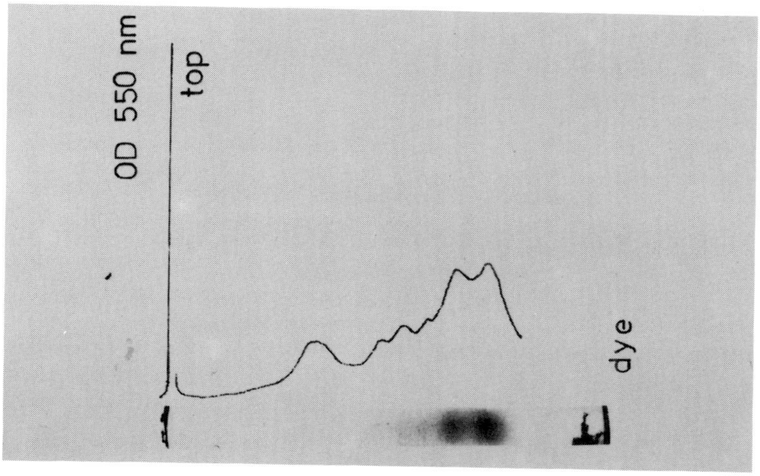

Fig. 2. Sodium dodecylsulfate-polyacrylamide gel electrophoresis of purified receptor protein from *Torpedo marmorata*.

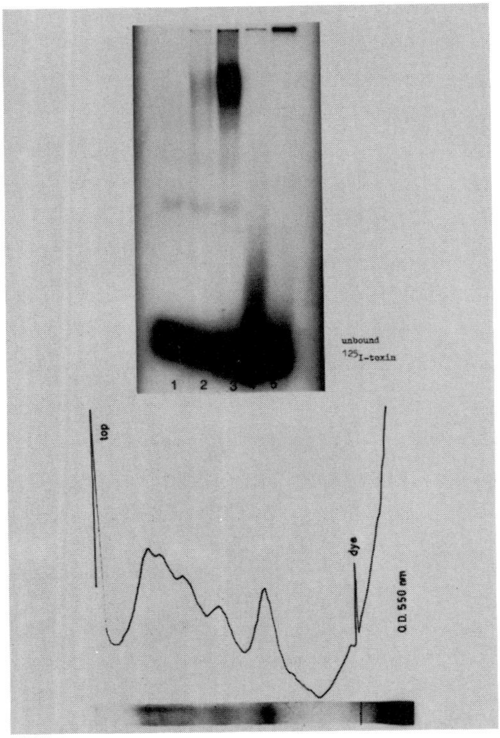

Fig. 3. Autoradiograph and densitometer scan of sodium dodecylsulfate-polyacrylamide gel electrophoresis of receptor-toxin complex cross-linked with suberimidate. The samples contained: (1) 100 µg receptor protein, 60 pmoles [^{125}I]-toxin, 0.1 mg/ml suberimidate, (2) 100 µg receptor protein, 60 pmoles [^{125}I]-toxin, 0.5 mg/ml suberimidate, (3) 100 µg receptor protein, 60 pmoles [^{125}I]-toxin, 1 mg/ml suberimidate, (4) controls: electrophoresis buffer, 60 pmoles [^{125}I]-toxin, 1 mg/ml suberimidate, (5) 100 µg pyruvate kinase, 60 pmoles [^{125}I]-toxin, 1 mg/ml suberimidate. All solutions contain 0.2 M triethanolamine, pH 8.5.

ly the dried gel was autoradiographed. Fig. 3 shows five bands. The bands correspond to molecular weights of 191 000, 167 000, 130 000, 82 000 and 53 000. Neither Naja naja toxin alone nor a mixture of toxin with control proteins such as

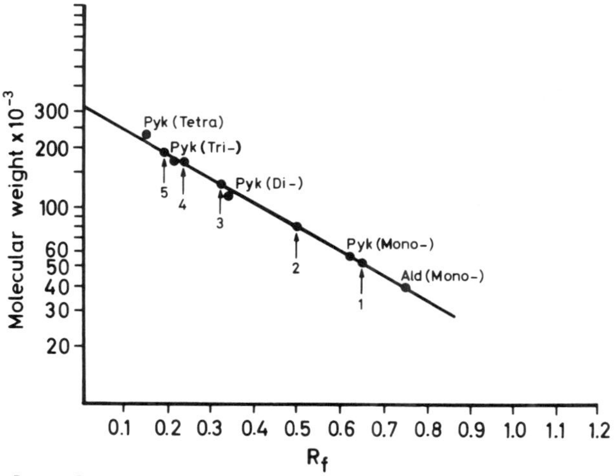

Fig. 4. Molecular weight determination of the receptor-toxin complex, Pyk, pyruvate kinase; Ald, aldolase. 1, 2, 3, 4, and 5 refer to the bands of the autoradiograph of Fig. 3.

pyruvate kinase yield significant bands under these conditions (Fig. 3, sample well 2, 4 and 5).

The results shown in Fig. 3 indicate that only one of the two low-molecular weight bands in the stained gel (Fig. 2) exhibits radioactivity. We interpret this to mean that only one of the two subunits binds the Naja naja toxin. The control experiment with pyruvate kinase indicates that only proteins which specifically bind the toxin show significant radioactivity under our conditions of cross-linking.

Since cross-linking with suberimidate requires amino groups in appropriate proximity there is the possibility that the second subunit, although binding the toxin, did not react. However, for the following reasons we consider this unlikely. In earlier experiments, the two subunits have been cross-linked to each other indicating that they both are reactive towards suberimidate [12]. Furthermore, the subunits of all proteins investigated by this method could be cross-linked probably due to the general abundance of amino groups.

According to the Ref. [21] the number of bands obtained by this method corresponds to the number of subunits in the molecule. This would mean that the Torpedo receptor protein, like the receptor from Electrophorus [12], appears to be composed of at least five subunits. None of the radioactively labelled bands (Fig. 4) can be due to impurities since our protein is homogeneous (Fig. 1). In contrast to coomassie blue stained gels the autoradiograph reveals only proteins that specifically bind toxin.

The molecular weights corresponding to the bands are difficult to evaluate because they include the toxin molecules and we have no way to determine how many toxin molecules are present in each band. The molecular weight of the low-molecular weight band corresponds to 53 000 Daltons. Subtracting 8 000 Daltons for the toxin from the molecular weight of this band, we obtain 45 000.

In conclusion, only one of the two subunits of the Torpedo receptor protein appears to bind Naja naja toxin. It remains to be determined if the second band corresponds to a degradation product of this subunit that has lost its binding capacity or if it has a different function.

Acknowledgements: We are grateful to Dr. R. Martin, Naples, for supplying Torpedo and to Ms. Jutta Birsner and Mr. Giampiero Bandini for technical assistance. This work was supported by the Deutsche Forschungsgemeinschaft (Sonderforschungsbereich 138: "Biologische Grenzflächen und Spezifität").

REFERENCES

1. Sund, H., and Weber, K.: The quaternary structure of proteins. Angew.Chem. 78, 217-232 (1966); Angew.Chem.Int.Ed. 5, 231-245 (1966).
2. Klotz, I.M., Langerman, N.R., and Darnall, D.W.: Quaternary structure of proteins. Ann.Rev.Biochem. 39, 25-62 (1970).
3. Müller-Hill, B.: The lac-repressor. Angew.Chem. 83, 195-207 (1971); Angew.Chem.Int.Ed. 10, 195-207 (1971).

4. Weber, K.: New structural model of E. coli aspartate transcarbamylase and the amino-acid sequence of the regulatory polypeptide chain. Nature 218, 1116-1119 (1968).

5. Seydoux, F., Malhotra, O.P., and Bernhard, S.A.: Half-site reactivity, Critical Reviews in Biochemistry 2, 227-257 (1974).

6. Kisselev, L.L., Favorova, O.O.: Aminoacyl-tRNA synthetases: Some recent results and achievements. Advances in Enzymology 40, 141-238 (1974).

7. Kosakowski, H.M., and Böck, A.: Substrate complexes of phenylalanyl-tRNA synthetase from Escherichia coli. Eur. J.Biochem. 24, 190-200 (1971).

8. Blanquet, S., Fayat, G., Waller, J.P., and Iwatsubo, M.: The mechanism of reaction of methionyl-tRNA synthetase from Escherichia coli. Interaction of the enzyme with ligands of the amino-acid-activation reaction. Eur.J.Biochem. 24, 461-469 (1972).

9. Schmidt, J., and Raftery, M.A.: Purification of acetylcholine receptors from Torpedo californica electroplax by affinity chromatography. Biochemistry 12, 852-856 (1973).

10. Biesecker, G.: Molecular properties of the cholinergic receptor purified from Electrophorus electricus. Biochemistry 12, 4403-4409 (1973).

11. Klett, R.P., Fulpius, B.W., Cooper, D., Smith, M., Reich, E., and Possani, L.D.: The acetylcholine receptor. I. Purification and characterization of a macromolecule isolated from Electrophorus electricus. J.Biol.Chem. 248, 6841-6853 (1973).

12. Hucho, F., and Changeux, J.P.: Molecular weight and quaternary structure of the cholinergic receptor protein extracted by detergents from Electrophorus electricus electric tissue. FEBS Letters 38, 11-15 (1973).

13. Meunier, J.C., Sealock, R., Olson, R., and Changeux, J. P.: Purification and properties of the cholinergic receptor protein from Electrophorus electricus electric tissue. Eur.J.Biochem. 45, 371-394 (1974).

14. Moody, T., Schmidt, J., and Raftery, M.A.: Binding of acetylcholine and related compounds to purified acetylcholine receptor from Torpedo californica electroplax. Biochem.Biophys.Res.Commun. 53, 761-772 (1973).

15. Chang, H.W.: Purification and characterization of acetylcholine receptor-I from Electrophorus electricus. Proc. Natl.Acad.Sci. USA 71, 2113-2117 (1974).

16. Eldefrawi, M.E., and Eldefrawi, A.T.: Purification and molecular properties of the acetylcholine receptor from Torpedo electroplax. Arch.Biochem.Biophys. 159, 362-373 (1973).

17. Weber, M., and Changeux, J.P.: Binding of Naja nigri-

collis $[^3H]\alpha$-toxin to membrane fragments from Electrophorus and Torpedo electric organs. I. Binding of the tritiated α-neurotoxin in the absence of effector. Mol. Pharmacology 10, 1-14 (1974).

18. Raftery, M.A., Schmidt, J., and Clark, D.G.: Specificity of α-bungarotoxin binding to Torpedo californica electroplax. Arch.Biochem.Biophys. 152, 882-886 (1972).

19. Karlsson, E., Heilbronn, E., and Widlund, L.: Isolation of the nicotinic acetylcholine receptor by biospecific chromatography on insolubilized Naja naja neurotoxin. FEBS Letters 28, 107-111 (1972).

20. Karlsson, E., Arnberg, H., and Eaker, D.: Isolation of the principal neurotoxin of two Naja naja subspecies. Eur.J.Biochem. 21, 1-16 (1971).

21. Davies, G.E., and Stark, G.R.: Use of dimethyl suberimidate, a cross-linking reagent in studying the subunit structure of oligomeric proteins. Proc.Natl.Acad.Sci. USA 66, 651-656 (1970).

Received August 17, 1974.

DISCUSSION

K. Weber: Just two comments. First, highly cross-linked proteins can show mobilities different from the value expected from their actual polypeptide-chain molecular weight. Second, non-protein material (i.e. carbohydrate side group and/or lipoprotein) can change the electrophoretic mobility, since the SDS normalization of normal proteins is not achieved, i.e. influences of charge, abnormal SDS binding and non-uniform conformation can be important.

Hucho: The calibration curve is composed of highly cross-linked proteins as well and we did not see large deviations in their electrophoretic mobility as compared to uncross-linked proteins. But since calibration was done with water-soluble enzymes we are well aware of the uncertainties of our molecular weight. Concerning the carbohydrate content there is a good check of its influence on the electrophoretic mobility: The R_f value has to be the same in gels of different concentrations. We did that check, and the molecular weights obtained were the same in 3.5, 5 and 7.5 % polyacrylamide gel.

Steinhardt: According to current ideas, the molecular weight of the SDS complexes should be about 2.4 times that of the uncombined protein (unless the latter contains a great deal of carbohydrate or lipid). Does such a factor prevail in your work? Reynolds has published results with a lipoprotein very recently.

Hucho: There are no indications for lipids but both bands stain for carbohydrate. As to the factor of 2.4, we did not evaluate this.

Kempfle: On SDS-gel electrophoresis you show two different peaks indicating a molecular weight of 45 000 and 54 000, respectively. On an autoradiogram with ^{125}I-labeled toxin-protein you find only one peak. Has this peak the lower (45 000) or the higher (54 000) molecular weight?

Hucho: The molecular weights you are mentioning refer to the receptor from Electrophorus. With the Torpedo receptor they are 37 000 and 49 000, respectively. We cannot tell which of the bands we labeled: The labeled band corresponds to a molecular weight of 53 000, the toxin has a molecular weight of 8 000, but we have no way to evaluate how many toxin molecules we cross-linked to the receptor.

Further Characterization of Purified Acetylcholine Receptor and its Incorporation into Phospholipid Vesicles

Mark G. McNamee, Cheryl L. Weill and *Arthur Karlin*

The acetylcholine receptor (ACHR) translates the binding of acetylcholine into an increase in cation permeability at post-synaptic membranes. The receptor protein can be solubilized from the membrane by non-ionic detergents and extensively purified by affinity chromatography. The isolated protein retains the specific binding properties attributed to the ACHR *in situ* and has been used to deduce many of the molecular properties of the ACHR. Progress in the purification and characterization of the ACHR has been reviewed recently (1). An important characteristic of purified ACHR is the specific activity, and some of our results comparing toxin binding with affinity labeling are presented.

A complete molecular description of receptor function must include its interactions with other membrane components. Considerable effort is now dedicated to the reconstitution of cation permeability control in membranes containing ACHR (2,3), and our initial efforts in this area are discussed.

PURIFICATION OF ACETYLCHOLINE RECEPTOR

Tissue from the main electric organ of *Electrophorus electricus* is homogenized in 1 mM EDTA-pH 7.4 to remove most soluble proteins. The pellet obtained by sedimentation at 54,000g for 45 min is homogenized in 1 M NaCl-2 mM $NaPO_4$-1 mM EDTA-pH 7.4 to solubilize 80-90% of the original acetylcholinesterase (ACHE) activity. The pellet is washed with 1 mM EDTA and then

Table 1. Affinity Column Purification of ACHR from Electrophorus Electricus.

Column Treatment	Protein		ACHE		ACHR	
	mg	%	pmoles	%	pmoles	%
Triton Extract	1145	100.0	2144	100.0	15300	100.0
Not Bound	1136	99.2	1036	48.3	0	0
Eluted with 50 mM NaCl	10.8	0.9	83	3.9	0	0
Eluted with 150 mM NaCl	0.78	0.1	468	21.8	1958	12.8
Eluted with 50 mM CARB + 100 mM NaCl	0.82	0.1	41	1.9	3813	25.0
Total Recovery	1148	100.3	1628	75.9	5771	37.8

Solubilized ACHR (186 ml) in 3% 'Triton X-100'-50 mM NaCl-10 mM $NaPO_4$-1 mM EDTA-3 mM NaN_3-pH 7.0 from 660g of the electric organ of Electrophorus electricus was passed through a 1.8 ml (5 mm x 9 cm) column of p-carboxytrimethylammonium-Affinose 401. The column was eluted succesively with 25 ml of 50 mM NaCl, 20 ml of 150 mM NaCl and 15 ml of 50 mM carbamylcholine (CARB) + 100 mM NaCl. All buffers contained 0.2% 'Triton X-100'-10 mM $NaPO_4$-1 mM EDTA-3 mM NaN_3-pH 7. Protein was determined by the Lowry (14) or Fluram (13) method; ACHE activity was measured by Ellman's procedure (19) and is expressed as pmoles of catalytic sites assuming a MW of 80,000 per catalytic site and 10 moles acetylthiocholine hydrolized per min per g enzyme; ACHR activity was determined by specific MBTA labeling (see text). CARB was removed by dialysis before assay for ACHE and ACHR.

extracted with 3% 'Triton X-100'-50 mM NaCl-10 mM $NaPO_4$-1 mM EDTA-3 mM NaN_3-pH 8 for 1 hr at 4°. The supernatant after centrifugation at 360,000g for 40 min is adjusted to pH 7 and contains ~0.7% of the initial ACHE and apparently all of the ACHR.

The above procedure differs from a previous isolation method (4) in that the tissue itself rather than a purified membrane preparation is used as the starting material. The specific activity of ACHR in the detergent extract is lower using the present procedure, but the same overall yield and specific activity of ACHR are obtained in the next step of purification.

Solubilized ACHR is purified by affinity chromatography using p-carboxyphenyltrimethylammonium, coupled by thioester linkage to Affinose 401 (Bio-Rad), as the adsorbing ligand and carbamylcholine as the eluting ligand (4). Table 1 provides a summary of a typical elution pattern from the affinity column. Although a significant fraction of the recovered ACHR is eluted in the high-salt buffer, the simultaneous removal of nearly all the remaining ACHE justifies the loss.

The ACHR can be further purified by centrifugation in a 5-20% sucrose density gradient containing 0.2% Triton X-100 . The ACHR moves as a symmetrical peak (Fig. 1) with an apparent $s_{20,w}$ value of 9.5 S, similar to independent observations (5,6). A 50% increase in specific activity is usually observed after centrifugation yielding ACHR with specific activity of 6-7 nmole MBTA sites/mg protein (see below). The ACHR is resolved from the remaining traces of ACHE, and less than .02% of the protein in the pooled ACHR fractions is active ACHE. The specific activity of ACHR is constant within experimental error across the main part of the peak, suggesting near homogeneity.

CHARACTERIZATION OF ACHR

Affinity Alkylation

The ACHR can be assayed at all stages of purification by affinity alkylation of dithiothreitol-reduced receptor with [^{3}H]MBTA(4-(N-maleimido)benzyl-[^{3}H]trimethylammonium iodide) using a modification of the "quick assay" procedure (4; manuscript in preparation). The portion of MBTA labeling that is

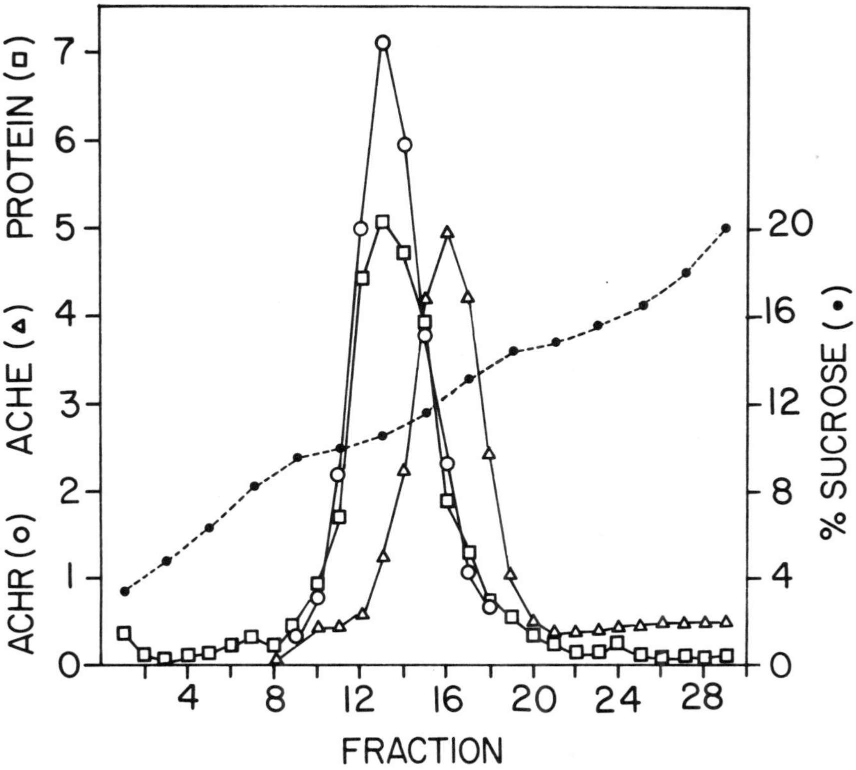

Figure 1. Sucrose density gradient centrifugation of ACHR.

Fractions from the affinity column eluted with 50 mM carbamylcholine + 100 mM NaCl (7.4 ml, 0.77 mg protein) were concentrated and dialyzed in a Schleicher and Schuell Collodion Bag Apparatus under reduced pressure at 0° against 150 mM NaCl-10 mM $NaPO_4$-1 mM EDTA-.2% 'Triton X-100'-pH 7. 0.45 ml of concentrated ACHR (specific activity 4.5nmoles/mg) was applied to the top of a 12 ml 5-20% sucrose gradient in the same buffer and centrifuged at 40,000 RPM for 14 hr at 5° in the SW41 rotor of an L2-65B Beckman ultracentrifuge. Fractions of 0.4 ml were collected with an ISCO Model 183 Fractionator and analyzed for % sucrose (●), protein (□,20μg/unit) (13), ACHE (△,0.2 p-moles catalytic sites/unit) (19), and ACHR (○,100 pmoles MBTA sites/unit). 93.2% of the protein, 100% of the ACHR and 20% of the ACHE were recovered (the remaining ACHE apparently aggregated and was not recovered from the bottom of the tube). The specific activities of ACHR for fractions 11-16, respectively, (nmoles/mg protein) were: 6.3, 5.7, 7.0, 6.3, 5.5, and 6.2.

blocked by Naja naja siamensis toxin 3 (7) is taken to be specific for ACHR; at least 95% of the total MBTA labeling of purified ACHR is blocked by toxin.

Toxin Binding

A [^{3}H]methyl derivative of N. n. siam. toxin 3 is prepared by reductive alkylation of native toxin with formaldehyde and [^{3}H]sodium borohydride (8,9). This derivative has 0.8 methyl-groups per molecule of toxin, a specific activity of 4 Ci/mmole and biological activity equivalent to native toxin (manuscript in preparation). The binding of [^{3}H]toxin to receptor is determined by incubating toxin and ACHR for 1 hr at 25° in 0.2% 'Triton X-100'-150 mM NaCl-10 mM $NaPO_4$-1 mM EDTA-pH 7 and then separating toxin-ACHR complex from free toxin on Bio-Gel P-30.

We have found that there are two toxin binding sites per MBTA binding site in purified ACHR. For example, a preparation of ACHR from the affinity column containing 4.9 nmoles MBTA sites/mg protein was found to bind 10.0 $\pm$ 0.2 nmoles [^{3}H] toxin/mg. Furthermore, toxin binding to ACHR, previously reacted with MBTA, was decreased only about 30%. This is in contrast to the greater than 95% blockage by toxin of MBTA labeling. There are clearly at least two classes of toxin binding sites, only one of which overlaps the MBTA binding site. The above results are consistent with reports that there are two α-bungarotoxin binding sites per small-ligand binding site in ACHR (10,11). The physiological significance of the other toxin site(s) is not known, and further studies on the relationship between affinity labeling and toxin binding are in progress.

Subunit Structure

ACHR, purified by affinity chromatography, gives rise by SDS-polyacrylamide gel electrophoresis to three polypeptide bands, as previously shown (4), corresponding to molecular weights of 40, 45, and 52,000 daltons (average values from several preparations). The relative mobilities and intensities of the three bands are the same before and after further purification of ACHR by sucrose density gradient centrifuation, even when samples are taken from different parts of the ACHR peak (Fig. 2). The band corresponding to ~40,000 daltons carries the site affinity alkylated by MBTA (4) and presumably contains at least one of the toxin binding sites. The functional significance of the other bands has not yet been ascertained. All three bands stain positively by the PAS procedure (12), but no direct determination of sugar content has been made.

The specific activity of the most highly purified ACHR preparations (7 nmoles MBTA sites/mg protein) corresponds to a minimum molecular weight of 140,000 daltons per MBTA binding site, a value close to the combined molecular weights of the three subunits (137,000 daltons). This agreement, although intriguing, may be fortuitous. Protein is routinely determined by the Fluram method (13), and for purified ACHR the Fluram and Lowry procedures (14) give different values for protein concentration, using bovine serum albumin as the standard. The Lowry value is 1.4 times the Fluram value and is consistent with amino acid analysis results. The specific activity of ACHR is correspondingly lower in terms of Lowry protein and gives a value of 200,000 daltons per MBTA binding site and presumably 100,000 daltons per toxin site. Estimates for the molecular weight of the ACHR complex have been obtained using a variety of techniques including sucrose density gradient centrifugation, gel filtration, and SDS-polyacrylamide gel electrophoresis after crosslinking of subunits (5,8,15), but any attempt to relate the results to specific combinations of subunits is recognized as speculative at the present time.

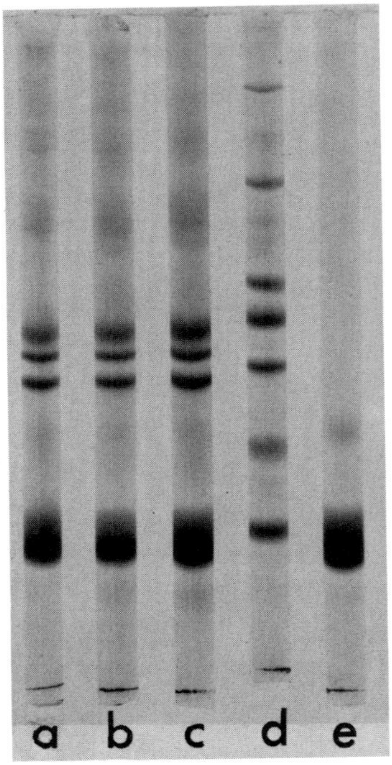

Figure 2. SDS-Polyacrylamide gel electrophoresis of purified ACHR.

To samples of ACHR from the affinity column (a) and from a subsequent sucrose density gradient centrifugation, pooled fractions 11-13 (b) and 14-16 (c) (see Fig. 1) was added 20µg of succinylated lysozyme as carrier. Protein was precipitated and washed with acetone to remove Triton X-100 and dissolved in 2% SDS-10 mM Tris acetate-1 mM EDTA-20 mM DTT-pH 8.0 at 50° for 1 hr. Succinylated lysozyme alone (e) shows no overlapping bands with ACHR and has no effect on the pattern obtained. The molecular weight standards (20) from the top to the bottom (d) are: myosin (220,000), β-galactosidase (130,000), bovine serum albumin (68,000), γ-globulin, heavy chain (50,000), aldolase (40,000), γ-globulin, light chain (23,500), and lysozyme (14,300).

INCORPORATION OF ACHR INTO PHOSPHOLIPID VESICLES

Reconstitution of ACHR function in model membrane systems is a principal objective of current receptor research since the mechanism by which the binding of acetylcholine and its congeners to the ACHR is transduced into a change in membrane cation permeability is not understood. It has yet to be demonstrated whether or not the ion permeability modulation properties attributed to the ACHR in situ and in native and reconstituted membrane vesicles (2,16) are an integral part of the ACHR complex isolated and purified by presently used methods (1).

We have incorporated purified ACHR into phospholipid vesicles by two methods, both of which are adaptations of methods used by Racker on a variety of membrane systems (17,18). The first involves sonication of detergent-depleted ACHR with phospholipids, and the second consists of slow dialysis of a cholate suspension of purified ACHR and phospholipid against detergent-free buffer.

In the first method, ACHR in 0.2% Triton X-100 is centrifuged into a detergent-free sucrose density gradient to remove excess Triton X-100 and then added to dried phospholipids and briefly sonicated. The co-sonicated mixture, in buffer containing 20% sucrose, is layered onto a 30-50% sucrose gradient and overlayed with a 0-15% sucrose gradient. During centrifugation the ACHR-phospholipid vesicles float to a lower density whereas ACHR sonicated without added lipids sediments to higher density, as expected (Fig. 3). Initial experiments were done using ACHR prelabeled with $[^3H]$MBTA.

In the second procedure, Triton X-100 associated with ACHR is exchanged for 0.2% sodium cholate by sucrose density gradient centrifugation. Phospholipid dispersed in cholate is then added (a mixture of egg phosphatidylcholine and purified phospholipids from electric organ tissue at a phospholipid/protein weight ratio of 200/1) and the mixture is dialyzed for 3 days at 4°. Nitrogen is continuously bubbled through the dialysis buffer (4000 volumes) containing 50 mM NaCl-50 mM KCl-10 mM Tris-50 μM $MgCl_2$-50 μM $CaCl_2$-100 μM EDTA-3 mM NaN_3-pH 7.4. A portion of the dialyzed mixture, containing <.001% cholate, is centrifuged in a sucrose density gradient under conditions

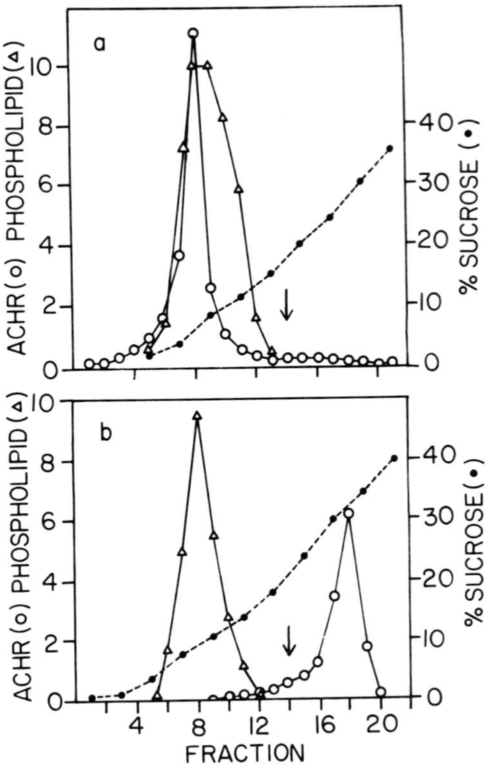

Figure 3. Incorporation of ACHR into phospholipid vesicles.

32 μg of ACHR was labeled with [^{3}H]MBTA by the "slow-labeling" procedure (4) and centrifuged in a 5-20% sucrose gradient (with no Triton X-100) for 12 hr at 40,000 RPM in an SW41 rotor at 5°. The detergent-depleted ACHR was added to 7 umoles dry egg phosphatidylcholine (Sigma) and 0.5μmoles purified eel phospholipids. The mixture was sonicated for 4 min in 30-second intervals at 0° using the microtip of a Bronwill Biosonik at power setting 60. 1.2 ml of the sonicated mixture in 20% sucrose-80 mM NaCl-10 mM Tris-pH 7.4 was layered on 4 ml of a 30-50% sucrose gradient in 50 mM NaCl-10 mM Tris-pH 7.4 and overlayed with 6 ml of a 0-15% sucrose gradient in the same buffer. The mixture was centrifuged for 15 hr at 40,000 RPM in an SW41 rotor at 5°. (a) Fractions of .55 ml were collected with an ISCO 183 Fractionator and analyzed for % sucrose (●), [^{3}H]MBTA (O, 1000 cpm/unit), and phospholipid (Δ, 1 μmole PO_4/unit) (21). (b) Samples containing only ACHR or only phospholipid were sonicated and centrifuged separately under identical conditions and analyzed as above. Units are the same as in (a) except that there are 2 μmoles PO_4/unit. The arrows indicate the initial sample position in the gradient.

similar to those in Fig. 3. The ACHR floats up in association with the phospholipid. Unlabeled ACHR is supplemented with a small amount of [^{3}H]MBTA-labeled ACHR, used as a marker. The unlabeled receptor activity is assayed throughout the procedure, and there is no loss of specific MBTA labeling capacity at any stage of the recombination. If the ACHR-phospholipid vesicles after dialysis are briefly sonicated in a bath sonicator with additional phospholipid and $^{22}Na^+$, non-leaky vesicles containing $^{22}Na^+$ can be produced; however, at this stage, there is a 45% loss of MBTA labeling capacity during 3 min of sonication at 25°. The optimal conditions for formation of ACHR-phospholipid vesicles showing permeability control are presently under investigation.

ACKNOWLEDGEMENTS

The technical assistance of Ms. Kang Liu is appreciated. This work was supported by research grants from the National Institute of Neurological Diseases and Stroke, and the National Science Foundation, and by a gift from the New York Heart Association, Inc.

REFERENCES

1. Karlin, A.: The Acetylcholine Receptor: Progress Report. Life Sciences 14, 1385-1415 (1974).
2. Hazelbauer, G. L., Changeux, J.-P.: Reconstitution of a Chemically Excitable Membrane. Proc. Nat. Acad. Sci. USA 71, 1479-1483 (1974).
3. Kemp, G., Dolly, J. O., Barnard, E. A., Wenner, C. E.: Reconstitution of a Partially Purified Endplate Acetylcholine Receptor Preparation in Lipid Bilayer Membranes. Biochem. Biophys. Res. Comm. 54, 607-613 (1973).
4. Karlin, A., Cowburn, D.: The Affinity-Labeling of Partially Purified Acetylcholine Receptor from Electric Tissue of *Electrophorus*. Proc. Nat. Acad. Sci. USA 70, 3636-3640 (1973).

5. Meunier, J. C., Olsen, R. W., Changeux, J.-P.: Studies on the Cholinergic Receptor Protein from Electrophorus Electricus. FEBS Letters 24, 63-68 (1972).

6. Raftery, M. A., Schmidt, J., Clark, D. G., Wolcott, R. C.: Demonstration of a Specific α-Bungarotoxin Binding Component in Electrophorus Electricus Electroplax Membranes. Biochem. Biophys. Res. Comm. 45, 1622-1629 (1971).

7. Karlsson, E., Arnberg, H., Eaker, D.: Isolation of the Principal Neurotoxins of Two Naja naja Subspecies. Eur. J. Biochem. 21, 1-16 (1971).

8. Biesecker, G.: Molecular Properties of the Cholinergic Receptor Purified from Electrophorus Electricus. Biochemistry 12, 4403-4409 (1973).

9. Rice, R. H., Means, G. E.: Radioactive Labeling of Proteins in Vitro. J. Biol. Chem. 246, 831-832 (1971).

10. Moody, T., Schmidt, J., Raftery, M. A.: Binding of Acetylcholine and Related Compounds to Purified Acetylcholine Receptor from Torpedo Californica Electroplax. Biochem. Biophys. Res. Comm. 53, 761-772 (1973).

11. Chang, H. W.: Purification and Characterization of Acetylcholine Receptor-I from Electrophorus Electricus. Proc. Nat. Acad. Sci. USA 71, 2113-2117 (1974).

12. Fairbanks, G., Steck, T. L., Wallach, D. F. H.: Electrophoretic Analysis of the Major Polypeptides of the Human Erythrocyte Membrane. Biochemistry 10, 2606-2616 (1971).

13. Udenfriend, S., Stein, S., Bohlen, P., Dairman, W., Leimgruber W., Weigele, M.: Flouresamine: A Reagent for Assay of Amino Acids, Peptides, Proteins, and Primary Amines in the Picomole Range. Science 178, 871-872 (1972).

14. Lowry, O. H., Rosebrough, N. J., Farr, A. L., Randall, R. J.: Protein Measurement with the Folin Phenol Reagent. J. Biol. Chem. 193, 265-275 (1951).

15. Hucho, F., Changeux, J.-P.: Molecular Weight and Quaternary Structure of the Cholinergic Receptor Protein Extracted from Electrophorus Electricus Electric Tissue. FEBS Letters 38, 11-15 (1973).

16. Kasai, M., Changeux, J.-P.: In Vitro Excitation of Purified Membrane Fragments by Cholinergic Agonists. I-III. J. Membrane Biol. 6, 1-80 (1971).

17. Racker, E.: A New Procedure for the Reconstitution of Biologically Active Phospholipid Vesicles. Biochem. Biophys. Res. Comm. 55, 224-230 (1973).

18. Racker, E.: Reconstitution of a Calcium Pump with Phospholipids and a Purified Ca^{++}-Adenosine Triphosphatase from Sarcoplasmic Reticulum. J. Biol. Chem. 247, 8198-8200 (1972).

19. Ellman, G. L., Courtney, K. D., Andres, V., Featherstone, R. M.: A New and Rapid Colorimetric Determination of Acetylcholinesterase Activity. Biochem. Pharm. 7,88-95 (1961).

20. Weber, K., Osborn, M.: The Reliability of Molecular Weight Determinations by Dodecyl Sulfate-Polyacrylamide Gel Electrophoresis. J. Biol. Chem. 244,4406-4412 (1969).

21. McClare, C. W. F.: An Accurate and Convenient Organic Phosphorus Assay. Anal. Biochem. 39,528-530 (1971).

Received July 25, 1974

DISCUSSION

Hucho: Do you know anything about the way in which the receptor is incorporated into the lipid vesicles? Is it adsorbed to the surface, is it wrapped by the lipid, or is it really incorporated into the bilayer? Do you have electron microscope pictures?

McNamee: We have no direct evidence that indicates how the receptor is associated with the lipid vesicles. We expect that the receptor is incorporated into the bilayer as demonstrated in a number of other systems where reconstitution was carried out by similar procedures. We have not yet obtained electron microscope pictures of our vesicles.

Helmreich: What was the composition of the lipids used for reconstitution? Did you add cholesterol?

McNamee: The main lipid used ($\sim$95 %) was phospatidyl choline from egg yolk purified by silicic acid chromatography. We also included a small amount of total phospholipids isolated and purified from eel electric tissue; these lipids were primarily phosphatidyl choline, phosphatidyl ethanolamine and phosphatidyl serine. In the experiments described here we did not add cholesterol but we have included it recently.

Structural and Functional Studies of an Acetylcholine Receptor

M. A. Raftery, J. Bode, R. Vandlen, D. Michaelson, J. Deutsch, T. Moody, M. J. Ross and R. M. Stroud

INTRODUCTION

Understanding of postsynaptic depolarization at the molecular level requires study of neuro-receptors at several physical levels of macromolecular order. Ideally, such a study should be carried out at the cellular level, the membrane level, at the level of solution chemistry of isolated receptors, and at the level of reconstitution of such isolated receptor macromolecules with the native phospholipids they encounter in their natural environment. Studies described in this communication deal with the biochemical and biophysical characterization of AcChR from *Torpedo californica* at the last three desirable levels.

Torpedo species are ideal for biochemical and biophysical studies of AcChR since the electroplax of these organisms contains high concentrations of such receptor macromolecules. It is also evident from recent studies in a number of laboratories that there are important species differences in acetylcholine receptors characterized thus far. Nevertheless, study of any one of these AcChR macromolecules in detail should facilitate our knowledge of the molecular mechanisms involved in postsynaptic depolarization.

STRUCTURAL STUDIES OF AcChR-RICH (acetylcholine receptor) AND AcChE-RICH (acetylcholine esterase) MEMBRANE PARTICLES

The fractionation, by sucrose density gradient centrifugation, of electroplax membrane particles from *Torpedo californica*

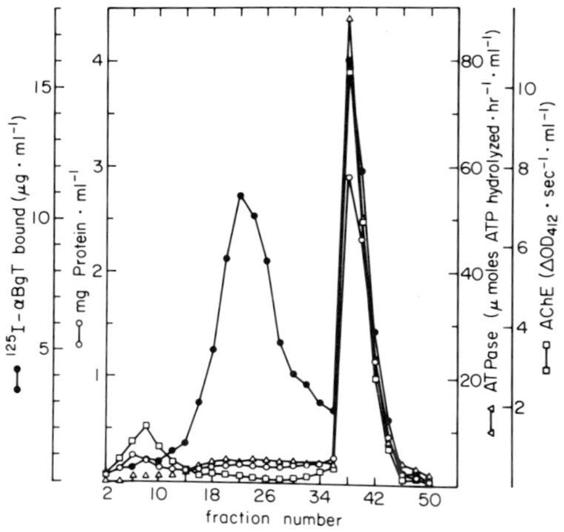

Fig. 1. Zonal centrifugation of <u>Torpedo californica</u> electroplax membrane fragments in a sucrose density gradient-low sucrose density at left.

(1,2) or <u>Electrophorus electricus</u> (3) have yielded such membrane particles highly enriched in AcChR activity, AcChE activity, or ATPase activity. Fig. 1 shows a large-scale separation of membrane particles from <u>Torpedo californica</u> electroplax (2). For this discussion, the particles of greatest interest are those enriched in AcChE activity (fractions 4 to 10) and AcChR activity (fractions 16 to 30). The degree of enrichment involved is best exemplified by the AcChR-rich particles; of the protein present in the AcChR-rich particles, 40 to 50 % of the total is represented by AcChR, on the basis of specific activity compared with isolated, purified AcChR from the same tissue. Negative staining of the two types of particles reveals some interesting differences. In Fig 2a, the results of negative staining of AcChE-rich particles is presented.

The most obvious feature is that the membranes are surrounded by a halo of subparticles 100 to 110 Å in diameter. It can be shown that most of these AcChE-rich particles contain AcChE activity. The simplest conclusion is that the subparticles represent AcChE oligomers associated with the membrane struc-

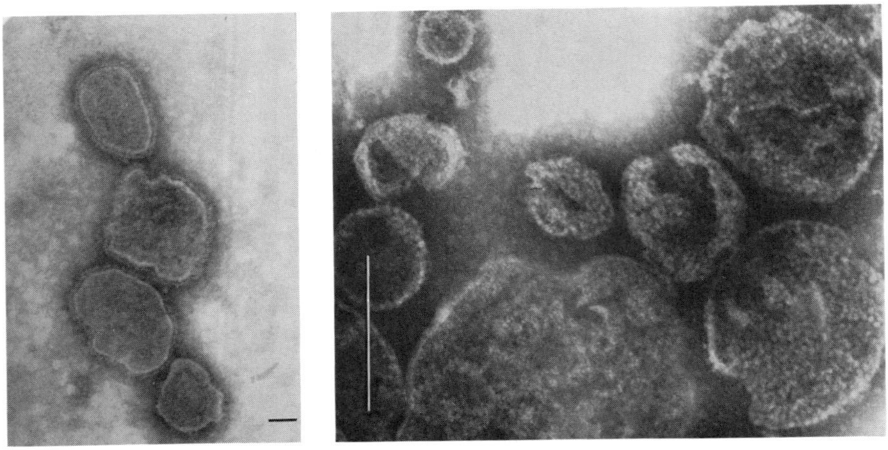

A B

Fig. 2. Negative staining of AcChE-rich membrane fragments (A, Scale 500 Å) and AcChR-rich membrane fragments (B, Scale 0,5 μ) with uranyl acetate.

ture. It is tempting to conclude that the AcChE oligomers are associated with these membrane structures via the 150 Å tail that has previously been observed by electron microscopic methods using isolated purified AcChE (4,5). Although it is attractive to suggest that the AcChE oligomers are associated by means of the 150 Å tail penetrating a membrane structure, this seems unlikely since AcChE can readily be extracted from electroplax membranes by high salt solutions. A more reasonable hypothesis is that the association is due to electrostatic bonding.

Negative staining procedures applied to the AcChR-rich particles (Fig. 2b) reveal in 80 % or more of the membrane particles a doughnut-like structure which covers the membrane surface. These structures, 80-90 Å in diameter, are presumed to represent AcChR molecules embedded in the membrane. Supportive evidence for this conclusion comes from the previous studies of Nichol and Potter (6) and Cartaud et al. (7), and more recent studies of Meunier et al. (8) on similar electron microscopic studies of isolated AcChR from Electrophorus elec-

tricus. It is not necessary, however, to conclude that the receptor oligomer as observed by such negative staining procedures is best represented by a rosette of protein subunits. Rather, depending on how the stain used penetrates a receptor molecule, various interpretations of the structure are possible, i.e., a cylinder of receptor subunits could equally well explain the results.

One of the more significant aspects of these results is that it seems unlikely that AcChE and AcChR are arranged in any kind of mosaic structure within or on the same membrane surface. It is more reasonable to conclude that AcChE and AcChR molecules occupy different patches on the postsynaptic membrane or, alternatively, that they are associated not as patches with the same membrane, but with different membrane structures. Therefore, at this point, it is not possible to give a detailed structural analysis of the postsynaptic cleft.

ISOLATION AND CHARACTERIZATION OF AcChR

AcChR from *Torpedo californica* can be quantitatively and rapidly solubilized by treatment of electroplax membranes with nonionic detergents (9). Affinity chromatography on quaternary ammonium resins, as previously described (10-12) yields in one step a highly purified preparation of AcChR from this tissue, as shown in Fig. 3. A summary of the quantitative aspects of this one-step separation is presented in

Table 1. Purification Table - AcChR

	Protein (mgs)	Activity (nmoles)	Specific Activity (nmoles/mg)
Homogenate	18,600	1950	0.11
Membrane Suspension	3250	1950	0.60
Triton Extract	1350	1950	1.45
Affinity Pool	95	1230	13

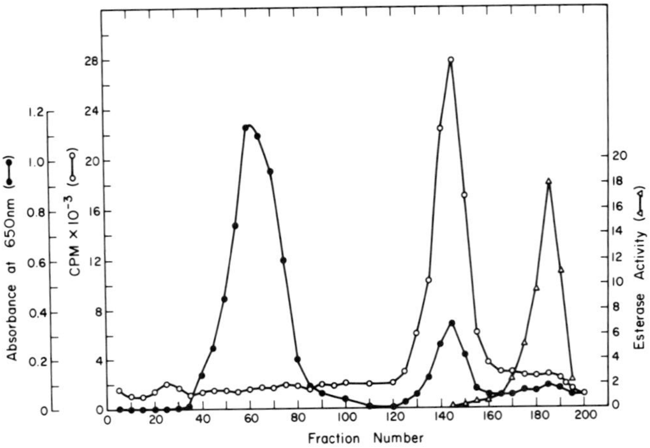

Fig. 3. Affinity chromatography of Triton-X-100 extracts according to Schmidt and Raftery (10,11).

Table 1. 100 mg of highly purified AcChR can be obtained in 63 % yield from 1 kg of original electroplax tissue. Gel electrophoresis patterns obtained from this purified material are presented in Fig. 4. Such electrophoresis carried out under non-denaturing conditions demonstrates that one major electrophoretic component is observed under the conditions used. Under denaturing conditions, with SDS and mercaptoethanol, a reasonably simple polypeptide composition was observed for isolated AcChR. The main component observed is a polypeptide of 40,000 M.W.; however, in addition, another component of 50,000 M.W. is always present, as is a minor component of approximately 65,000 M.W. The amounts of the two minor components, compared with the major one of 40,000 M.W., seem to vary from preparation to preparation; they may, therefore, represent tenacious impurities or integral components of a molecular aggregate which are, however, readily dissociable. Therefore, at the present time, it is difficult to define in exact terms the subunit composition of isolated AcChR. It appears unlikely that the polypeptide components observed are the result of proteolytic degradation of subunits of higher molecular weight since it has been shown (2) that AcChR-rich

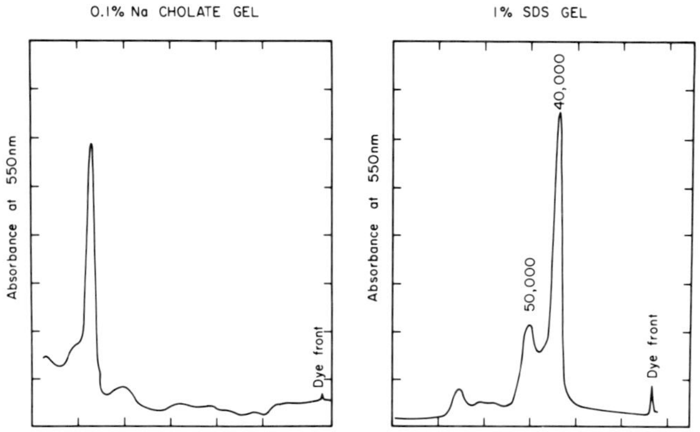

Fig. 4. Gel electrophoresis of purified AcChR under non-denaturing (left) and denaturing (right) conditions.

membrane particles contain the same polypeptide components as the isolated purified material. We consider it unlikely that degradation has occurred at the membrane isolation stage in a quantitative fashion.

Sucrose density gradient centrifugation of the isolated purified material shows a major component (Fig. 5) which has the same SDS gel pattern as the unfractionated material. The small amount of material more dense than the major component

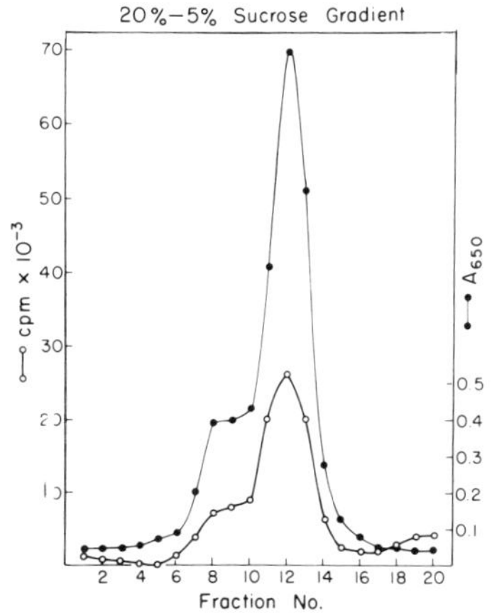

Fig. 5. Sucrose density gradient centrifugation of purified AcChR.

(Fig. 5). has the same SDS gel pattern and is most likely an aggregate of the major component.

Further characterization of the isolated purified receptor has been conducted. It is a glycoprotein containing approximately 5 % neutral sugars: mannose, galactose, and glucose in the ratio of 8 to 2 to 1. In addition, from amino acid analysis, N-acetyl-D-glucosamine has been identified as another carbohydrate constituent.

Because of the occurrence of sugar residues, it is difficult to define the subunit stoichiometry of the isolated material based on staining of SDS gels, since all bands that are observed - 40,000, 50,000 and 65,000 M.W.- have a positive reaction for carbohydrate moities. Nevertheless, cross-linking experiments with dimethyl-suberimidate indicate that the isolated AcChR from Torpedo californica has a molecular weight of approximately 250,000 in agreement with recent studies (13, 14) of isolated purified AcChR from Electrophorus electricus. Nevertheless, at the present time it is not possible to make definite conclusions regarding the exact subunit structure.

LIGAND BINDING TO AcChR

We have previously shown the number of cholinergic ligand binding sites for acetylcholine and its analogs appear to be equal to half the number of α-Bgt binding sites (15,16). Some of the more instructive results obtained in our laboratory for such a half-site relationship stems from our studies of the fluorescent cholinergic analog bis-(3-aminopyridinium) 1,10-decane diiodide (DAP). Fig. 6 shows a double reciprocal plot for the binding of DAP to isolated purified AcChR and the competition of acetylcholine with such binding. The pattern observed is of the competitive type. Displacement of bound DAP from AcChR by other cholinergic ligands can be followed by the generation of fluorescence due to free DAP. (DAP

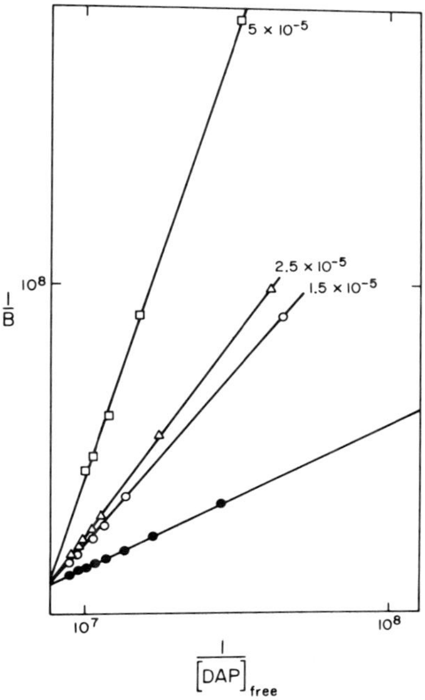

Fig. 6. Double reciprocal plot of DAP binding to purified AcChR and its competition with various concentrations of acetylcholine.

fluorescence is completely quenched in DAP-AcChR complexes). Figure 7 shows the titration of DAP from its receptor-associated state by various cholinergic ligands. α-Bgt can completely displace DAP from the bound state, as indicated in Fig. 7 by "α-Bgt level" and decamethonium and hexamethonium clearly displace DAP from its bound state in a simple titratable manner. However, carbamylcholine, even in very high concentrations, cannot fully release DAP to the α-Bgt level. Such an observation aroused the suspicion that carbamylcholine could displace one of the two pyridinium moieties of DAP without necessarily resulting in complete dissociation of DAP from its receptor-bound state. This suspicion was confirmed by centrifugation of AcChR in the presence of both DAP and acetylcholine or carbamylcholine. The results presented in Fig. 8 show that following centrifugation to pellet AcChR the

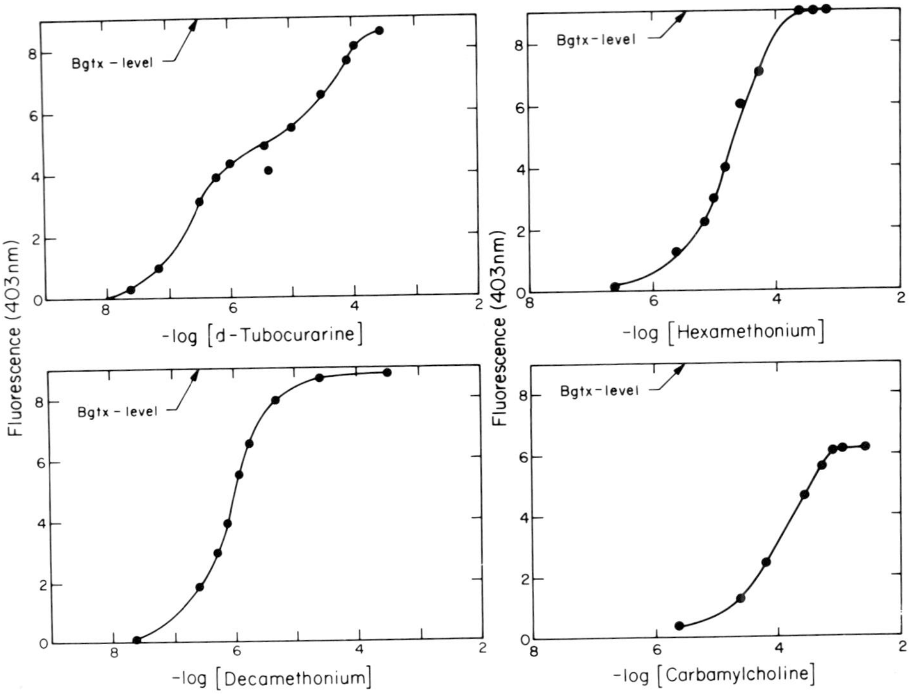

Fig. 7. Displacement of DAP from its AcChR-bound state by α-bungarotoxin (α-Bgt) and by various cholinergic analogs.

amount of fluorescence in the supernatant of such centrifugations was the same as that observed in the total sample, for DAP displacement by ligands such as decamethonium. However, with acetylcholine it was observed that the fluorescence generated was pelleted with the AcChR. The logical conclusion is that acetylcholine or carbamylcholine displace one of the two aminopyridinium moieties of DAP without DAP dissociating from the receptor-bound state. These results have given rise to a model of cholinergic ligand binding to isolated AcChR as shown in Fig. 9. In this model bis-quaternary ligands such as DAP, decamethonium and hexamethonium associate with two binding subsites on the AcChR surface. Depolarizing agonists such as acetylcholine or carbamylcholine bind to one of the two subsites (subsite A). As shown in the model, the other subsite appears to preferentially complex inorganic cations,

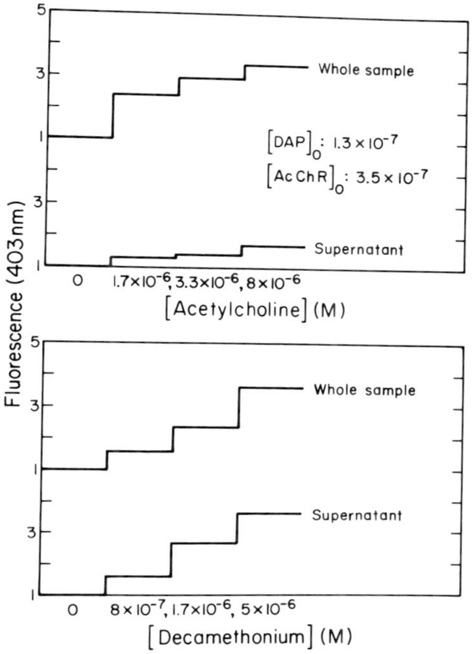

Fig. 8. Comparison of DAP liberated from its AcChR-bound state by decamethonium and acetylcholine before and after centrifugation.

monovalent or divalent, with high affinity. Such cations are competitive for the binding of bis-quaternary ligands such as DAP (16), decamethonium or hexamethonium (17). Additionally, the binding of d-tubocurarine is competitively inhibited by cations such as sodium (Fig. 10). On the other hand, the binding of acetylcholine is not inhibited by such cations (Fig. 10). Thus, of the two binding subsites for bis-quaternary ligands, one subsite is the specific neurotransmitter recognition site; the other binding subsite appears to preferentially complex inorganic cations.

These conclusions are substantiated by recent studies (18) of the alkylation of AcChR by trialkyloxonium salts. Treatment of purified AcChR with trimethyloxonium fluoroborate results in abolition of DAP binding, drastic reduction in the number

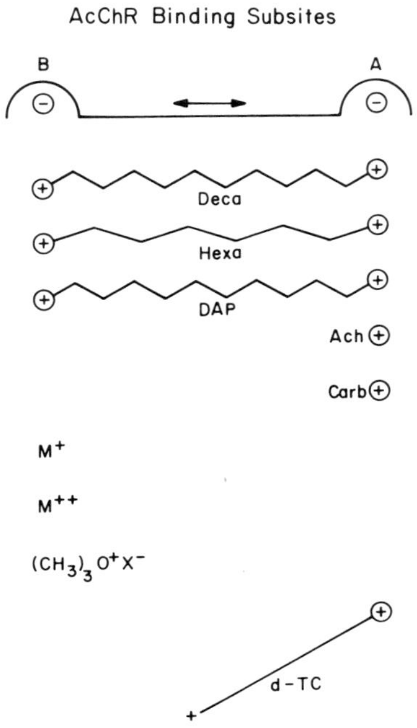

Fig. 9. Schematic model of ligand binding subsites on AcChR.

of binding sites for decamethonium without effect on the number of acetylcholine binding sites. Additionally, the dissociation constant for decamethonium is unaffected, although the number of binding sites is decreased; neither the dissociation constant nor the number of binding sites for acetylcholine are affected. Furthermore, in the presence of high concentrations of sodium chloride, the specific inactivation is inhibited. All of these results support the model presented in Fig. 9 of two binding subsites on the receptor surface.

An important question relating to neurotransmitter binding to AcChR is whether or not such binding is cooperative. It has been suggested (19,20) that the cooperativity observed in membrane depolarization as a function of agonist concentration is perhaps related to cooperative binding. We have observed that the binding of acetylcholine or of carbamylcholine to

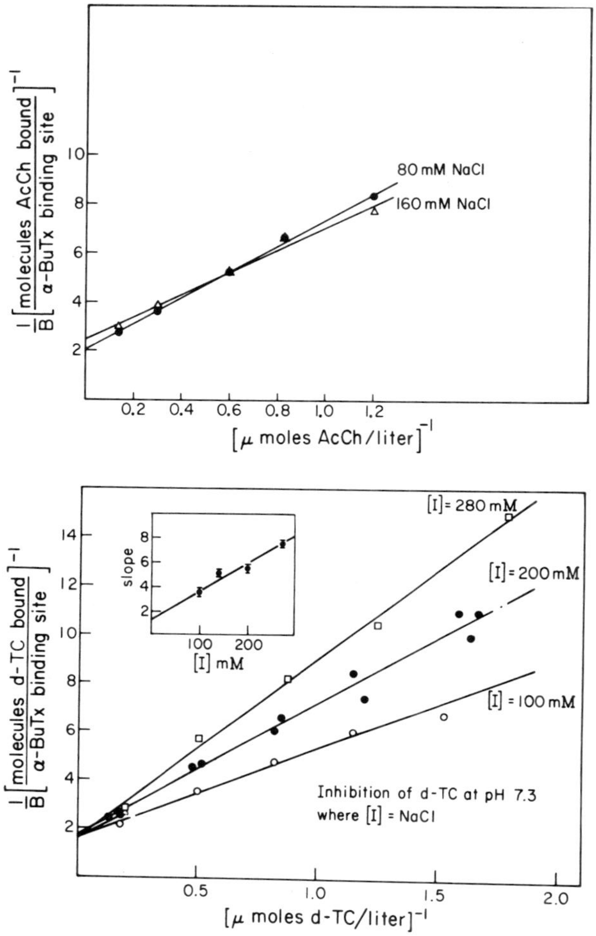

Fig. 10. Double-reciprocal plots of the antagonism between NaCl and acetylcholine (top) and d-tubocurarine binding to AcChR.

isolated purified AcChR and to AcChR-rich membrane fragments from Torpedo californica is not cooperative. Results shown in Fig. 11 emphasize the half of the site's phenomenon previously discussed for DAP; these are also observed for acetylcholine. It is important to point out that positive cooperativity in ligand binding is not necessary to explain positive cooperativity in physiological response (21,22).

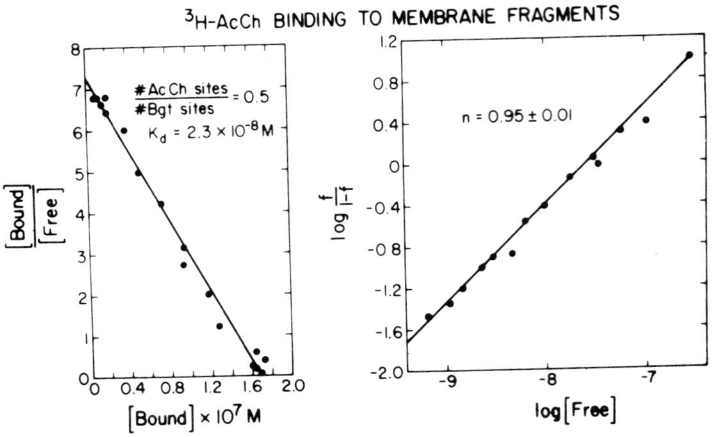

Fig. 11. Scatchard (left) and Hill (right) plots of acetylcholine binding to AcChR-rich membrane fragments.

The clearest demonstration of a half-site relationship for the binding of cholinergic ligands compared with α-Bgt binding is shown by the results given in Fig. 12 where the release of DAP from its AcChR-bound state is monitored by fluorescence increase upon addition of α-Bgt to saturating concentrations. However, it is obvious that essentially no release of DAP occurs until approximately half the concentration of α-Bgt necessary to saturate has been added. That the α-Bgt is forming a complex can be demonstrated by use of ^{125}I-α-Bgt to monitor complex formation. These results clearly show that both α-Bgt and DAP can occupy separate binding sites on the purified AcChR simultaneously. The observed half-site relationship could result from a number of causes such as (a) negative cooperativity induced by ligand binding (23); (b) preexistent asymmetry in AcChR (24); or (c) two classes of binding site, perhaps on different polypeptide chains. At the present time, insufficient evidence is available to clearly distinguish the contributions made by any or all of these effects. The fact that saturation of AcChR binding sites by DAP slows down the kinetics of toxin association with all of its binding sites (25), suggests that ne-

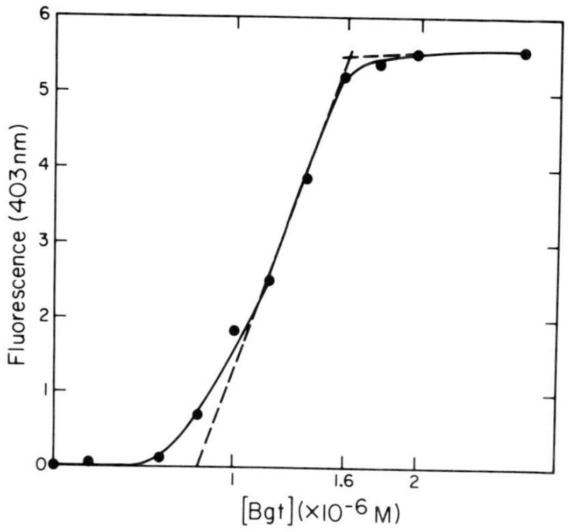

Fig. 12 Titration of AcChR-bound DAP as a function of added α-Bgt

gative cooperativity, induced by DAP binding, may well make a contribution to the half-site phenomenon which is observed. On the other hand, there is also some evidence for preexistent asymmetry in the AcChR based on studies of ligand binding to the AcChR species half saturated with α-Bgt, i.e., toxin complexed with the binding sites to which ligands do not bind (26).

The results discussed immediately above indicate that ligands such as DAP may well play some role in inducing conformation changes in the AcChR.

We have also studied the binding of the fluorescent dye 1-(5-dimethylaminonaphthalene-1-sulfonamide) ethane-3-trimethylammonium iodide (Dns-chol) originally synthesized by Weber et al. (1971) (27) and first used to study binding to a proteolipid isolated from electroplax (28). More recently, Dns-chol has been used by Cohen and Changeux (1973) (29) and by Cohen et al. (1974) (30) to determine its ability to bind to AcChR-rich fragments isolated from the electric organ of

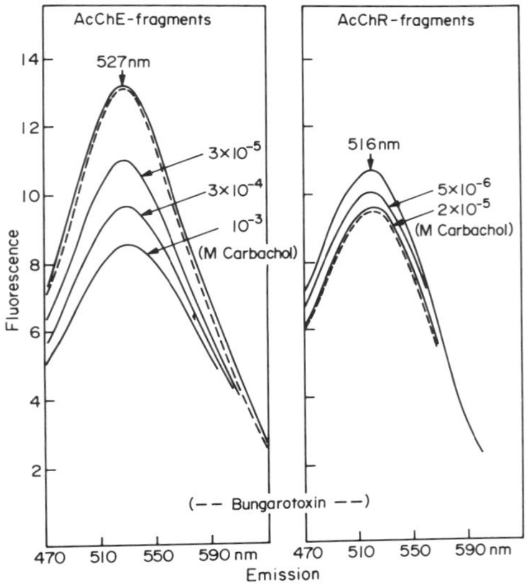

Fig. 13. Fluorescence spectra Dns-chol bound to AcChE-rich (left) and AcChR-rich (right) membrane fragments and the effects of α-Bgt and carbachol on emission intensity performed as described in (29).

Torpedo marmorata. We have studied the binding of the Dns-chol to AcChE-rich fragments and to AcChR-rich fragments from Torpedo californica and the results of these studies are shown in Fig. 13. For AcChE-rich fragments excitation at 287 nm leads to emission with a maximum at 527 nm, as shown on the left side of Fig. 13. Addition of α-Bgt does not result in any decrease in the emission intensity, as shown by the dashed line, as expected, since the AcChE-rich fragments do not bind α-Bgt. On the other hand, addition of carbamylcholine at various concentrations reduced the fluorescence intensity without any shift in the wavelength of maximum emission. The maximum emission with AcChR-rich fragments is 516 nm. As shown on the right-hand side of Fig. 13, addition of α-Bgt to saturation causes a reduction in the intensity at this wavelength. Additionally, carbamylcholine at saturating concentrations, reduces the emission intensity to almost the same level as α-Bgt at saturation. There is no change in the wavelength of maximum emission upon addition of the agonist carbamyl-

Table 2. I_{50} Values for AcChR - Purified and Membrane-Bound

	Purified AcChR 5 mM Tris	AcChR Membranes 20 mM NaCl-5 mM Tris	Ratio
Acetylcholine	2.5×10^{-6}	6×10^{-9}	420
Carbamylcholine	4.5×10^{-5}	4×10^{-8}	1100
Nicotine	8×10^{-5}	1×10^{-7}	800
Phenyltrimethyl-ammonium	8×10^{-5}	2×10^{-7}	400
Choline	1.7×10^{-3}	1×10^{-5}	170
Decamethonium	2×10^{-6}	9×10^{-9}	220
Hexamethonium	2×10^{-6}	5×10^{-7}	4.0
d-Tubocurarine	1×10^{-7}	3×10^{-8}	3.3
Gallamine	2×10^{-7}	8×10^{-8}	2.5
DAP	1×10^{-6}	2×10^{-8}	50

choline to AcChR-rich fragments from *Torpedo californica*. Also, when isolated purified AcChR was first denatured by heating, the same amount of fluorescence intensity was observed at 516 nm as was found with AcChR poisoned with α-Bgt. Therefore, we conclude that the binding of Dns-chol to isolated purified AcChR or to AcChR-rich fragments from *Torpedo californica* is largely nonspecific.

Dissociation constants can be measured directly on purified membrane fragments or purified AcChR by equilibrium dialysis or by centrifugation methods. In addition, other methods have been used in our laboratory to measure inhibition constants for the same ligands. The first of these involved reduction of the initial rate of ^{125}I-α-Bgt association with receptor in the presence of added ligand (15,25,31).

Table 2 summarizes some of the pertinent results obtained by inhibition of the initial rate of ^{125}I-α-Bgt association with purified receptor and purified membrane fragments. Inhibition constants vary over three orders of magnitude. The most significant difference observed is in the binding of agonists, acetylcholine, carbamylcholine, nicotine, and decamethonium, where all of these bind considerably more tightly to the receptor in its native membrane environment than they do to the purified material. On the other hand, the antagonists, hexamethonium and d-tubocurarine, display no significant difference in their affinities for the receptor in the purified and membrane-bound states. Effects opposite to these, i.e., agonists binding more strongly to purified receptor than to membrane-bound receptor have been observed by Meunier and Changeux (1973) (32) in Electrophorus electroplax AcChR and these authors have suggested the release of a membrane constraint, yielding a relaxed receptor, as the origin of such effects. It is evident from the data presented in Table 2 that the change undergone by the receptor upon solubilization appears to be restricted to a rather precise region. The change is localized at the acetylcholine-binding subsite and it is not possible yet to say whether quaternary structural changes in the acetylcholine receptor oligomer accompany these tertiary structural changes that are observed. In addition, it is also possible that surface charge, especially negative charges of the membrane particles exert some effect upon the binding constants for the positively charged ligands. The structural change in the receptor is likely to be a very subtle one as evidenced by the greater than 1 order of magnitude difference observed in the ratios for acetylcholine and carbamylcholine (Table 2).

RECONSTITUTION

Purified AcChR from Torpedo californica can be exchanged from Triton X-100 solution to sodium cholate solution following adsorption on DEAE-cellulose. Elution in sodium cholate can be achieved by use of high concentrations of NaCl. Following the removal of most of the cholate, addition of the AcChR preparation to sonicated Torpedo californica phospholipids, followed by extensive dialysis, results in a re-association of the AcChR with phospholipids. By sucrose density gradient centrifugation it can be demonstrated that approximately 70 % of the protein forms a lipoprotein complex. Addition of $^{22}NaCl$ to this preparation results in equilibration of the ^{22}Na between inside and outside of closed structures which are formed. The enclosed volume of such reconstituted vesicles can be obtained following rapid adsorption onto DEAE-cellulose columns and immediate wash-away of external sodium. Rupture of the absorbed vesicles with solutions containing Triton X-100 releases the internal ^{22}Na, which can be washed off the DEAE columns, thus giving an estimate of the internal volume of the vesicles. Figure 14 shows the rate of leakage of ^{22}Na from such vesicles as a function of time. Such leakage is reasonably rapid and it can be shown (33) that this rapid leakage rate is also observed in Torpedo californica phospholipid vesicles prepared without the addition of AcChR. Thus com-

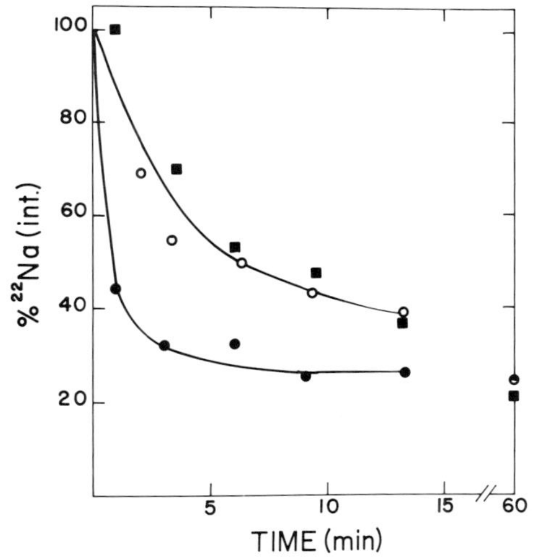

Fig. 14. ^{22}Na efflux from reconstituted AcChR-lipid vesicles: ○-○, flux of vesicles without additives; ●-●, flux in presence of 100 μM carbamylcholine; ■-■, flux in presence of 100 μM carbamylcholine following prior saturation with α-Bgt.

pared to vesicles prepared from synthetic phospholipids, the naturally occurring lipids of the electroplax form leaky vesicles. However, addition of carbamylcholine to our reconstituted vesicles results in a dramatic increase in the rate of efflux of internal ^{22}Na. That this is a specific cholinergic agonist-induced excitability of the reformed membrane is clearly shown (Fig. 14) by the demonstration of complete abolition of the enhancement by carbamylcholine when the system is incubated with α-Bgt prior to addition of carbamylcholine.

One of the most significant experiments described in this communication is that upon re-association with Torpedo phospholipids, isolated purified AcChR reforms a chemically excitable membrane as shown in Fig. 14. The significance of this finding is that the isolated protein, in addition to containing the specific neurotransmitter recognition site, i.e., the binding subsite for acetylcholine, also apparently contains the molecular features which are necessary for ion-translocation upon association with the neurotransmitter. This represents the first demonstration of such an effect using isolated purified material; it obviously allows the study of this important postsynaptic macromolecule in reconstituted systems so that in the future detailed molecular insight may be obtained into the molecular mechanism of postsynaptic depolarization.

X-RAY DIFFRACTION

X-ray diffraction from oriented fragments of AcChR-rich membranes yields information about the organization of receptor subunits in the membrane. Samples for X-ray diffraction were prepared by ultracentrifugation at 100-200 x 10^3g for 10-40 hrs. into specially constructed cells. Under certain conditions, the diffraction pattern shown in Fig. 15 was obtained from the hydrated pelleted specimen. This photograph shows that the membrane fragments are well organized into almost

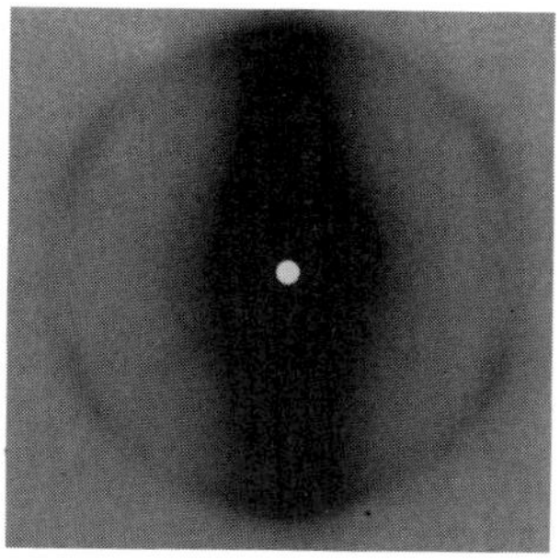

Fig. 15. Low-angle X-ray diffraction from a fully hydrated, pelleted sample taken at 2°C. The normal to the membrane surfaces is vertical. The off-axial maxima have a d spacing of 37 Å.

parallel sheets. The off-axis reflections (spacing 37 Å) indicate that the protein molecules within the membrane sheets have been organized into three-dimensional crystalline domains. The radial width of this family of diffraction maxima gives information about the ordered domain sizes while the angular dispersion of intensity shows the degree to which the membrane-protein sheets are parallel to each other.

These data indicate that the proteins are assembled into domains of about 700 Å across. The angular distribution shows peaks of almost gaussian distribution about their centers with angular half-width of $\pm$ 12°. These maxima lie on layer lines with a vertical spacing of 74 Å, which is indicative of a true repeat distance in the protein domain of 74 Å perpendicular to the plane of the membranes. The second order, 37 Å maximum on the meridian shows that the vertical distance between subunits is 37 Å. The meridian also shows a broad pillar of continuous diffraction with three orders of a 110 Å repeat superimposed on it.These meridional maxima are weak and fairly sharp, and belong to a second family which may represent the multilayer repeat between lipid bilayers. Another family of four or five sharp but weak diffraction maxima intersect the equator, but have a large, almost continuous angular disper-

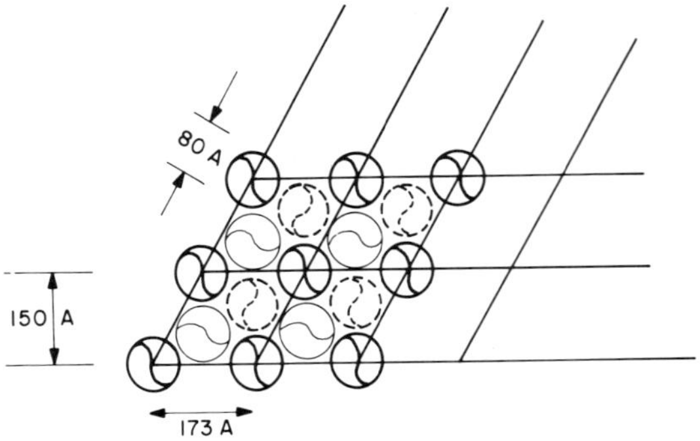

Fig. 16. A hexagonal lattice of cell side 173 Å is suggested by the equatorial diffraction maxima.

sion. This family may represent multiple orders of a lattice, possibly either hexagonal or close-packed cubic with a first-order spacing of 150 Å (Fig. 16). The existence of a diffuse 4.6 Å maximum in all hydrated specimens examined at 2°C shows that the Torpedo lipids in the bilayers are above the phase transition; this demonstrates that the crystalline diffraction almost certainly arises from the ordering of proteins and not phospolipids.

On drying in nitrogen, the sample used in Fig. 15 lost the crystalline order, the pellet diminished in size considerably, and the meridian showed a single order of 57 Å spacing, which is close to, though slightly larger than the multilayer spacing found in dried phospholipid bilayers (34).

Samples have been prepared which, though they show no evidence of crystallinity, still show a clear, continuous diffraction pattern along the meridian (perpendicular to the membrane) with a first broad maximum at about 78 Å and a second at about 34 Å.

The diameter of particles seen in negatively stained electron micrographs is about 80 Å. With the assumptions that 80 Å is close to the true particle diameter, and that subunits have a molecular weight of about 40,000 daltons, an estimate of the number of subunits per 37 Å layer can be made. The volume of one layer would be about 150-180 x 10^3 $Å^3$. The volume occupied by a protein of molecular weight 40,000 will depend on the hydration state and the closeness of contact between subunits (35). It is expected to lie in the range between 50,000 $Å^3$ for pure hydrated protein to about 100,000 $Å^3$ if the packing is more like that found in protein crystals.

Taken in concert with the approximate molecular weight of the multisubunit receptor of 240,000, the X-ray and electron microscope evidence suggests a structure built up of three layers each containing two monomers of 40,000 molecular weight. The exact mode of packing requires further investigation; however, a model such as that shown in Fig. 17 is reasonably consistent with the observed X-ray diffraction data. The important features of such a model are that the structure located in the membrane is polar, and is presumably limited in size by the presence of the bilayer. In such a model there would be preexistent structural asymmetry between the three pairs of subunits (in the top, middle and bottom of the bilayer). This is broadly consistent with electrophysiological data that cholinergic ligands are effective only on the synap-

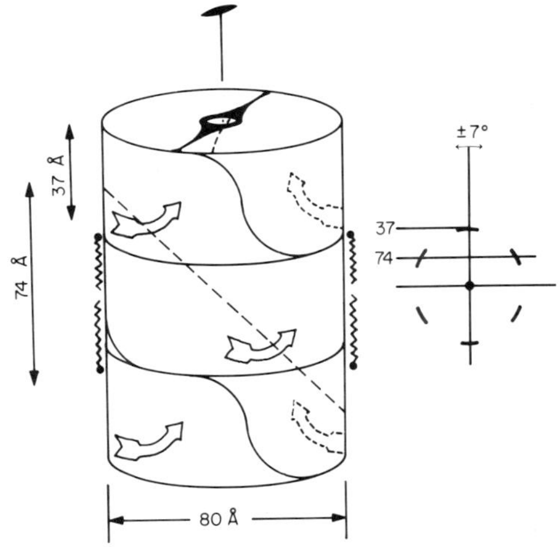

Fig. 17. A tentative model for the AcChR which embodies a 74 Å true repeat and a 37 Å separation between subunits perpendicular to the membrane layers.

tic side of the membrane (36).

The possibility of different subunit types, or different conformational states of the subunits embedded in the bilayer cannot be ruled out, and the model at this stage must be considered only as one of several possibilities.

ACKNOWLEDGEMENTS

The research presented here was supported by USPHS Grants NS-10294 and GM-19984 and by a grant from the Sloan Foundation. M.A. Raftery and R.M. Stroud gratefully acknowledge the National Institute of Health for receipt of Research Career Development awards. Contribution number 4954.

REFERENCES

1. Duguid, J.R. and Raftery, M.A. (1973). Biochemistry 12, 3593.

2. Michaelson, D., Vandlen, R., Bode, J., Moody, T., Schmidt, J. and Raftery, M.A. (1974, submitted). Arch. Biochem. Biophys.

3. Duguid, J.R. and Raftery, M.A. (1973). Arch. Biochem. Biophys. 159, 512.

4. Dudai, Y., Herzberg, M. and Silman, I. (1973). Proc. Natl. Acad. Sci. 70, 247.

5. Rieger, F., Bons, S., Massoulie, J. and Cartaud, J. (1973). Eur. J. Biochem. 34, 539.

6. Nichol, A. and Potter, L.T. (1973). Brain Res. 57, 508.

7. Cartaud, J., Benedetti, E.L., Cohen, J., Meunier, J.-C. and Changeux, J.-P. (1973). FEBS Lett. 33, 109.

8. Meunier, J.-C., Sealock, R., Olsen, R. and Changeux, J.-P. (1974). Eur. J. Biochem. 45, 371.

9. Raftery, M.A., Schmidt, J. and Clark, D.G. (1972). Arch. Biochem. Biophys. 152, 882.

10. Schmidt, J. and Raftery, M.A. (1972). Biochem.Biophys. Res. Commun. 49, 572.

11. Schmidt, J. and Raftery, M.A. (1973). Biochemistry 12, 4106.

12. Vandlen, R., Schmidt, J. and Raftery, M.A. (1974, in press). Proc. Am. Chem. Soc. Affinity Chromat., Los Angeles.

13. Biesecker, G. (1973). Biochemistry, 12, 4403.

14. Hucho, F. and Changeux, J.-P. (1973). FEBS Lett. 38, 11.

15. Moody, T., Schmidt, J. and Raftery, M.A. (1973). Biochem. Biophys. Res. Commun. 53, 761.

16. Martinez-Carrion, M. and Raftery, M.A. (1973). Biochem. Biophys. Res. Commun. 55, 1156.

17. Moody, T. and Raftery, M.A. (1974, in preparation).

18. Chao, Y. and Raftery, M.A. (1974, in press). Biochem. Biophys. Res. Commun.

19. Changeux, J.-P. and Podleski, T.R. (1968). Proc. Natl. Acad. Sci. 59, 944.

20. Changeux, J.-P., Thiery, J., Tung, T. and Kittel, C. (1967). Proc. Natl. Acad. Sci. 57, 331.

21. Levitzki, A. (1974). J. Theor. Biol. 44, 367.

22. Levitzki, A., Segel, L.A. and Steer, M.L. (1974, in press).

23. Conway, A. and Koshland, D.E., Jr. (1968). Biochemistry 7, 4011.

24. MacQuarrie, R.A. and Bernhard, S. (1971). Biochemistry 10, 2456.

25. Deutsch, J. and Raftery, M.A. (1974 , in preparation).

26. Bode, J. and Raftery, M.A. (1974, submitted). Biochemistry

27. Weber, G., Borris, D., DeRobertis, E., Barrantes, F., LaTorre, J., and DeCarlin, M. (1971). Mol. Pharmacol. 7, 530.

28. DeRobertis, E. (1971). Science 171, 963.

29. Cohen, J. and Changeux, J.-P. (1973). Biochemistry 12, 4855.

30. Cohen, J., Weber, M. and Changeux, J.-P. (1974). C.R. Acad. Sci., Paris, D. 278, 1269.

31. Schmidt, J. and Raftery, M.A. (1974, in press). J. Neurochem.

32. Meunier, J.-C. and Changeux, J.-P. (1973). FEBS Lett. 32, 143.

33. Michaelson, D.M. and Raftery, M.A. (1974, in press). Proc. Natl. Acad. Sci.

34. Levine, Y.K. and Wilkins, M.H.F. (1971). Nature New Biology, 230, 69.

35. Mathews, B.W. (1968). J. Mol. Biol., 33, 491.

36. Del Castillo, J. and Katz, B. (1955). J. Physiol. (London), 128, 157.

Received September 1, 1974

DISCUSSION

Veeger: Have you any data concerning the degree of fluorescence polarization of the protein-bound bis-(3-aminopyridinium)1,10-decane diiodide, partially replaced by carbamylcholine. I would be interested to see whether it has some degrees of freedom. It is possible that the increase of fluorescence is due to the removal of a quenching group by substrate binding from the vicinity of the chromophore, thus quenching not being due to the acetylcholine site at all.

Raftery: So far we have not done any fluorescence polarization work. The subsite model presented in Fig. 9 is the simplest interpretation of all the binding data which we have to date. If we were, however, to involve a third site, say for acetylcholine, then it is unlikely that the data for ligands such as decamethonium would be so clean e.g. it is an agonist and therefore should interact with the third site. However, its K_I for inhibition of bis-(3-aminopyridinium)1,10-decane diiodide binding agrees precisely with its K_d, determined directly, so that we do not at present see the necessity to make the picture more complex.

Changeux: In your model of the receptor protein incorporated into the membrane you postulate that all identical dimers are piled on top of each other. Why does the pile stop at three units and does not make a huge indefinite helix. In other words, what limits the growth of the macromolecular assembly.

Raftery: We do not necessarily postulate identical subunits in the stack of dimers. It is possible that the middle pair is structurally or conformationally different and this alone would limit the size of the receptor assembly.

Changeux: A comment relevant to the fluorescence experiments you have done with *Torpedo californica* membrane fragments and Dns-choline. You do not see the wandering of the shift observed with *Torpedo marmorata* membranes and the same ligand. Obviously, one cause for the difference presented might be the species difference. However, with your membrane fragments containing the receptor the relative amount of emitted light which decreases in the presence of α-toxin is rather small (may be 10-20 % instead of the 70 % observed by Cohen and Changeux). Several explanations might be given for this difference. One is that there exists a relatively large amount of non-specific binding which indeed takes place with *Torpedo* when the receptor-rich membrane fragments have a low specific activity. My question is, therefore: What is the specific activity of the membrane fragments at the end of the fluorescence experiment?

Raftery: The specific activity of the AcChR-rich fragments is the same at the end of the fluorescence experiments as it was at the beginning in various preparations used. These contained from 20-40 % of their total protein content represented by receptor. In contrast to the results of Cohen and Changeux using *Torpedo marmorata* we also find, as you say, only about 20 % fluorescence intensity decrease on addition of toxin or ligands which represents tremendous non-specific binding. Similarly, with our purified receptor we find only about 30 % decrease upon saturation with toxin or ligands. Indeed, boiled denatured *Torpedo californica* AcChR binds the same amount of Dns-choline non-specifically as it does before such drastic treatment. Our interpretation is that the AcChR from the two species are different in some subtle manner. This also appears to be clear from differences in affinity for bis-quaternary ligands as well as from differences in the ratio of toxin sites to ligand binding sites.

Sund: According to your Fig. 9 you suggest that decamethonium as well as hexamethonium are bound to each subsite. The length of both molecules is very different and, therefore, I assume if both subsites fit to decamethonium the system does not fit to hexamethonium, and vice versa. You found that the number of binding sites for decamethonium and hexamethonium are the same. Could it be possible that hexamethonium or decamethonium binds only to one of the two subsites? If this is the case, how can you explain that according to your table 2 the I_{50} values are the same for both?

Raftery: The model we present in Fig. 9 is a simplified version for reasons of clarity. Either decamethonium or hexamethonium could clearly prefer a given conformation in a flexible receptor merely driving a conformational equilibrium, or alternatively, interacting strongly with only one negative subsite, while the rest of the free energy of binding comes from hydrophobic interactions involving the methylene groups. Another ligand, d-tubocurarine is shown in Fig. 9 to bind to subsite A and not to subsite B. This is an oversimplification since we know (see Fig. 10) that its binding is inhibited by NaCl. However, the K_I for NaCl is ~70 mM which is a high value compared with those obtained for NaCl inhibiting decamethonium, hexamethonium, bis-(3-aminopyridinium)1.10-decane diiode or α-bungarotoxin. We take this to mean that d-tubocurarine can bind in a number of ways, one of them involving subsite B.

Helmreich: I would like to come back to the question raised by Dr. Changeux. If you envision the receptor assembly to be inserted in the membrane as a cylinder with one face communicating with the outside and the other face reaching into the hydrophobic membrane core, you might be able to show that the sugar residues are reactive in the membrane from the outside. Can one do that?

Raftery: Yes, using concanavalin A which binds to our receptor (presumably to terminal mannose residues) we should be able to provide another demonstration of polarity. One further comment is appropriate. It is not necessary to have a phospholipid bilayer associated with the central pair of subunits, as is hinted at in Fig. 17. It is also possible that the lipids interact with the bottom pair of subunits only; these could expose a hydrophobic surface (covered in the top identical pair by their interaction with the central pair), thus facilitating penetration of a lipid bilayer and giving rise to the polarity we observe in our reconstitution experiments, i.e. 95 ± 5 % of toxin binding sites on the outside of the receptor-lipid vesicles (cf. Ref. 33).

Hucho: The doughnuts shown by the electron microscope appear to consist of more than just two subunits as would follow from your model. Would you please comment on that?

Raftery: First, it is not possible to fit six spherical protein subunits of molecular weight about 40 000 into an 80 Å circle. Secondly it is not necessary to interpret the electron micrographs as representing a circle of subunits, unless they were elongated. Depending on how the stain penetrates the subunit assembly we propose and, depending on the tilt of the assembly with respect to the perpendicular axis, one could visualize any number of subunits (up to the total present) by negative staining.

Cohen: It is interesting that there appear to be slight but clear functional differences between the receptor-rich membranes purified from two fish as similar as Torpedo californica and Torpedo marmorata. Apparently, the affinity of the membrane-bound receptor of Torpedo californica for bis-onium (decamethonium, hexamethonium) are two orders of magnitude greater than that of Torpedo marmorata (M. Weber and J.P. Changeux, Mol. Pharmacology 10, 1-40 [1974]). However, the affinities for mono-quaternary amines (acetylcholine, carbamylcholine, nicotine) are the same for the two preparations. A second and perhaps related difference is the nature of interaction between Dns-choline and receptor sites and secondary binding sites (cf. J.B. Cohen and J.P. Changeux, Biochemistry 12, 4855 [1973]). What are the conditions you have used for membrane isolation and for the ligand interaction studies? Do you feel the differences reflect differences between isolation procedures or between species?

Raftery: We have used differing conditions of ionic strength for membrane isolation and consistently get the same results. Our interpretation is that the differences observed are species related (see comments relating to J.P. Changeux' question).

Section V. Antibodies, Drugs and Metabolites

Chairmen: *E. Antonini* and *M. Raftery*

Antibody Combining Sites as a Model for Molecular Recognition

I. Pecht

I. INTRODUCTION

The combining sites of antibodies present a rather instructive example of molecular recognition. The immunoglobulins constitute a group of protein molecules which are very similar in general structure, yet exhibit practically an unlimited range of different specificities. Sequence and protein chemistry studies of immunoglobulins revealed the following basic picture of these proteins [1,2]: Two pairs of polypeptide chains of different lengths (light 22000 - 25000 Daltons and heavy 50000 - 55000 Daltons) are joined together by disulphide bonds. Each chain has regions of homology which are 110 to 120 residues long of constant sequence within a given class. Four such homology regions are present in the heavy chain and two in the light. The amino acid sequence of the N-terminal region of each of these chains is variable and contains the residues responsible for the specificity of the combining site formed by that part of the immunoglobulin molecule.

The resolution by X-ray crystallography of the three-dimensional structure of immunoglobulins, particularly of the antigen-combining site located in the variable regions [3-5] together with results obtained by other physico-chemical approaches [6-8], have made it clear that the interaction between these sites and their respective ligands are determined by the same forces which are operative in other binding sites of proteins, notably in enzymes. The specificity and affinity of the sites result from an appropriate spatial arrangement of amino acid residues within the site to fit the complementary groups of the hapten. In solution, the antibody (Ab)-hapten (H) (or antigen) complex (AbH) exists in a dynamic equilibrium. The elementary steps by which the complex is formed and dissociates have been followed by fast kinetic methods, mainly by chemical relaxation temperature-jump technique [8-10]. Noteworthy is the observation that in all cases hitherto examined, this equilibrium has been found to fit a single-step mechanism expressed by the following equation: $Ab + H \underset{k_{21}}{\overset{k_{12}}{\rightleftharpoons}} AbH$. This is surprising in view of the fact that the antigen binding is only one major function of the immunoglogulin (for a review, cf. ref. 7). The other functions (e.g. binding of complement) are triggered by the antigen binding and are localized at the Fc region of the protein. Thus an extra step would be expected by which the "signal" of binding is transmitted from the combining site in the Fv region to the Fc region. By studying the circular polarization of the emission of the tryptophan residues of antibodies it has recently been found that upon antigen binding, conformational changes are indeed taking place in both the Fab and Fc regions of the immunoglobulin molecules [11]. Moreover, it was found that such structural transitions are not induced by binding small haptens [11]. It is because of this difference that the recognition steps of small ligands by the antibody combining site may be studied in a very detailed manner without interference from secondary steps [12].

The interaction between combining sites of antibodies and their respective ligands were probed by both kinetic and fluorometric techniques. From the chemical relaxation data, the specific rates of binding and dissociation of different ligands to the sites were determined. The induction of a circularly polarized component in the emission of fluorescent probes covalently attached to the haptens was used as criterion for immobilization of these side chains upon interaction with the contact residues of the combining site and therefore served as a probe for the dimensions of the site.

II. HOMOGENEOUS COMBINING SITES ON IMMUNOGLOBULINS: KINETIC MAPPING

Homogeneous combining sites with affinity towards three different types of haptens have been investigated. These were sites of the following IgA myeloma proteins produced by plasmacytoma cells of mice [13]:

1. MOPC 315, specific for 2,4 dinitrophenyl derivatives.
2. HOPC 8 and TEPC 15, specific for phosphoryl choline.
3. XRPC 24, specific for galactose oligomers.

In temperature-jump experiments all three different systems of combining sites and their ligands exhibit only a single relaxation time which decreased linearly with the increase in the sum of free concentrations of combining sites and respective ligands. This suggests that a single-step binding equilibrium is operative for all these binding processes. Thus an analysis of the interaction between each of these sites and a series of systematically varied derivatives of its ligand may be carried out in terms of their specific rates of binding, k_{12}, and dissociation, k_{21}. The combining site of MOPC 315 is the one most extensively investigated by this method. The structural features of more than 50 different derivatives of the 2,4-dinitrophenyl group were correlated with their specific rates of interaction with the protein combining site. These belonged to the following structurally distinct series:

1. Nitrophenyl derivatives differing in number and location of nitro groups or in the nature of ring substituent.
2. Derivatives of 2,4-dinitroaniline substituted at the amino-nitrogen by straight or branched alkyl chains, including alkyl carboxylates and alkyl ammonium groups.
3. Bis-substituted 2,4-dinitroaniline derivatives by alkyl groups on the amine.
4. α-N-DNP derivatives of glycine and alanine and its dipeptide, including the different diastereoisomers of alanine and $(alanine)_2$.

5. $(lysyl)_n$ ε-N-DNP lysine derivatives where n varied from 0 to 8.

Tables 1 and 2 summarize data obtained for the interaction of MOPC 315 and some of the above-mentioned ligands. From these data it is seen that, though some variation in the k_{12} values is observed, the main source for difference in the affinity of the ligands stems in the variations of the specific rates of dissociation, namely in their reciprocal residence time within the combining site. Thus it is the value of k_{21} which reflects the complementarity between a certain hapten and the combining site.

2,4-dinitroaniline derivatives where one or both amino hydrogens were substituted by alkyl groups showed the following trend in k_{21} values: From 927 sec^{-1} decreasing to 373 sec^{-1} and to 121 sec^{-1}, then rising to 555 sec^{-1} for the unsubstituted, monomethyl, dimethyl and diethyl 2,4-dinitroaniline, respectively. This sensitivity to subtle changes in the ligand structure illustrates clearly the specificity of the combining site as well as the resolving capacity of the method. A closer analysis of the k_{21} values presented in Table 1 enabled us to characterize the combining site of MOPC 315 in terms of its different subsites of interaction. Each subsite was defined as a space element within the site, located at a certain distance from the region where the DNP ring is anchored and in which interactions of a specific nature (electrostatic, hydrogen bonding, hydrophobic etc.) take place between contact amino acid residues of the subsite and the hapten. The assumption that different ligands are positioned in the site in such a way that the DNP ring always occupies the same region is supported by a variety of other results obtained by spectroscopic and affinity-labeling studies [6,14]. Using the DNP subsite as the reference point (S_1), location of further subsites are given in terms of distances from the geometric center of the DNP ring.

Table 1. Specific rates of interaction and equilibrium constants (K) of different 2,4-dinitroaniline derivatives and MOPC 315

No.	R*	$k_{12} \times 10^{-8}$ ($M^{-1}sec^{-1}$)	k_{21} (sec^{-1})	k_{12}/k_{21} 10^{-6}(M^{-1})	$K \times 10^{-6}$ (M^{-1})
1	CH_3	3.5 ± 0.2	370 ± 30	0.94	
2	$(CH_2)_3CH_3$	2.5 ± 0.2	79 ± 10	3.20	3.4
3	$(CH_2)_5CH_3$	1.2 ± 0.1	123 ± 18	1.0	1.1
4	CH_2COO^-	1.7 ± 0.2	1340 ± 100	0.13	0.2
5	$CH_2CH_2COO^-$	1.8 ± 0.3	285 ± 50	0.64	
6	$CH_2CH_2NH_3^+$	0.7 ± 0.1	735 ± 65	0.089	0.13
7	$CH(CH_3)_2$	3.7 ± 0.2	108 ± 9	3.4	
8	$C(CH_3)_3$	3.6 ± 0.3	58 ± 6	6.1	
9	CH_2phenyl	1.8 ± 0.2	47 ± 9	3.8	
10	$(CH_2)_3COO^-$	3.8 ± 0.2	45 ± 9	8.4	
11	$(CH_2)_3NH_3^+$	0.8 ± 0.1	850 ± 100	0.098	
12	$(CH_2)_5COO^-$	1.2 ± 0.1	60 ± 10	2.0	
13	$(CH_2)_6NH_3^+$	0.9 ± 0.1	170 ± 10	7.5	
14	cyclohexyl	2.6 ± 0.2	116 ± 10	2.2	

*

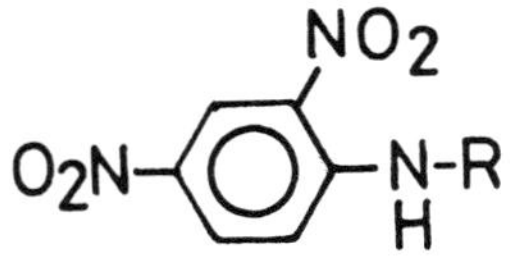

The k_{21} values in Table 1 for the series of aliphatic derivatives show pronounced low values (i.e. longer residence time in the site) for those having bulky, hydrophobic groups adjacent to the amino group 1 at the aromatic ring. For example, the iso-propyl (No. 7) and tert-butyl (No. 8) derivatives show a four- to six-fold decrease as compared with the methyl derivatives (No. 1). The same trend has already been indicated when the mono and bis substitution on the amine were compared. Noteworthy is the finding that excessive bulkiness may cause steric hindrance and therefore an increase in k_{21}, e.g. for the N,N-diethyl-2,4-dinitroaniline it reaches a value of 555 sec^{-1} compared with 121 sec^{-1} for the dimethyl derivative. These data help in assessing a limit to the width of the combining site at this point as ~ 6Å and clearly indicate the presence of a hydrophobic subsite extending up to ~ 8Å from the center of the DNP ring. The behaviour of the electrostatically charged derivatives is instructive both in corroborating the conclusions drawn about the extent of the hydrophobic subsite as well as for the identification of a further subsite where electrostatic interactions prevail. Thus a rather dramatic decrease is oberserved in k_{21} as the carboxylate is moved away from the ring (No. 4,5). Already increasing its distance by one methylene group (to ~6Å) reduces k_{21} by a factor 5. That this effect is due to the repulsion of the hydrophilic, charged group from the hydrophobic subsite is supported by the fact that a positively charged derivative of the same size (No. 6) is also being repelled from this region in the combining site. It is only when the charged group is as far as to be outside the hydrophobic subsite and when it is negative in sign, that a minimum value in k_{21} is reached. The presence of a positively charged subsite at this region is again corroborated by the behaviour of the positively charged derivatives. The k_{21} values for these haptens increase first (No. 6,11) and only when significantly distant from the DNP (~10Å), their effect becomes less pronounced (No. 13). Furthermore, when the positively charged groups are as distant as in the latter ligand, their specific rates of dissociation are rather similar to those of the uncharged aliphatic side chains (No. 3). That this is due to the fact that the tail of the ligand may protrude outside the combining site is also born out by the data obtained with oligo-lysyl-ε-N-DNP lysine derivatives which will be described below. Further evidence for the subsites assignment along with its refinement was achieved by analysing the data obtained for the interaction between α-N-DNP oligopeptides of different stereoisomeric structure[8] where the same chemical residues are present, yet in different spatial arrangements.

In Table 2 the kinetic data for the interaction between the oligo-lysyl-ε-N-DNP lysine are summarized [15]. It is rather interesting that, as the number of lysyl residues increases, only small changes are found in the k_{21} values and it is the specific rate of binding, k_{12}, which decreases in a marked fashion. It is worth noting at this point that a significant decrease is also observed in the k_{12} values for all DNP derivatives carrying a net positive charge (Table 1, Nos. 6, 11, 13). This effect of slowing down the binding of positively charged ligands is interpreted as a result of <u>extra</u> site interaction between the negatively charged protein and positive ligands. Two different experimental approaches support this suggestion:

1. Raising the ionic strength causes an enhancement of k_{12}.
2. Substituting alanyl residues in place of two or three lysyls in the penta- and hexa-lysyl derivatives, also causes a pronounced increase in k_{12}.

There are several general conclusions which evolve from these results; the first is the importance of examining the variations in the equilibrium constants in terms of the changes in <u>both</u> specific rates of binding and dissociation. The specificity in terms of recognition is expressed throughout in k_{21}. This is still the case even when the ligand is carrying a relatively large side chain, e.g. in the oligo-lysyl series, as it is only the first lysyl residue which is encompassed by the combining site. The significant changes in k_{12} clearly indicate the complexity of the binding step between the biopolymer and the ligand which involves several encounters and non-specific interactions leading ultimately to the specific binding.

Table 2. Specific rates of interaction between the IgA-MOPC 315 and a series of oligo-lysyl-ε-DNP lysine derivatives

$(lys)_n$-ε-N-DNP lysine	$k_{12} \times 10^{-7}$ (M^{-1} sec^{-1})	k_{21} (sec^{-1})	$K = k_{12}/k_{21} \times 10^{-6}$ (M^{-1})
n = 0	14.0 ± 1.4	53 ± 5	2.6
1	8.7 ± 0.8	35 ± 4	2.5
2	8.1 ± 0.6	27 ± 4	3.0
3	8.0 ± 0.6	32 ± 6	2.5
4	7.0 ± 0.7	31 ± 6	2.3
5	4.8 ± 0.8	29 ± 6	1.7
6	3.7 ± 0.6	29 ± 4	1.3
7	3.3 ± 0.4	31 ± 5	1.1
8	3.3 ± 0.3	32 ± 5	1.0
α-N-acetyl-ε-N-DNP lysine	12.0 ± 1	51 ± 7	2.4

III. INDUCED CIRCULAR POLARIZATION OF FLUORESCENCE - SENSING THE SITE'S DIMENSIONS

The linear and circular polarization of the fluorescence emitted by complexes formed between MOPC 315 and a series of specially designed fluorescent DNP derivatives has been measured [16]. These ligands have the general structure: 2,4-DNP-NH$(CH_2)_n$NH-A, where n was varied from 2 to 6 and A = 9-aminoacridine. The binding of these ligands to the protein MOPC 315 was determined by following the quenching of the protein emission and its affinity was found to vary from $6 \times 10^5 M^{-1}$ for the n = 2 derivative to $1 \times 10^6 M^{-1}$ for the n = 6 derivative. Fig. 1 shows the emission spectrum of the above-mentioned complexes along with the circularly polarized emission spectra expressed in terms of g_{em} - the emission anisotropy factor [16] which is a measure for the degree of chirality of the exited state of the emitter or of its environ-

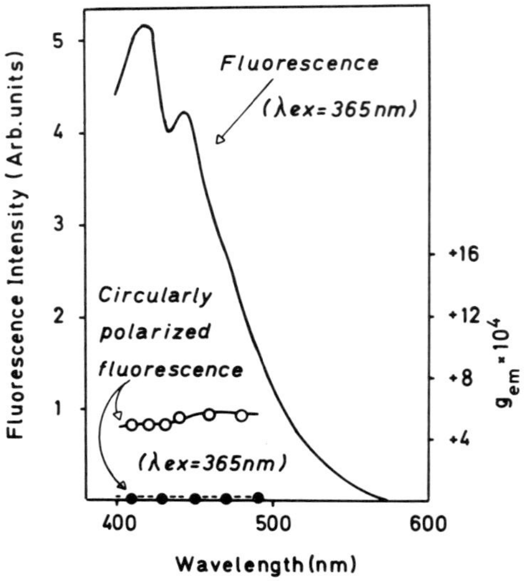

Fig. 1. Emission and circularly polarized emission spectra of the complexes formed between MOPC 315 and the series of ligands having the general structure DNP-NH-$(CH_2)_n$-HN-acridine. (-o- for n = 2; -●- for n > 2).

ment. Only the complex formed between the shortest derivative (n = 2) and MOPC 315 was found to exhibit a circularly polarized component in its emission [16]. This finding means that only when the 9-aminoacridine group is as close as ca.7.5 Å to the center of the DNP anchoring point (~9 Å from the 4-nitro-group end) it begins to sense an asymmetric environment caused by the contact residues of the site. This gives an independent criterion for the dimensions of the combining site: (a) The width at the above-mentioned distance from the DNP ring subsite (S_1) reaches that of the acridine ring (i. e. ~9 Å) as compared with narrower dimensions close to S_1 (~6 Å as judged from the k_{21} values for the series of bis-substituted 2,4-dinitroaniline). (b) The depth (or length, in case the ligand is bound in a lateral fashion) required to accommodate, at least in part, the 9-aminoacridine in the shortest derivative of that series is about 11 Å. These dimensions are in good agreement with those estimated on the basis of the kinetic mapping and also with very recent estimates made from

EPR measurements of spin-labelled DNP derivatives [14]. Both our measurements of the circular polarization of the fluorophore-labelled DNP derivatives and those with the spin-labelled compounds bring support to the suggestion that the DNP ring is held quite rigidly in its subsite.

IV. COMBINING SITES OF DIFFERENT SPECIFICITIES

Homogeneous combining of the two chemically rather different types of ligands are presently investigated by the methods described above. These are mouse myeloma proteins with affinities for phosphoryl choline (HOPC 8 and TEPC 15) [17] and galactose oligomers (XRPC 24) [18]. Except for having distincly different specificities, the first type of site is of interest because recently the structure of a related phosphoryl choline-binding mouse myeloma protein has been resolved by X-ray crystallography [3] and the galactose binding site is the first homogeneous site for a hapten of repeating structural units.

In chemical relaxation measurements of the equilibrium of these two immunoglobulins with their respective ligands, again a single concentration-dependent relaxation time was observed implying a single-step binding equilibrium to prevail also with this site. An illustration for this case is given in Fig. 2, where the concentration dependence of the relaxation times for the intact and Fab fragment of HOPC 8 and phosphoryl choline are shown [19].

The specific rates of binding and dissociation are $k_{12} = 4.1 \pm 0.5 \times 10^7\ M^{-1}\ sec^{-1}$; $k_{21} = 1.1 \pm 0.2 \times 10^2\ sec^{-1}$ for the intact HOPC 8 and $k_{12} = 4.0 \pm 0.3 \times 10^7\ M^{-1}\ sec^{-1}$; $k_{21} = 0.82 \pm 0.15 \times 10^2\ sec^{-1}$ for the Fab fragment. The k_{12}/k_{21} ratio is in good agreement with the equilibrium constants determined by other methods ($5.5 \times 10^5\ M^{-1}$) [20]. Both specific rates of binding and dissociation fall into the range observed for ligand binding to the MOPC 315. The k_{12} value seems somewhat

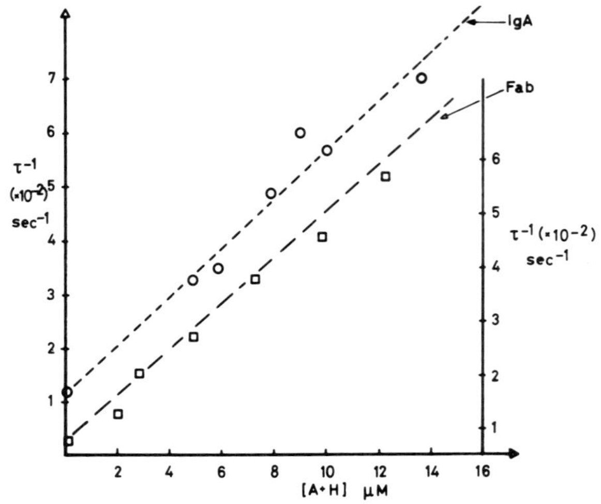

Fig. 2. A plot of the reciprocal of the observed relaxation times as a function of the sum of free sites of HOPC 8 and phosphoryl choline (at pH 8.0 in 0.02 M Tris buffer and 0.15 N NaCl at 25°C following the intrinsic protein fluorescence changes occurring on ligand binding).

low for a rate of binding of such a small ligand and it may be a result of carrying a net electrostatic charge. However the limited amount of data do not allow a more detailed assignment.

The specific rate of binding of tetragalactose to XRPC 24 is found to be significantly lower even than that of the latter ($k_{12} = 4.9 \times 10^5\ M^{-1}\ sec^{-1}$; $k_{21} = 1.3\ sec^{-1}$). It is very similar to the values obtained for the interaction between normally induced, heterogeneous anti-dextran antibodies and oligoisomaltose derivatives [21].

In Table 3 the results obtained for the binding of the oligoalanyl peptides to fractions of anti-polyalanine antibodies are summarized [23]. The antibodies absorbed on an affinity column, were fractionated by elution with alanine oligomers of different lengths [24]. To enable the spectroscopic detection of the interaction of the peptides with the antibodies, use has been made of the non-radiative energy transfer from the

Table 3. Specific rates of interaction of alanine peptides with fractionated anti-polyalanine antibodies

Hapten	Antibody eluted by	k_{12} ($M^{-1}sec^{-1}$)	k_{21} (sec^{-1})	k_{12}/k_{21} (M^{-1})
D(Ala)-R*	$D(Ala)_2$	4.6×10^5	3.1	1.5×10^5
	$D(Ala)_3$	5.7×10^6	2.0	2.9×10^6
	$D(Ala)_4$	3.0×10^6	1.5	2.0×10^6
$D(Ala)_2$-R*	$D(Ala)_2$	4.1×10^6	1.2	3.4×10^6
	$D(Ala)_3$	4.3×10^6	1.8	2.4×10^6
	$D(Ala)_4$	1.7×10^6	3.8	4.5×10^4

*R = HN-$(CH_2)_2$-NH-dansyl

antibodies tryptophanes to a dansyl group (dansyl = dimethyl-amino-naphthalene-sulphonyl) attached covalently to the alanine oligo-peptides [23]. As in the case of the antibodies with specificity for oligosaccharides, these data show a pronounced slowing down of the specific rates of binding. The variation in the specific rates for the two investigated fractions are limited and do not allow a clear correlation between ligand size and specific eluate. The fluorescence properties of the dansyl residue bound via its alanyl side chain to the anti-alanine antibodies are not affected by the combining site. This is in contradistinction to the effect of binding the dansyl to anti-dansyl antibodies where pronounced changes in quantum yields, emission maximum and the induction of a large circularly polarized component are observed [22]. This difference implies that the fluorophore is in former case protruding out of the site.

All kinetic data available to date indicate that the interaction between haptens and their respective site is rather like a "rigid key-rigid lock" interaction. The question of how the second functional role of the immunoglobulin, namely complement binding, is being triggered was mentioned earlier along with the suggestion that the size of the ligand may be the decisive factor for this process. It is therefore rather intriguing that the observed rates of binding of larger, oligomeric haptens are consistently lower by two to three orders of magnitude than those of smaller and more compact ligands. Whether this conjecture is really meaningful requires further experimental data.

I am indebted to my coworkers D. Haselkorn and A. Licht for permission to use unpublished results and to Dr. R.A. Dwek for making available preprints of his papers. The support of the Volkswagen foundation is gratefully acknowledged.

REFERENCES

1. Dorrington, K.J., and Tanford, C.: Adv. Immunol. 12, 333-357 (1970).
2. Kabat, E.A., and Wu, T.T.: Ann. N.Y. Acad. Sci. 190, 382-391 (1971).
3. Padlan, E.A., Segal, D.M., Spande, T.F., Davies, D.R., Rudikoff, S., and Potter, M.: Nature New Biol. 245, 165-167 (1973).
4. Amzel, L.M., Poljak, R.J., Saul, F., Varga, J.M., and Richards, F.F.: Proc. Natl. Acad. Sci. USA 71, 1427-1430 (1974).
5. Schiffer, M., Girling, K.L., Ely, K.R., and Edmundson, A.B.: Biochem. 12, 4629-4631 (1973).
6. Haimovich, J., Eisen, H.N., and Givol, D.: Ann. N.Y. Acad. Sci. 190, 352-359 (1972).
7. Metzger, H.: Adv. Immunol. 18, 169-207 (1974).
8. Haselkorn, D., Friedman, S., Givol, D., and Pecht, I.: Biochem. 13, 2210-2221 (1974).
9. Day, L.A., Sturtevant, J.U., and Singer, S.J.: Ann. N.Y. Acad. Sci. 103, 619-626 (1963).
10. Froese, A.: Immunochem. 5, 253-258, (1968).
11. Givol, D., Pecht, I., Hochman,J., Schlessinger, J., and Steinberg, I.Z.: Progress in Immunology, 2, (1974).
12. Pecht, I., Haselkorn, D., and Friendman, S.: FEBS Letters 24, 331-334 (1972).
13. Potter, M.: Physiol. Rev. 52, 631-719 (1972).

14. Dwek, R.A., Knott, J.C.A., Marsh, D., Press, E.M., Price, N.C., and White, A.I.: Europ. J. Biochem. in press.

15. Haselkorn, D., Yaron, A., Sela, M., and Pecht, I.: to be published.

16. Hartmann, U., Schlessinger, J., and Pecht, I.: Israel J. Med. Sci. in press.

17. Potter, M., and Leon, M.A.: Science 162, 369 (1968).

18. Jolley, U.E., Rudikoff, S., Potter, M., and Glaudemans, C.P.J.: Biochemistry 12, 3039-3044 (1973).

19. Hartmann, U., Walker, M., and Pecht, I.: unpublished.

20. Pollet, R., and Edelhoch, H.: J. Biol. Chem. 248, 5447 (1973).

21. Kabat, E.A., and Pecht, I.: unpublished.

22. Schlessinger, J., Steinberg, I.Z., and Pecht, I.: J. Mol. Biol. 87, 725-732 (1974).

23. Licht, A., and Pecht, I.: unpublished.

24. Schechter, I.: Nature 228, 639-641 (1970).

Received August 9, 1974.

DISCUSSION

Wagner: The binding to antibodies of the different DNP-acridine derivatives could also be monitored by measuring the polarization of the acridine fluorescence. Has this been done and are the results consistent with those of the other methods?

Pecht: The binding of the DNP-acridine derivatives to the MOPC 315 combining site leads to a significant change in the linear polarization of the emission of the acridine group. The difference in this polarization is, however, not as well resolved for the complexes formed with the different lengths of derivatives as it is when the circular polarization is measured. As stated in the abstract the structural implications of these circular-polarization measurements are in excellent agreement with those obtained by other methods.

Woody: If I remember correctly, the binding of small hapten does not lead to complement binding. It would be interesting to look at the circularly polarized luminescence of small hapten complexes with Ab to determine whether the conformational change you have demonstrated with large haptens is directly correlated with that leading to complement binding.

Pecht: In several myeloma proteins that were investigated by that method, phosphoryl choline which was shown to be the specific ligand did not induce any changes in g_{em}. The interesting reason for this behavior may obviously be their small dimensions and we are currently involved in investigation of systematically varying sizes of ligands, for example, oligo-alanine derivatives with anti-alanine antibodies.

Woody: Your observation of a vanishing CD as opposed to a substantial CPL in the anti-dansyl system does not necessarily imply a translational motion of the hapten. It could be primarily a torsional difference between the ground and excited states.

Pecht: The observed differences in chirality of the bound dansyl (to their respective antibodies) between ground and excited state may, in fact, be due to either torsional or translational motion of the ligand inside the combining site. We cannot differentiate between the two possibilities and they may both be involved in this heterogeneous population of sites.

von Hippel: In a well-characterized, simple (one tryptophan) system, how much of a change do you get in g_{em} in going to complete denaturation? What is the sensitivity of the instrument (how large must g_{em} be to be seen - e.g. for tryptophan)?

Pecht: In the case of protein denaturation the tryptophan g_{em} value will completely vanish. One illustrative example is that of the single tryptophan of azurin where the single trp emits at the unusual low wavelength of 308 nm. The g_{em} value of the native protein is of the order of 5×10^{-4} and decreases to zero upon denaturation by 6 M guanidinium-HCl. An interesting case is partial denaturation by acid: in this case, the g_{em} value decreases first at the red end of the band, i.e. that part of the molecule population, that has been partially denatured, emits at higher wavelengths and contributes significantly less to the circularly polarized emission.

The detection sensitivity per tryptophan is of the order of $\pm 5 \times 10^{-5}$.

Veeger: Have you measured the excitation wavelength dependence of circularly polarized tryptophan emission? We have recently got some success with energy transfer studies by exciting at the red edge of the tryptophan band, helping to exclude some of the possibilities.

Pecht: We did not measure excitation wavelength dependence in the case of tryptophan emission. In the case of dansyl emission this has been done and the results are rather interesting and informative (cf. Schlessinger, Steinberg and Pecht, J. Mol. Biol. 87, 1974).

Veeger: Is it possible to obtain information about the nature of the micro-fluctuations in the dansyl-CPL emission, by using tryptophan-sensitised energy transfer?

Pecht: No. This is impossible in the first place because of the heterogeneity of the antibody sites.

Antibiotic Action on Enzymes Involved in Peptidoglycan Synthesis

Harold R. Perkins, Jean-Marie Ghuysen, Jean-Marie Frère and *Manuel Nieto*

Peptidoglycan is a ubiquitous component of the envelopes of all bacteria that have rigid cell walls. It consists of unbranched glycan chains made up of alternating units of N-acetylglucosamine and N-acetyl (or rarely N-glycolyl) muramic acid, linked β-1,4 as in the simpler chitin. Muramic acid itself is glucosamine substituted on the 3-position by a residue of D-lactic acid in ether linkage. The resultant carboxyl groups are substituted by short specific peptide side chains consisting of alternating L and D amino acid residues, the first being L-alanine or rarely glycine or L-serine and the second D-glutamic acid. In the completed cell wall the substituted glycan chains are cross-linked by means of tri-functional amino acid residues in the primary peptide side-chain, occasionally *via* the D-glutamic acid residue but more often through a dibasic amino acid (generally diaminopimelic acid or lysine) in the third position, and sometimes with the intervention of an intercalating peptide chain (6,19). The mechanism by which this cross-linking occurs, and its inhibition, are the main topics of this paper.

In the biosynthesis of the glycan, the primary peptide side-chains are first attached to UDP-N-acetylmuramic acid by sequential addition of the appropriate amino acid at the carboxyl terminus. After the first three residues (L,D,L) a dipeptide of D-alanine is added to complete the pentapeptide side-chain of the glycan precursor. This precursor [e.g. UDP-N-acetylmuramyl-L-alanyl-D-isoglutamyl-(L)-meso-diaminopimelyl-D-alanyl-D-alanine (UDP-MurNAc-pentapeptide) in Gram negatives, Bacilli and *Streptomyces* strain R39 (see below)] is then linked to undecaprenyl phosphate with elimination of UMP and formation of a pyrophosphate linkage. The next step in glycan synthesis consists of the substitution at the muramic acid C(4) position of β-N-acetylglucosaminine from UDPGlc-NAc, UDP being eliminated. At this stage, where appropriate, extra amino acids that will ultimately form the cross-bridges are added. The disaccharide-peptide units still linked at their reducing ends to undecaprenyl pyrophosphate are then polymerized to yield the primary glycan chains by a mechanism in which growth occurs at the reducing end (22). The newly-formed glycan chain is then linked to the pre-existing peptidoglycan network (13,23) by a transpeptidation reaction in which the sub-terminal D-alanine residue on one side-chain becomes attached to an amino group on another, either directly or *via* cross-bridging amino acids, and the terminal D-alanine is eliminated (Fig. 1). On side-chains where the *D*-ala-*D*-Ala OH dipeptide is not used to form a cross-link, the terminal D-alanine residue is often removed by a *D*,*D*-carboxypeptidase, and thereafter the next *D*-alanine residue is removed by an *L*,*D*-carboxypeptidase.

It was proposed in 1965 (20,24) and is now generally accepted, that it is the formation of the inter-chain cross-links by the transpeptidation reaction that is sensitive to β-lactam antibiotics. Further, just as the *D*,*D*-transpeptidases are often sensitive to penicillin, so too are the analogous *D*,*D*-carboxypeptidases (9). Many strains of *Streptomyces* excrete soluble *D*,*D*-carboxypeptidases and in 1972 we showed that some of these purified enzymes could perform *in vitro* transpeptidation reactions that were penicillin-sensitive (18). These transpeptidation reactions were related to those that occur in the cross-linking of peptidoglycan, and later work showed that the enzymes would join peptide fragments together *in vitro* to make products identical with the cross-link occurring in the cell walls of the parent organism (8). Much of what follows describes the

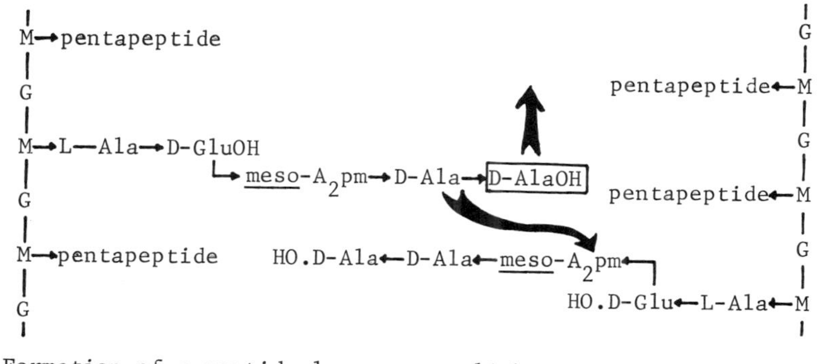

Fig. 1. Formation of a peptidoglycan cross-linkage

interaction of D,D-transpeptidases with their substrates (the carboxyl-donor peptides and the acceptors with free amino groups) and with β-lactam antibiotics.

INTERACTION OF TRANSPEPTIDASES WITH DONORS AND ACCEPTORS

The D,D-carboxypeptidase-transpeptidases under discussion are those of *Streptomyces* strains R61 and R39. These enzymes have been obtained in a high state of purity (2,5), and their specificities for various peptides acting as carboxypeptidase substrates have been examined in detail (11,12). In each case the C-terminal sequence giving the highest activity is R.D-Ala-D-Ala but the preceding residue also has a considerable effect on the enzyme kinetics. Thus for the R61 enzyme, Km = 12 mM and V_{max} = 5160 μmol/mg protein/h when Ac_2-L-Lys-D-Ala-D-Ala is the substrate, whereas the values are 15 mM and 23 μmol/mg protein/h with α-Ac-L-Lys-D-Ala-D-Ala, i.e. with a free ε-amino group (11). The corresponding values for the R39 enzyme are 0.8 mM and 1200 μmol/mg protein/h with the diacetylated substrate and 0.2 mM and 2200 μmol/mg protein/h when there is a free ε-amino group (12). Although not every carboxypeptidase substrate has been tested, it seems probable that all these substrates will function as donors in the transpeptidase reaction when the mixture has been supplemented with a suitable acceptor. The range of substrates that will function as acceptors seems to be related to the type of cross-linkage that exists in the peptidoglycan of the organism producing the enzyme. Thus in strain R61 the cross-link is formed between the penultimate D-alanine of one primary side-chain and a glycine residue that is attached to the ε-amino group of the L,L-diaminopimelic acid belonging to another side-chain. Correspondingly, glycine and many peptides with an N-terminal glycine residue, will act as acceptors in transpeptidation reactions, although other amino compounds can also function (17). In strain R39, on the other hand, the peptidoglycan linkage is directly from a D-alanine residue to the amino group at the D-centre of *meso*-diaminopimelic acid and in this case suitable transpeptidation acceptors must have an amino group in α-position to the carboxyl group of a D-amino acid or glycine (17).

Whenever these soluble *Streptomyces* enzymes perform transpeptidation, concomitant hydrolysis of the substrate by carboxypeptidase action also occurs, and the kinetics are therefore necessarily complex (1,3). Nevertheless, in the presence of saturating concentrations of donor, it is possible to determine from Lineweaver-Burk plots an apparent K_m for the acceptor. The values obtained with the R61 enzyme were about 5 mM for meso-diaminopimelic acid as acceptor and 3 mM for the peptide Gly-L-Ala. Thus the 'affinity' of this enzyme for donor and acceptor is about the same. When complex acceptors related to peptidoglycan fragments were used, it became apparent that certain features of those substances exerted a profound effect on the relative amount of transpeptidation and donor hydrolysis that occurred (7,8,17). Thus, for instance, when increasing concentrations of the tetrapeptide L-Ala-D-isoGln-(L)-meso-A_2pm-D-Ala were provided as acceptor for the R39 enzyme, with Ac_2-L-Lys-D-Ala-D-Ala as donor, then transpeptidation rose to a maximum at an acceptor concentration of about 0.8 mM and at higher concentrations both transpeptidation and hydrolysis reactions were progressively inhibited until eventually the tripeptide donor present in the reaction mixtures remained unused (7). This particular phenomenon was dependent on the α-amide group on the glutamic acid residue, since high concentrations of the same tetrapeptide, non-amidated, inhibited overall attack on the donor by limiting its hydrolysis, but did not decrease the amount of transpeptidation product formed. More extensive studies (8) showed that when the donor site of the enzyme was saturated, the rate of the total reaction (hydrolysis and transpeptidation) was the same as the rate of hydrolysis alone when no acceptor was added. Thus the enzyme had the same turnover number for D-alanine release whether or not acceptor was present. The concentration of amidated tetrapeptide required to inhibit transpeptidation as well as hydrolysis was lower when the enzyme was less saturated by donor. Thus, as indicated above, a high concentration of amidated tetrapeptide acceptor can 'freeze' the enzyme so that both substrates remain completely unattacked. These and similar results led to the conclusion that mechanisms able to change the acceptor and donor concentrations or to alter some particular structural feature of these peptides are important for control of the activity of the exocellular D,D-carboxypeptidase-transpeptidase of strain R39. Similar mechanisms could well be involved in the control of the peptide cross-linking system *in vivo* (8).

INTERACTION OF TRANSPEPTIDASES WITH β-LACTAM ANTIBIOTICS

As indicated, the transpeptidation reactions are sensitive to penicillin. Tipper and Strominger (20) proposed that penicillin functioned as an analogue of a terminal acyl-D-Ala-D-Ala sequence, which having arrived at the active site of a transpeptidase, permanently inactivated the enzyme by penicilloylation of the protein. Recent work with the purified D,D-carboxypeptidase-transpeptidases has now begun to throw some new light on how these enzymes interact with β-lactam antibiotics.

When a purified preparation of the R61 enzyme was mixed with benzylpenicillin, the circular dichroism spectrum of the enzyme in the near u.v. was extensively altered (Fig. 2) (14). The peptide region in the far u.v. remained unchanged. The interaction between the enzyme and the antibiotic showed high affinity and no further change in the c.d. of the enzyme occurred after the addition of one mol of penicillin for each mol of enzyme (38,000 mol wt). Examination of the fluorescence of the enzyme proved more convenient for studying the interaction with penicillin. Excitation was at 273 nm and emission was observed at 320 nm. When one molar proportion of benzylpenicillin was added to 7.5 x 10^{-7} M enzyme in 10 mM sodium phosphate buffer, pH 7.0, at 25°C the fluorescent emission decreased over a period of some 10 min, the final quenching being 25-30% (Fig. 3). By making successive additions of penicillin to a sample of enzyme and allowing 10 min after each addition for completion of the interaction, the quenching of fluorescence was used to follow penicillin binding. The result once more was that the enzyme bound one molar proportion of antibiotic and further additions did not affect the fluorescence (14). The inactivation by penicillin was evidently reversible, because fully inactivated enzyme recovered all its activity during overnight dialysis against 10 mM buffer, pH 7. The binding of penicillin or

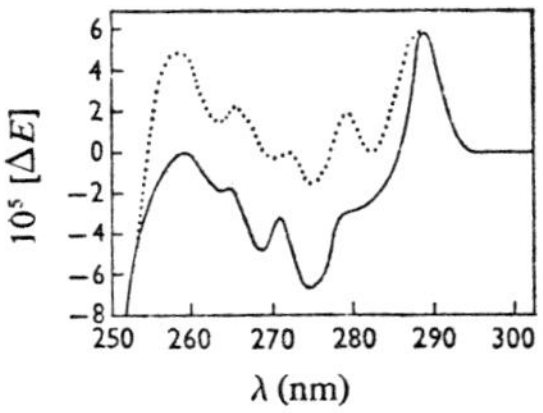

Fig. 2. Circular dichroism of the R61 enzyme with (......) and without (———) penicillin (14).

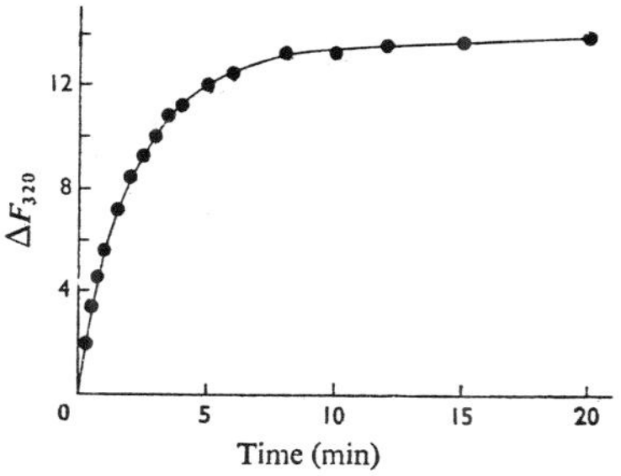

Fig. 3. Time course of the binding of sodium benzylpenicillin to the R61 enzyme. The ordinate represents quenching of fluorescence (14).

its release were unaffected by the presence of thiol compounds, which contrasts with the behaviour of an enzyme from *Bacillus subtilis* (10).

The forward rate constant for the association of penicillin and R61 enzyme was calculated from second-order kinetic plots during the first 3-4 min after mixing. The value measured at 25°C in 10 mM sodium phosphate buffer pH 7 was $k_f = 1.8 \times 10^4$ litre. mol^{-1}. s^{-1}, and in the presence of 57 mM Ac_2-L-Lys-D-Ala-D-Ala (a donor substrate at a concentration about five times its K_m value) $k_f = 5.7 \times 10^3$ litre. mol^{-1}. s^{-1}. This relatively small change in the reaction rate could not be ascribed to competition between antibiotic and substrate, however, because the non-substrate, non-inhibitor peptide Ac_2-L-Lys-D-Ala produced a similar effect ($k_f = 4.4 \times 10^3$ litre. mol^{-1}. s^{-1}) (14). It was also observed that penicillin binding, as measured by fluorescence quenching, could occur with enzyme that had been made enzymically inactive by either urea or guanidinium chloride.

From the titration of penicillin with the R61 enzyme, as followed by quenching of fluorescence, an association constant could be calculated. Hence, with the k_f determined as described above, the rate constant for the dissociation of the complex could be deduced - in 10 mM sodium phosphate buffer it was 0.8×10^{-4} s^{-1}, corresponding to a half-life of 2.4 h at 25°C (14). This is comparable to a value of 1.35 h observed at 37°C, in experiments in which an excess of penicillinase was present throughout and enzyme recovery was followed by measuring its carboxypeptidase activity (unpublished work).

The purified enzyme from strain R39 was also inactivated by β-lactam antibiotics (12). The binding of penicillin to this enzyme could not be studied by the use of fluorescence or c.d. because neither parameter was affected. Instead, binding was examined by measuring the loss of carboxypeptidase activity as more penicillin was added. Again it was found that one mol of penicillin was needed to inactivate one mol of enzyme (53,300 mol wt) (4). A sample of the enzyme was also inactivated by addition of an excess of ^{14}C-benzylpenicillin and filtered on a Sephadex column to remove unfixed antibiotic. At 37°C and in pH 7.7 buffer (0.1 M Tris, 0.2 M NaCl, 0.05 M $MgCl_2$) in which the untreated enzyme was completely stable, the radioactive penicillin was released and enzymic activity recovered at almost exactly the same rate (half-lives 73 and 65 h, respectively). The kinetic characteristics of the recovered enzyme, both as carboxypeptidase and as transpeptidase, were unchanged and furthermore the regenerated enzyme reacted with fresh penicillin exactly as before.

The nature of the radioactive compounds released from the ^{14}C-benzylpenicillin-R39 enzyme complex was also examined (4). It was found that the main compound released in Tris-NaCl-$MgCl_2$ buffer was not benzylpenicillin or benzylpenicilloic acid, nor was it the same as the compound that arose when penicillin alone was incubated under the same conditions, in which case the main product was probably a benzylpenicilloyl-Tris compound. When the ^{14}C-penicillin-enzyme was dissociated in phosphate buffer (10 mM, pH 7) release of radioactivity was more rapid (half-life at 37°C 11-14 h) but enzymic activity did not recover; the enzyme is, in any case, unstable in this buffer. Under these conditions the main product of penicillin release seems to be benzylpenicilloic acid, as judged by both chromatography and electrophoresis, but the identification was not certain.

It did not appear that SH groups were involved in either enzyme activity or penicillin binding, since neither iodoacetate nor 5,5'-dithio-*bis*-(2-nitrobenzoic acid) had any effect.

The reaction of the R39 enzyme with β-lactam antibiotics was also studied with two cephalosporins (4). The chromogenic cephalosporin 87-312 (3-[2,4-dinitrostyryl]-[6R-7R]-7-[2-thienylacetamido]-ceph-3-em-4-carboxylic acid, E-isomer) (15) immediately changed colour when added to the enzyme, even at $0^{\circ}C$. The new absorption spectrum was found to be almost identical with that of cephalosporin 87-312 that had been hydrolysed by β-lactamase (i.e. the absorption maximum had shifted from 386 to 482 nm). Thus the reaction between the enzyme and the antibiotic could be followed by the loss of enzymic activity, on the one hand, and the change in absorption maximum on the other. As the activity disappeared there was a parallel change in the absorption spectrum of the antibiotic (Fig. 4). According to the published extinction coefficient for cephalosporin 87-312, the end point of the enzyme-antibiotic reaction occurred when 1.27 molar proportions of cephalosporin had been added, compared with a value of 0.92 for benzylpenicillin and the same 90% pure enzyme preparation. It seems most likely that the true stoicheiometry is in fact 1:1. This view was supported by observations with cephaloridine, another cephalosporin that gives a change of absorption spectrum when its β-lactam ring is opened (16). A titration with R39 enzyme similar to that described for cephalosporin 87-312 produced a molar ratio of antibiotic:enzyme of 0.92. The rate constants for the dissociation of the cephalosporin-enzyme complexes, based on recovery of enzyme activity, were also calculated for the decomposition in Tris-NaCl-$MgCl_2$ buffer at $37^{\circ}C$. The half-lives were 130 h for cephalosporin 87-312 and 300 h for cephaloridine, compared with 65 h for benzylpenicillin (4).

Parallel experiments performed with the R61 enzyme showed that the spectrum of cephalosporin 87-312 bound to this enzyme was not the same as that of the β-lactamase product (Frère, Leyh-Bouille, Ghuysen & Perkins, unpublished work). In this case the latter spectrum only appeared after the breakdown of the complex accompanied by restoration of enzymic activity. As with the R39 enzyme, the complex with R61 enzyme appeared to require 1.25 mol of cephalosporin 87-312.

There seemed no doubt that the cephalosporin 87-312 and benzylpenicillin were competing for the same site on both enzymes, because experiments showed that their reactions with the protein were mutually exclusive.

The above results suggest that although both enzymes caused the ultimate destruction of penicillin, they did not do so by a simple opening of the β-lactam ring, since benzylpenicilloic acid was not the product, at least in Tris-NaCl-$MgCl_2$ buffer. Since the regenerated enzyme had also recovered

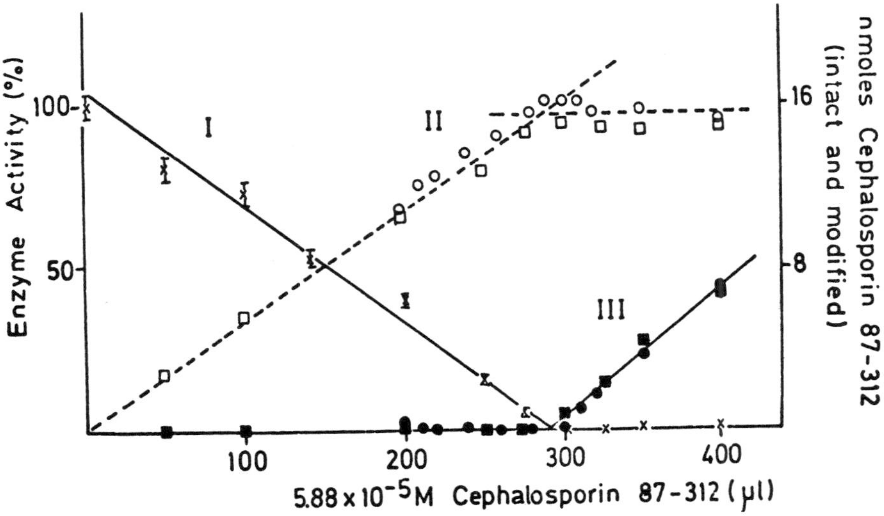

Fig. 4. Titration of R39 enzyme with cephalosporin 87-312. I: Residual enzyme activity, II: modified antibiotic concentration deduced from E_{482}, III: unchanged antibiotic concentration deduced from E_{386}(4).

its sensitivity to penicillin, it was evident that the released penicillin had not carried with it some labile part of the protein. What bond is formed between β-lactam antibiotics and the soluble D,D-carboxypeptidase-transpeptidases of Streptomyces species remains unknown at present, except that no sulphydryl group seems to be involved. This situation is different from that with the D,D-carboxypeptidase of B. subtilis, where benzylpenicillin apparently formed a thioester bond (21). The main difference between the enzymes from strains R61 and R39 appears to be that the penicillin complex of the former is much more unstable, with a half-life of 1.35 h compared with 65 h for the latter. As with the R39 enzyme, the 'penicillin' released from its complex with the R61 enzyme was neither benzylpenicillin nor benzylpenicilloic acid (unpublished work). It is now thought (P.M. Blumberg, personal communication) that the DD-carboxypeptidases of B. subtilis and B. stearothermophilus also convert penicillin to a product that is not penicilloic acid.

At present it is not possible to say what site on the enzymes is occupied by β-lactam antibiotics. Penicillins prevent enzyme action, either by stopping the initial binding of the donor peptide or by preventing the ensuing chemical reaction that leads either to hydrolysis or to transpeptidation. Until there is an independent method of measuring donor binding, these possibilities cannot be clearly distinguished. With antibiotics like cephaloridine that are firmly bound, it should prove possible to isolate fragments of enzyme retaining the modified antibiotic, thus eventually defining the nature of the binding site.

Acknowledgements

The work in Liège was supported in part by the Fonds National de la Recherche Scientifique, the Fonds de la Recherche Fondamentale Collective, Brussels, Belgium (contract No. 1000) and by the Institut pour l'Encouragement de la Recherche Scientifique dans l'Industrie et l'Agriculture, Brussels, Belgium (contract No. 1699). J-M.F. is Chargé de Recherches du Fonds National de la Recherche Scientifique, Brussels, Belgium.

1. Frère, J-M.: Enzymic mechanisms involving concomitant transfer and hydrolysis reactions. Biochem. J. 135, 469-481 (1973).
2. Frère, J-M., Ghuysen, J-M., Perkins, H.R. & Nieto, M.: Molecular weight and amino acid composition of the exocellular DD-carboxypeptidase-transpeptidase of Streptomyces R61. Biochem. J. 135, 463-468 (1973).
3. Frère, J-M., Ghuysen, J-M., Perkins, H.R. & Nieto, M.: Kinetics of concomitant transfer and hydrolysis reactions catalysed by the exocellular DD-carboxypeptidase-transpeptidase of Streptomyces R61. Biochem. J. 135, 483-492 (1973).
4. Frère, J-M., Ghuysen, J-M., Reynolds, P.E., Moreno, R. & Perkins, H.R.: Binding of β-lactam antibiotics to the exocellular DD-carboxypeptidase-transpeptidase of Streptomyces R39. Biochem. J. (in press).
5. Frère, J-M., Moreno, R., Ghuysen, J-M. & Perkins, H.R.: Molecular weight, amino acid composition and physico-chemical properties of the exocellular DD-carboxypeptidase-transpeptidase of Streptomyces R39. Biochem. J. (in press).
6. Ghuysen, J-M.: Use of bacteriolytic enzymes in determination of wall structure and their role in cell metabolism. Bact. Rev. 32, 425-464 (1968).

7. Ghuysen, J-M., Leyh-Bouille, M., Campbell, J.N., Moreno, R., Frère, J-M., Duez, C., Nieto, M. & Perkins, H.R.: Structure of the wall peptidoglycan of Streptomyces R39 and the specificity profile of its exocellular DD-carboxypeptidase-transpeptidase for peptide acceptors. Biochemistry 12, 1243-1251 (1973).

8. Ghuysen, J-M., Reynolds, P.E., Perkins, H.R., Frère, J-M. & Moreno, R.: Effects of donor and acceptor peptides on concomitant hydrolysis and transfer reactions catalysed by the exocellular DD-carboxypeptidase-transpeptidase from Streptomyces R39. Biochemistry 13, 2539-2547 (1974).

9. Izaki, K., Matsuhashi, M. & Strominger, J.L.: Glycopeptide transpeptidase and D-alanine carboxypeptidase : penicillin sensitive enzymatic reactions. Proc. Natl. Acad. Sci., U.S. 55, 656-663 (1966).

10. Lawrence, P.J. & Strominger, J.L.: Biosynthesis of the peptidoglycan of bacterial cell walls. The reversible fixation of radioactive penicillin G to the D-alanine carboxypeptidase of Bacillus subtilis. J. Biol. Chem. 245, 3660-3666 (1970).

11. Leyh-Bouille, M., Coyette, J., Ghuysen, J-M., Idczak, J., Perkins, H.R. & Nieto, M.: Penicillin-sensitive DD-carboxypeptidase from Streptomyces strain R61. Biochemistry 10, 2163-2170 (1971).

12. Leyh-Bouille, M., Nakel, M., Frère, J-M., Johnson, K., Ghuysen, J-M., Nieto, M. & Perkins, H.R.: Penicillin-sensitive DD-carboxypeptidases from Streptomyces strains R39 and K11. Biochemistry 11, 1290-1298 (1972).

13. Mirelman, D., Bracha, R. & Sharon, N.: Role of the penicillin-sensitive transpeptidation reaction in attachment of newly synthesized peptidoglycan to cell walls of Micrococcus luteus. Proc. Natl. Acad. Sci. U.S. 69, 3355-3359 (1972).

14. Nieto, M., Perkins, H.R., Frère, J-M. & Ghuysen, J-M.: Fluorescence and circular dichroism studies on the Streptomyces R61 DD-carboxypeptidase and transpeptidase. Penicillin binding by the enzyme. Biochem. J. 135, 493-505 (1973).

15. O'Callaghan, C.H., Morris, A., Kirby, S.M. & Shingler, A.H.: Novel methods for detection of β-lactamases by using a chromogenic cephalosporin substrate. Antimicrob. Ag. Chemother. 1, 283-288 (1972).

16. O'Callaghan, C.H., Muggleton, P.W. & Ross, G.W.: Effects of β-lactamase from Gram-negative organisms on cephalosporins and penicillins. Antimicrob. Ag. Chemother. 1968 pp. 57-63 (1969).

17. Perkins, H.R., Nieto, M., Frère, J-M., Leyh-Bouille, M. & Ghuysen, J-M.: Streptomyces DD-carboxypeptidases as transpeptidases. The specificity for amino compounds acting as carboxyl acceptors. Biochem. J. 131, 707-718 (1973).

18. Pollock, J.J., Ghuysen, J-M., Linder, R., Salton, M.R.J., Perkins, H.R., Nieto, M., Leyh-Bouille, M., Frère, J-M. & Johnson, K.: Transpeptidase activity of Streptomyces D-alanyl-D carboxypeptidases. Proc. Natl. Acad. Sci. U.S. 69, 662-666 (1972).

19. Schleifer, K.H. & Kandler, O.: Peptidoglycan types of bacterial walls and their taxonomic implications. Bact. Rev. 36, 407-477 (1972).

20. Tipper, D.J. & Strominger, J.L.: Mechanism of action of penicillins : a proposal based on their structural similarity to acyl-D-alanyl-D-alanine. Proc. Natl. Acad. Sci. U.S. 54, 1133-1141 (1965).

21. Umbreit, J.N. & Strominger, J.L.: D-Alanine carboxypeptidase from Bacillus subtilis membranes. Interaction with penicillins and cephalosporins. J. Biol. Chem. 248, 6767-6771 (1973).

22. Ward, J.B. & Perkins, H.R.: The direction of glycan synthesis in a bacterial peptidoglycan. Biochem. J. 135, 721-728 (1973).

23. Ward, J.B. & Perkins, H.R.: Peptidoglycan biosynthesis by preparations from Bacillus licheniformis : cross-linking of newly synthesized chains to preformed cell wall. Biochem. J. 139, 781-784 (1974).

24. Wise, E.M. & Park, J.T.: Penicillin : its basic site of action as an inhibitor of a peptide cross-linking reaction in cell wall mucopeptide synthesis. Proc. Natl. Acad. Sci. U.S. 54, 75-81 (1965).

Received August 3, 1974

DISCUSSION

Sund: It seems to me that the experimental data of your fluorescence titration follow a sigmoidal shaped curve rather than a normal hyperbolic curve.

Perkins: You may be right, but I think one would need to carry out experiments in the light of more recent results using a more sensitive method of following the association. The change in fluorescence at low molar proportions of penicillin was rather small to measure accurately.

Veeger: Have you compared enzyme activities in buffers of equal ionic strength, in order to elucidate specific effects of polyvalent ions on your enzyme?

Perkins: We did not use phosphate buffer at high ionic strength because the enzyme is particularly sensitive to that anion, and rapidly loses its activity.

Problems in Studying Protein Binding with Regard to the Interaction of Steroids with a Specific Binding Globulin

R. A. Lutz, U.-W. Wiegand and *H. G. Weder*

INTRODUCTION

The study of protein structure may often be aided by investigating the interactions of low-molecular ligands with the protein in question. This is illustrated by various papers of this symposium for enzymes or receptor proteins. Specific interactions manifest themselves by high binding constants. But a weak binding between ligand and biopolymer may also occur. In these cases, the study of interactions characterizes the behaviour of the ligand. Drugs and metabolites of exogenous and endogenous compounds are no exception and therefore are also capable of binding to proteins with high or low specifity. Low specific binding occurs mostly with albumin, whereas some globulins are able to interact with drugs and metabolites in a highly specific manner.

Protein binding is of extreme importance for endogenous and exogenous compounds. The free concentration of the ligand in blood and tissue depends on the degree of binding. This greatly influences the distribution, the rate of metabolism and excretion in the organism. For exogenous and endogenous compounds the binding affinity is of importance for another reason. The mutual displacement of ligands depends on the relative values of the binding constants. The displacement of endogenous compounds has significant physiological consequences, whereas for exogenous ligands pharmacological and toxicological implications occur. Contraceptives increase the binding of testosterone to its binding protein and a reduced concentration of the free and active hormone results. The simultaneous administration of oral hypoglycemic agents and anticoagulants may enhance hypoglycemia or increase the anticoagulant effect.

To get information about such physiological or pharmacological effects due to protein binding it is necessary to study the mechanism of binding. For this purpose one should be able to determine the binding constants, the degree of binding and the binding capacity as accurately as possible.

The most primary aim in studying binding is to try to reproduce in vivo conditions. The design of binding experiments should therefore take this into account. An important limiting parameter in this respect is the concentration. The well-known separation methods like gelfiltration, ultracentrifugation and dialysis have the advantage that it is possible to work at corresponding in vivo concentrations, provided a sensitive analytical determination of the ligand exists. In addition, temperature, pH and ionic strength should be controlled. In contrast to the above mentioned methods, NMR-spectroscopy and calorimetry are limited in their sensitivity.

In general the binding proteins have more than one single set of binding sites. In these cases, some problems may arise in evaluating the binding parameters from a sensitive plot like the Scatchard plot. Some of the problems encountered with Scatchard plots and their evaluation will therefore be discussed in the first part of this study. As an application of the method developed in the first part, the binding of sex hormones and their metabolites to plasma proteins was then investigated. The respective results and some of their physiologically important conclusions are discussed in the remainder of this paper.

RESULTS AND DISCUSSION

Evaluation of Binding Parameters from Scatchard Plots:

To get accurate binding parameters for two or more binding sites from a Scatchard plot it is necessary to perform a suitable fitting procedure. We have proposed a numerical fitting procedure (1) with a fit criterion described geometrically as follows: Along an experimental Scatchard curve for which the general binding equation is valid one has to construct a band of constant width d at a distance d/2 on both sides of the curve. The band is chosen in such a manner that all experimental points are within this band. Through iterations one gets the best fitting curve for which d is a minimum.

As an example of this fitting procedure we wish to present its application in the case of a steroid hormone metabolite interacting with plasma proteins. In figure 1 the raw data of such a dialysis experiment are shown. From this figure it is evident that the variations of duplicate points tend to fall on

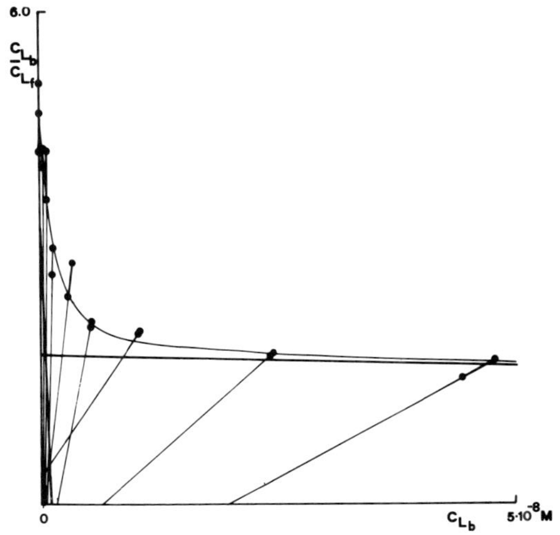

Figure 1: Scatchard plot of the interaction of 5α-dihydrotestosterone with 2% human pregnancy plasma in 1/15 M phosphate-buffer pH 7.4 at 25°C.

a line going through the origin. This line is the locus of the points having the same free ligand concentration. It is no coincidence that the free ligand concentration is actually the only directly measured quantity with our equilibrium dialysis system. The error in measuring this quantity is therefore given by the experimental setup. In the Scatchard plot all other parameters are calculated from this one quantity and a small error in it will be greatly magnified in these other parameters. When performing fitting procedures this fact has to be taken into account. It implies a very complicated weighting procedure which has never been carried out completely. With our fit criterion this problem is partially circumvented. One of the advantages of this method is the possibility of determining the bandwidth as a function of one parameter. Practical experience showed, however, that this width may be kept fixed. In our routine procedure it is either measured normally, e.g. perpendicular to the curve (see figure 2) or radially along the line of constant free ligand concentration (see figure 3).

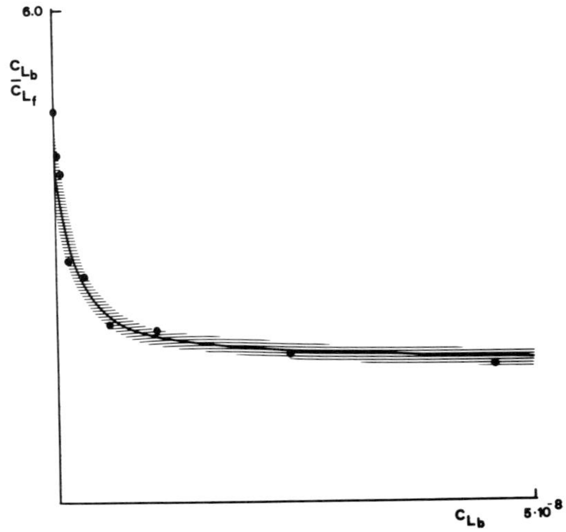

Figure 2: Scatchard plot of the mean experimental points of figure 1. The computer-optimized band is shown where the bandwidth has been measured normally.

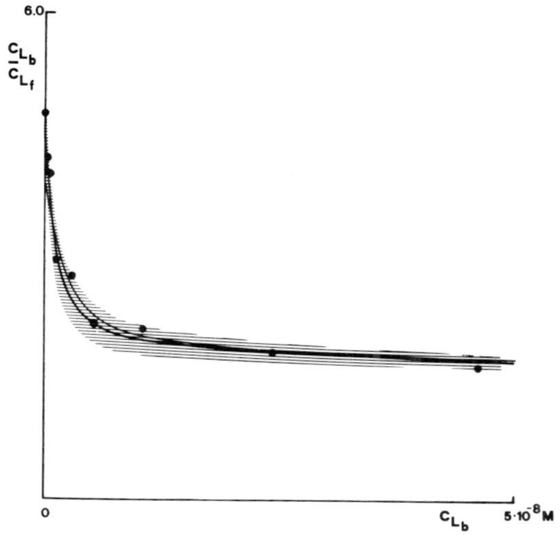

Figure 3: Scatchard plot of the mean experimental points of figure 1. The computer-optimized band is shown where the band-width has been measured radially along the line of constant free ligand concentration.

Another advantage of this fitting procedure is the fact that negative binding parameters as they may be obtained by other procedures are excluded in a very elegant way. Binding parameters with no physical meaning are therefore excluded. The disadvantage, however, is that a single bad point determines the bandwidth and causes a loss in sensitivity. For the high-affinity binding sites accurate low saturation points are extremely important. If we have binding constants of different orders of magnitude, the sites with the highest affinity may be saturated for relatively low values of the total saturation fraction. The information on this site may obviously come mainly from the low saturation range. The question then arises how to overcome the inherent poor experimental accuracy at low saturation. With a given experimental accuracy the only remedy is repeated measurements under the same conditions which is possible with our dialysis techniques (2,3). Table 1 shows that the number of the points is less important than the

Number of points	saturation range with respect to the high affinity site	association constants	
		$K_1 \pm$ S.D. $\times 10^8$ M^{-1}	$K_2 \pm$ S.D. $\times 10^4$ M^{-1}
9	0.02 - 7.0		
7	0.05 - 3.0	4.633±0.003	1.146±1.244
6	0.1 - 3.0		
5	0.2 - 3.0		
9	0.03 -10.0		
7	0.1 - 5.0		
6	0.2 - 5.0	10.50±5.70	
5	0.3 - 5.0		
4	0.8 - 5.0		

Table 1: Variations of the binding parameters as a function of the number of points and the saturation range considered. The points in the upper part are computer-calculated points refed into the computer, so that only rounding errors were introduced. The points in the lower part are means of actual experimental data of a dialysis experiment of 5α-dihydrotestosterone with 2% pregnancy plasma in 1/15 M phosphate-buffer pH 7.5 at 22°C.

quality. Reducing the number of high-quality points has relatively little influence on the high-affinity association constant. Repeated measurement of the same point should therefore be carried out rather than measuring a large number of individual points. If the binding system consists, as in our case, of a high affinity binding site and a non-specific binding site, the Scatchard plot has a severe drawback: The non-specific binding sites having a large binding capacity are measured only at a very low saturation range so that the error may become very large. This is very clearly demonstrated in table 1 where the low-affinity binding site is varying over 100%, even with theoretical computer-calculated points when the excess of low-affinity binding sites is very large (a couple of thousands in our example). Specific and non-specific binding sites should therefore be measured separately.

The Binding of Sex Hormones to Human Plasma Proteins:

Research done in the last few years shows that the binding of sex hormones or its metabolites to plasma proteins is of considerable importance. For instance, increased concentration of testosterone does not necessarily lead to masculinisation if the protein binding is also increased. The best example to illustrate this is pregnancy where no virilisation of the mother takes place. On the other hand, pathological conditions exist with increased binding but normal hormonal levels: Testicular feminisation and gynacomastia may partially be caused by this mechanism.

The following numbers may demonstrate the biological importance of the sex hormone binding: Only 2% of all the testosterone are freely circulating in human plasma (4). The major fraction (mean: 65.8%) is bound to a specific binding protein, the human sex hormone-binding globulin (HSBG, 5). The remainder (30.7%) is bound to albumin with low affinity and to transcortin (1.5%), the corticosteroid-binding protein. This HSBG has been our main interest as it has a high specifity and affinity for testosterone and dihydrotestosterone.

The following conclusions could be drawn from displacement experiments, in which the percent of displaced labelled testosterone by testosterone metabolites has been studied (6). The binding takes place mainly with 5α-androstane derivatives with polar groups in ring A and in the 17β-OH-position. 19-nor-derivatives are bound less stongly with the exception of 19-nor-androst-5-ene-3β, 17β-diol, which is bound more strongly than its androstene analogue. Estradiol is also bound, but with less affinity than testosterone. An unhindered 17β-hydroxy-function seems to be the prerequesite for binding. If this group is hindered sterically by a 17α-aliphatic group or if it is replaced by a 17α-hydroxy group, the binding activity is decreased considerably. A polar group such as an oxygen atom seems to be important in ring A, but it is not critical whether this group is in axial or in equatorial position or whether it is a keto function. Additional polar groups such as hydroxy functions or double bonds all tend to lower the binding affinity. Dihydrotestosterone is the metabolite which is most strongly bound. It is also the most potent androgen for sex hormone responsive tissues. These displacement experiments permit the comparison of the relative binding affinities of the individual steroids, but they do not give an indication about the magnitude of this affinity. In order to get this information the association constant has to be determined. Table 2 lists the association constants of the major sex hormone metabolites as determined in our laboratory.

Our main interest was to study the nature of the binding forces involved. We therefore measured the association constants as

Temperature °C	dihydrotestosterone M^{-1}	testosterone (7) M^{-1}	estradiol M^{-1}
5	-	$3{,}1\cdot10^{9}$	$1{,}0\cdot10^{9}$
25	$2{,}0\cdot10^{9}$	$1{,}2\cdot10^{9}$	$7{,}2\cdot10^{8}$
37	$3{,}0\cdot10^{9}$	$7{,}2\cdot10^{8}$	$3{,}4\cdot10^{8}$
45	$3{,}6\cdot10^{9}$	-	-

Table 2: Association constants of sex hormones with the human sex hormone-binding globulin (HSBG). The experiments have been carried out as described (7).

a function of temperature. Increasing the temperature from 5 to 37°C lowers the binding constant by a factor of about 4 in the case of testosterone (see table 2). Using the van't Hoff equation the enthalpy change can be calculated and was found to be -7.7 kcal/mole. From this result and the free-energy change, the entropy change was calculated to be 15.8 entropy units. This large and positive entropy change indicates that hydrophobic interactions due to the very lipophilic nucleus are among the binding forces. The protein, therefore, must have a large hydrophobic area at the binding site(s). Van der Waal's or hydrogen bonding forces also have to be involved as the association constant is decreasing with increasing temperature. Is is known that intact sulfhydryl groups are necessary for the binding activity of HSBG (8). Substances which react with SH-groups markedly lower this activity. It would, therefore, be possible that these groups are necessary for these dipole interactions.

The fact that only 17β-hydroxy-andostane derivatives are bound shows that the conformation of the ligand and, therefore, also of the binding sites on the protein are of importance. This conformation may be maintained by disulfide bridges as these are necessary for the binding activity (8,9). Recent experiments show that in the case of estradiol the binding constant is also decreasing with raising temperature. But in contrast to testosterone, raising the temperature from 4 to 37°C decreases the association constant by a factor of about 3 (see table 2). In the case of dihydrotestosterone, on the other hand, there is a slight increase of the association constants with raising temperature and the entropy change is more than twice that of the testosterone. It seems that the strong binding of dihydrotestosterone to HSBG is to a large extent due to the strong hydrophobic interaction. In the view of these results and as it is assumed that only the free hormone is fully biologically active, the function of HSBG might be the protection of the

	dihydrotestosterone	testosterone	estradiol
ΔG^o	$-12.7\ kcal \cdot M^{-1}$	$-12.4\ kcal \cdot M^{-1}$	$-12.1\ kcal \cdot M^{-1}$
ΔH^o	$+5\ kcal \cdot M^{-1}$	$-7.7\ kcal \cdot M^{-1}$	$-5.4\ kcal \cdot M^{-1}$
ΔS^o	>40 e.u.	+15.8 e.u.	+22.5 e.u.

Table 3: Free energy, enthalpy and entropy change of the binding of steroids to HSBG.

target organs from too high hormonal concentrations. Dihydrotestosterone, being in some respects the most active of all the androgens, should therefore be bound most strongly. This is consistent with our results and the data of tables 2 and 3.

Although there is still much to be learned about the hormonal regulation in the human body, there is evidence of the involvement of HSBG in this regulation, as there is a marked increase of the binding capacity during pregnancy. Estrogen administration has the same effect, while the administration of androgens lowers it. It has therefore been proposed that the HSBG amplifies the estrogen action by influencing the free testosterone and its free active metabolites (10,11). An increase in HSBG due to estrogens increases the binding of testosterone or dihydrotestosterone, so that less free androgens are available for the target organs. An increase in testosterone, on the other hand, produces a decrease in HSBG, so that more testosterone and metabolites are available for the end organ. This seems a very interesting hypothesis demonstrating the importance of the binding proteins in hormonal regulation. The results so far obtained are only a small step in further understanding how nature brings about the hormonal regulation. But the fact that the sex hormone-binding globulin does not have a wide distribution in animals (9) opens another question:

How is this regulation being done in those animals where a protein comparable to HSBG is missing? Other proteins seem to take over, at least partially the function of HSBG, but these proteins do not have the high binding affinities of HSBG and therefore cannot have the same profound effects on hormonal activity as HSBG.

Literature:

1) Weder, H.G., Schildknecht, J., Lutz, R.A. & Kesselring, P.: Determination of Binding Parameters from Scatchard Plots: Theoretical and Practical Considerations, Eur.J.Biochem. 42, 475-481 (1974).
2) Weder, H.G., Schildknecht, J. & Kesselring, P.: A new equilibrium dialyzing system, Amer. Lab. No. 10, 15-21 (1971).
3) Wiegand, U.-W., Lutz, R.A. & Weder, H.G.: Microdialysis: new tools for dynamic and equilibrium binding studies, to be published in FEBS Letters, 1974.
4) Vermeulen, A. & Verdonck, L.: Studies on the Binding of Testosterone to Human Plasma, Steroids 11, 609-635 (1968).
5) Mercier, C., Alfsen, A. & Baulieu, E.E.: A testosterone binding globulin in: Proc.sec.Symp. on steroid hormones, Ghent 1965, Excerpta Medica Foundation, 1966, p.212.
6) Corvol, P. & Bardin, C.W.: Species Distribution of Testosterone-Binding Globulin, Biol. Reprod. 8, 277-282 (1973).
7) Lutz, R.A., Lutz-Ewan, L. & Weder, H.G.: Further Studies on the Temperature Dependence of the Binding of Testosterone to Human Pregnancy Plasma Proteins, Steroids, 21, 423-431 (1973).
8) Pearlman, W.H., Fong, I.F.F. & Tou, J.H.: A Further Study of a Testosterone-binding Component of Human Pregnancy Serum, J.Biol. Chem. 244, 1373-1380 (1969).
9) Rosner, W., Toppel, S. & Smith R.N.: Testosterone-Estradiol-Binding Globulin of Human Plasma: Denaturation and Protection, Biochem. Biophys. Acta 351, 92-98 (1974).
10) Burke, C.W. & Anderson, D.C.: Interrelationships of unbound testosterone and oestradiol in human serum at 37°C, and a biological role for sex-hormone-binding globulin, J.Endoctrinol. 53, R26 - R27 (1972).
11) Burke, C.W. & Anderson, D.C.: Sex-Hormone-Binding Globulin is an Oestrogen Amplifier, Nature 240, 38-40 (1972).

Acknowledgement: This work was supported by the Swiss National Foundation for Scientific Research.

Received July 30, 1974

DISCUSSION

Wagner: Is it proven that the three hormones you worked with bind to the same site? Is the binding mutually competitive?

Lutz: Yes, this has been shown by numerous competition experiments in our laboratory as in the works of Murphy, Steeno, Vermeulen and others (Murphy, B.E.P., J.Biochem. 46, 299-302 [1968]; Steeno, O., Heyns, W., Van Baelen, H. and de Moor, P., Ann. Endocr. (Paris) 29, (Suppl.), 141-148 [1968]; Vermeulen, A. and Verdonck, L., Steroids 11, 609-635 [1968]).

Sund: Would you please explain your figure 1 in more detail, especially with regard to the various straight lines? With regard to your figures 1 - 3, is there no intercept on the abscissa?

Lutz: In figure 1 straight lines are drawn through the individual points corresponding to the same total ligand concentrations. As already stated, these lines tend to go near the origin. This fact underlines my statement that the variation of the individual points tends to occur along the line of constant free ligand concentration. This line has the equation

$$Y = {}^{1}/_{C_{Lf}} \cdot x$$

and, therefore, goes through the origin. The other two straight lines represent the partial plots of the high- and low-affinity binding sites, respectively. The intercepts of these lines with the abscissa represent the binding capacities. The intercept on the abscissa of the low-affinity binding site is, of course, way out to the right of figure 1. These partial plots have been omitted on the following two figures as the computer directly gives the numerical value of the binding capacities. This last part should answer your second question.

Steinhardt: Great refinement in calculating binding data makes one question whether the data warrant it. Was equilibrium truly reached (tested from both sides)? Was ligand lost to the membrane? Does statistical binding with the assumption of identical binding constants (and no conformational transition) really prevail? Data on other multiple-binding proteins, such as serum albumin, show that even a formal application of the Scatchard plot procedure often fails. In the present experiments binding occurred to two proteins. Were their concentrations known? Was the maximum binding per globulin molecule known? If not, the binding constants are assumed identical.

Lutz: Equilibrium was reached. We test this for all our binding experiments by dialyzing the ligand against buffer alone. The decrease and increase in radioactivity in each of the cell halves is measured until it reaches the same constant amount in both of them. This indicates that equilibrium is reached. The dialysis time, the time until equilibrium is reached, is generally below four hours with the equipment used. The ligand lost to the membrane or the cell wall is measured by calculating the difference in radioactivity added and recovered. The amount of ligand lost in this way is generally below 5 %. If the dialysis time and the amount of ligand lost is above the stated value with our equipment, different membranes should be used.

It is very well known to us that the Scatchard plot may only be applied if the assumptions it is based on are valid. Fortunately, the computer does not accept experimental data if these assumptions are not fulfilled. There are several indications in the computer output for the user to check if anything went wrong. In fact, we have the experience that the computer is quite critical in accepting experimental points. The computer-calculated Scatchard curve is therefore never forced on the results. In the present work the good experimental results have all been accepted by the computer. Furthermore, most of the research groups in this field have been using the Scatchard-type plot for evaluation and no deviations from the assumptions have been observed so far. But as there is still some controversy about the molecular weight of this globulin, the binding per molecule of globulin cannot be used. Instead of the saturation fraction in the original Scatchard equation, the concentration of the bound ligand has to be used. Theoretically, this should have no effect on the association constant as both sides of the original Scatchard equation are multiplied by the protein concentration. But, of course, the dimensions on the coordinates do change. For instance, the binding capacities (dimension: moles/liter) are obtained on the abscissa instead of the number of binding sites (without dimension). This binding capacity is a measure for the concentration of the binding sites on the protein or the maximal concentration of bound ligand. I hope that this answers your question.

Veeger: Are you implying with your conclusions that if I remove my sex hormone-binding globulin from my blood, that my virility will increase?

Lutz: Yes! It is not generally accepted, or at least there is strong evidence, that mainly the free hormone is the active hormone (Vermeulen, A. and Verdonck, L., Steroids 11, 609-635 [1968]; Vermeulen, A., Stoica, T. and Verdonck, L., J.Clin. Endocr. 33, 759-767 [1971]; Vermeulen, A. and Verdonck, L., J. Steroid Biochem. 3, 421-426 [1972]). Removing sex hormone-binding globulin would, therefore, increase the free hormone concentration and with this the active portion of the hormone. Of course, there also would be increased metabolism and increased negative feedback through the brain stem, but I do not want to go too much into details here. Anyway, until these other regulatory mechanisms come into play, there would most probably be an outburst in your virility whatever that may cause. Furthermore, the "androgen-estrogen-balance" will greatly be disturbed.

The Interaction of Low-Molecular Compounds with Serum Albumin as Measured by Optical Methods. Circular Dichroism of Bilirubin-Human Serum Albumin Complexes in the Presence of Alcohols in Aqueous Solution

G. Blauer and *E. Lavie*

SUMMARY

Measurements of circular dichroism (CD) in the visible region were carried out on bilirubin (2.5×10^{-5} M)-human serum albumin (5×10^{-5} M) complexes in aqueous solution (pH 7.4; 27°C) in the presence of various water-soluble alcohols. The two large and proximate CD bands of opposite sign observed in water both changed their sign at concentrations of 1-pentanol close to 0.2 M. These observed inversions required increasing concentrations of straight-chain aliphatic alcohols as the chain length of the alcohol decreased. Secondary or tertiary alcohols were less effective than the corresponding primary normal alcohols in effecting the changes in optical activity, while ethylene glycol or methanol, up to about 5 M, did not affect the CD spectra considerably.

Hydrophobic interactions between the alcohols and the protein are considered to be important for the observed changes. Among possible contributions to the optical activity changes observed, differences in the mode of coupling between electric transition dipole moments of dissymmetrically bound bilirubin apparently monitor conformational changes, which occur in the bound bile pigment as a result of perturbations by the alcohols at the binding site. These perturbations could also be due to conformational changes in the protein carrier caused by its interaction with alcohol molecules. Thus, the optical activity of a bound protein ligand may serve as a sensitive indicator of protein interactions with other ligands.

INTRODUCTION

Protein-ligand interactions, which are of fundamental importance in life processes, can be investigated in various ways [e.g., 1,2]. Among the physicochemical methods which have been applied, such as X-ray diffraction, nuclear magnetic resonance, electron paramagnetic resonance, fluorescence, etc. optical measurements play an important role. Whenever applicable, optical activity (CD and ORD) data have been used as a very sensitive tool for investigation of protein-ligand interactions [3]. The present report may serve to illustrate some applications of this method to physiologically important complexes of bile pigments with proteins.

Complexes of bilirubin with HSA have been shown to exhibit very large rotatory strengths in the visible-region transitions of the bound bile pigments [4,5]. At excess protein, both the sign and magnitude of the observed Cotton effects depend on the pH of the aqueous solution of the complex [4,5], on the mode of preparation and treatment of the albumin [4], on the presence of various electrolytes in the system, including fatty acids [6,7], and on the presence of certain short-chain aliphatic alcohols [8]. The high sensitivity of optical activity measurements in these systems is also demonstrated by the large differences obtained for complexes of bilirubin with serum albumins from different species [9]. In the absence of protein, no measurable optical activity has been observed for free bilirubin in aqueous solution [10].

The present report deals with the effect of various lower aliphatic alcohols on some optical properties of the bilirubin-HSA complex at neutral pH, using charcoal-treated

---

Abbreviations: HSA, human serum albumin; CD, circular dichroism; ORD, optical rotatory dispersion.

albumin. Some of the primary normal alcohols cause large changes in the visible-region CD spectra of the bound bilirubin, the efficiency increasing with the hydrocarbon chain length and decreasing for the corresponding secondary alcohols. Relatively low concentrations of some alcohols are sufficient to effect inversions in the sign of the large CD bands. Thus, the optical activity of a specific protein-ligand complex may be used to monitor in a sensitive way interactions of the protein with other low-molecular ligands.

MATERIALS AND METHODS

Bilirubin was obtained from Sigma. On thin-layer chromatography only traces of impurities (isomers of bilirubin) could be detected [11]. HSA, crystallized (Lot 31), was obtained from Pentex and was treated with charcoal [12]. Except for ethylene glycol, which was redistilled, all commercial alcohols used were of analytical grade: methanol (Frutarom,Haifa); ethanol, 1-propanol, 1-butanol, 2-butanol, 1-pentanol (Merck); 2-propanol (Mallinckrodt); 1-hexanol (B.D.H.); glycerol (Frutarom); tert.butyl alcohol (Fluka). The refractive indices n_D^{20} of all alcohols used were measured and agreed with published values. Other substances used were also of analytical grade.

CD measurements were carried out on a Cary Model 60 recording spectropolarimeter with a Model 6002 accessory. The absorbance of the solutions measured did not exceed about 1.3. Light-absorption measurements were performed on a Cary Model 14 recording spectrophotometer. Viscosity measurements were carried out in capillary viscometers at 27.00±0.05°C. See [4] for other experimental details and preparation of the complexes. The alcohols were always added as the last component of the mixture.

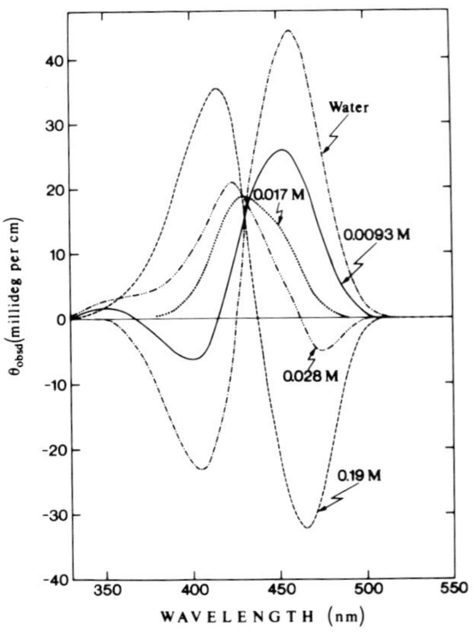

Fig.1. CD spectra of the bilirubin-HSA complex in aqueous solution at different concentrations of 1-pentanol. Bilirubin, 2.5×10^{-5} M; HSA, 5.0×10^{-5} M; NaCl, 0.1 M; Tris-HCl, 0.02 M; pH 7.4±0.1; temp., 27.0±0.5°C. The curve designated "Water" was obtained without added alcohol. Measurements were started about 20 min after preparation of the solutions. In some cases, a decrease in the absolute ellipticity values was recorded after an additional hour. θ is the observed ellipticity.

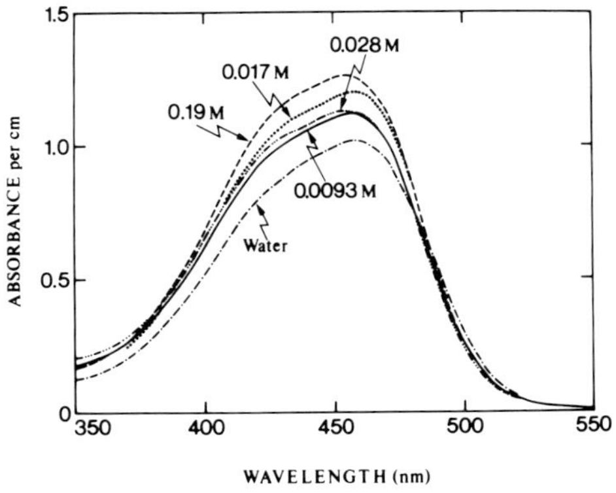

Fig.2. Light-absorption spectra of the bilirubin-HSA systems as given for Fig.1. Temp., 24±1°C. Measurements were carried out within about one hour from the time of preparation of the solutions.

RESULTS AND DISCUSSION

Fig.1 shows the effect on the CD spectrum in the visible region obtained by increasing the concentrations of 1-pentanol in an aqueous solution of the bilirubin-HSA complex at neutral pH. It appears that both observed main CD bands change their sign above about 0.02 M 1-pentanol, as compared with the aqueous solution in the absence of the alcohol. At concentrations of 1-pentanol between zero and 0.2 M, intermediate spectra are obtained with only a positive band at 0.017 M 1-pentanol. An "isoelliptic" point appears to be formed near 430 nm, although some curves intersect slightly above it. Analogous results were obtained for 1-propanol [8]. Only small changes in the corresponding ellipticity values of the extrema are observed upon increasing the HSA concentration to $2x10^{-3}$ M in the presence of 0.19 M 1-pentanol. The results indicate that over 90% of the bilirubin is bound to the protein under the conditions given for Fig.1. Any free bilirubin present would not contribute significantly to the observed optical activity [10]. Experiments at higher concentrations of 1-pentanol were precluded by the insolubility of the alcohol in water. It may be noted that both main CD band extrema show significantly higher ellipticity values than those reported in previous work [4]. This is probably due to the different HSA preparations used.

The corresponding light-absorption spectra are shown in Fig.2. A hyperchromic trend is apparent for the complexes at higher alcohol concentrations. Obviously, these spectra, obtained in the wavelength range 350 to 550 nm, are much less sensitive to changes in the concentration of 1-pentanol, as compared with the CD spectra.

Results obtained with other lower alcohols are summarized in Fig.3. The ellipticity values at 457 nm (observed long-wavelength extremum in water with no alcohols added) have been

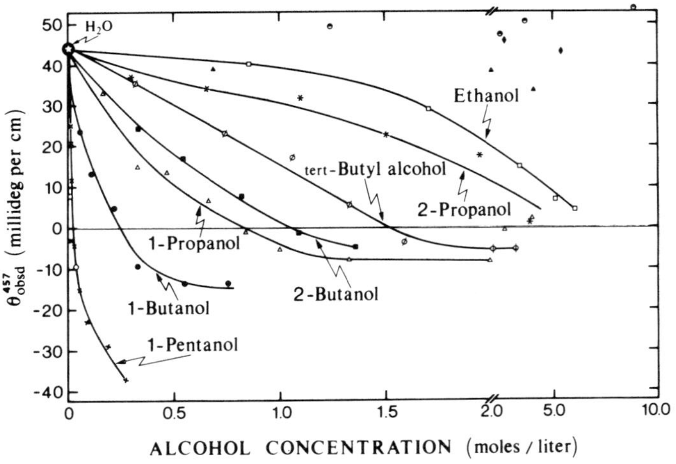

Fig.3. Observed ellipticities at 457 nm of the bilirubin-HSA complex in aqueous solution in the presence of various concentrations of different aliphatic alcohols. Conditions as given for Fig.1. In several cases, measurements at higher alcohol concentrations were limited by their solubility in water. The point designated "H_2O" refers to the value obtained in aqueous solution in the absence of alcohols. (◓) Methanol; (○) 1-hexanol; (▲) glycerol; (⧫) ethylene glycol.

taken as a measure of the changes occurring with increasing alcohol concentration. It appears that the efficiency of change increases in the order: methanol < ethanol < 2-propanol < tert.butyl alcohol < 2-butanol < 1-propanol < 1-butanol < 1-pentanol < 1-hexanol. Also, the changes observed appear to level off at smaller absolute values of θ, the shorter the hydrocarbon chain. At concentrations of 1-propanol above about 2 M, the ellipticity seems to revert again to positive values. In the concentration range given, the multifunctional alcohols ethylene glycol and glycerol do not appreciably affect the ellipticity at 457 nm, while methanol induces even higher values.

Increasing the HSA concentration to 1.2×10^{-4} M at constant alcohol concentrations did not cause further appreciable changes in the observed ellipticity of the above alcohol systems. Preliminary experiments carried out with either 0.84 M 1-propanol or 0.19 M 1-pentanol at molar ratios bilirubin/HSA larger than one showed considerably less change in the CD spectrum than indicated in Fig.3. Further analysis of these systems is in progress.

Ellipticity measurements (conditions as given for Fig.1) of the complex in the far-ultraviolet region in the presence of 0.8 M 1-propanol or 0.04 M 1-hexanol did not show significant differences from the ellipticity values at 222 nm,obtained in the absence of these alcohols.

Evidently, an increase in the length of the unbranched hydrocarbon chain of the primary alcohols increases their effectiveness to induce the changes in ellipticity observed in the visible region. This would suggest that hydrophobic interactions between the alcohols and the protein play an important role (see, for example, [2,13-16] and refs. cited therein). In isomeric alcohols with either a branched hydrocarbon chain (tert.butyl alcohol) or a secondary hydroxyl group (e.g., 2-butanol), the hydrophobicity apparently decreases (see [15]) and the observed changes in optical activity are largely reduced, as compared with the corresponding straight-chain primary alcohols. Further substitution by the polar hydroxyl groups, such as in the case of the multifunctional alcohols, up to concentrations of about 5 M, abolishes these effects, leaving the spectrum of the complex largely unchanged.

While the binding of the alcohols to HSA appears to be determined to a considerable extent by hydrophobic interactions, there is insufficient evidence so far as to the degree of cooperativity of these interactions [15] in the present case. Resolution of intermediate CD spectra in the presence of 1-propanol [8] indicated a requirement of at least five Gaussian curves, four of which could be attributed to an equilibrium between two different complexes, one formed in the absence of alcohol and the other at higher alcohol concentrations. On the other hand, transition curves obtained for various alcohols (Fig.3) do not suggest cooperative interactions. It may be noted that most of the transitions described for other proteins in the presence of alcohols [15,16] occur at significantly higher concentrations of the latter than in the present system.

As in our previous work (e.g., [4]), possible effects of microheterogeneity or of the thiol content of the albumin on the optical activity of the bilirubin complex were not investigated. It has been shown in the case of 1-propanol [8] that fractionated HSA monomer gave practically the same results as the unfractionated protein, which contains about 10% by weight of dimer. Reversibility of the effects of 1-propanol on the CD spectra of the bilirubin-HSA complex has been reported earlier [8]. Analogous experiments carried out with 1-pentanol by dilution of the complex, at constant ionic strength, from 0.19 M to 0.009 M 1-pentanol again resulted in the same inverted CD spectrum as obtained for a solution prepared directly to the same composition. Therefore, reversibility is also demonstrated with regard to the CD effect of 1-pentanol.

Previous interpretations of the inversion in the sign and of other changes in the Cotton effects observed for bilirubin-HSA complexes in aqueous solution (at different pH values [4], electrolyte concentrations [7] and upon addition of fatty acids [7]) included coupling between dissymmetrically oriented electric transition dipole moments of the two dipyrromethene chromophores of the bound bilirubin [4]. At a sufficient excess of HSA, as in the present case, practically only one specific binding site for bilirubin is considered to be occupied [5,17]. Under these conditions, intramolecular optical interactions may contribute to the optical activity, provided that the two bilirubin halves change their relative positions by rotation around the center methylene bridge and that the molecule assumes a dissymmetric conformation upon binding to a specific site on the protein. Sterically opposite positions of the two dipyrromethene chromophores could then lead to an observed inversion in the sign of the two main and proximate CD bands in the visible region.

It appears that subtle changes at the binding site, which could perhaps be coupled with conformational changes in the protein, would be monitored in a sensitive way by changes in the CD spectrum of the bound bilirubin. Apparently, the various alcohols affect the mode of binding of the bilirubin by the protein either directly at the binding site or indirectly, by combining with the protein and changing its conformation. Analogous considerations would apply if part of the optical activity observed were due to electric dipole coupling between bilirubin and protein transitions or to formation of skewed chromophore conformations of the bound bilirubin leading to inherent dissymmetry of the chromophore.

A number of preliminary viscosity measurements of aqueous HSA solutions (NaCl, 0.1 M; Tris, 0.02 M; pH 7.4; 27°C) indicated an increase by a factor of about 3 in the intrinsic viscosity in 0.84 M 1-propanol as compared with HSA in the absence of the alcohol (about 4 $ml \cdot g^{-1}$). Inclusion of 0.19 M 1-pentanol raised the intrinsic viscosity of the HSA solution to about 30 $ml \cdot g^{-1}$, indicating a substantial change in the shape of the protein. A possible correlation between the CD spectral changes in the bilirubin complex and the conformation of the protein at various concentrations of different alcohols is now under investigation. It is conceivable that subtle conformational changes, which would not be reflected in hydrodynamic properties of the protein carrier, are monitored by the optical activity of the specifically bound bilirubin.

ACKNOWLEDGEMENTS

The skillful technical assistance of Mrs.J.Silfen is appreciated. This work was supported by a grant (No.11 1772) from the Stiftung Volkswagenwerk.

REFERENCES

1. Edsall, J.T., Wyman, J.: Biophysical Chemistry, Vol.1. Academic Press, New York, 1958.

2. Steinhardt, J., Reynolds, J.A.: Multiple Equilibria in Proteins. Academic Press, New York-London,1969.

3. Blauer, G.: Optical activity of conjugated proteins. In: Structure and Bonding, Vol.18. Springer-Verlag, Heidelberg, 1974, pp.69-129.

4. Blauer, G., Harmatz, D., Snir, J.: Optical properties of bilirubin-serum albumin complexes in aqueous solution. I. Dependence on pH. Biochim.Biophys. Acta 278, 68-88 (1972).

5. Beaven, G.H., d'Albis, A., Gratzer, W.B.: The interaction of bilirubin with human serum albumin. Eur.J. Biochem. 33, 500-509 (1973).

6. Woolley,III, P.V., Hunter, M.J.: Binding and circular dichroism data on bilirubin-albumin in the presence of oleate and salicylate. Arch.Biochem. Biophys. 140, 197-209 (1970).

7. Blauer, G., Harmatz, D.: Optical properties of bilirubin-serum albumin complexes in aqueous solution. II. Effects of electrolytes and of concentration. Biochim.Biophys.Acta 278, 89-100 (1972).

8. Blauer, G., Lavie, E.: The effect of lower aliphatic alcohols on the circular dichroism of the complex bilirubin-human serum albumin in aqueous solution. FEBS Letters 41, 143-146 (1974).

9. Blauer, G., Harmatz, D.: Conformational studies on complexes of biopolymers with small molecules. Circular dichroism of complexes of bilirubin with serum albumin from different species. In: Jerusalem Symp.Quantum Chem.Biochem. (Bergmann, E.D., Pullman, B.,Eds.), Vol.V, pp.709-714. Isr. Acad.Sci.Human., Jerusalem, 1973.

10. Blauer, G., King, T.E.: Interactions of bilirubin with bovine serum albumin in aqueous solution. J. Biol.Chem. 245, 372-381 (1970).

11. McDonagh, A.F., Assisi, F.: Commercial bilirubin: a trinity of isomers. FEBS Letters 18, 315-317 (1971).

12. Chen, R.F.: Removal of fatty acids from serum albumin by charcoal treatment. J.Biol.Chem. 242, 173-181 (1967).

13. Jirgensons, B.: Effects of n-propyl alcohol and detergents on the optical rotatory dispersion of α-chymotrypsinogen, β-casein, histone fraction F1, and soybean trypsin inhibitor. J.Biol.Chem. 242, 912-918 (1967).

14. Sun, S.F.: Viscosity and hydrogen ion equilibrium studies on the effects of organic solvents on the conformation of bovine serum albumin in aqueous solutions. Biochim.Biophys.Acta 181, 473-476 (1969).

15. Herskovits, T.T., Gadegbeku, B., Jaillet, H.: On the structural stability and solvent denaturation of proteins. I. Denaturation by the alcohols and glycols. J.Biol.Chem. 245, 2588-2598 (1970).

16. Kaminsky, L.S., Yong, F.C., King, T.E.: Circular dichroism studies of the perturbations of cytochrome c by alcohols. J.Biol.Chem. 247, 1354-1359 (1972).

17. Jacobsen, J.: Binding of bilirubin to human serum albumin. Determination of the dissociation constants. FEBS Letters 5, 112-114 (1969).

Received 12 July 1974

DISCUSSION

Maass: In order to correlate the conformational changes in the bilirubin with tertiary structure changes of the protein did you observe any CD changes in the UV-bands of the protein with pH in the absence of bilirubin?

Blauer: There were no significant differences in the far ultraviolet ellipticity of the protein near 222 nm between pH 4 and 10, nor upon inclusion of either 0.04 M 1-hexanol or 0.8 M 1-propanol at neutral pH. This indicates that the secondary structure remains largely unchanged under these conditions. Measurements in the region of aromatic amino acids were not sufficiently reproducible.

Hemmerich: Stimulated by Dr. Blauer's finding of the surprisingly high CD of the bilirubin-BSA complex, we have tried to simulate this behavior in a "chemical" (i.e. low-molecular weight) system in order to decide, whether the induced chirality of bilirubin was protein-specific or not. Indeed, we found (Ph.D.-thesis O. Mayer) that the protein effect can be simulated quantitatively in petrol ether solutions of bilirubin containing α-phenethylamine as low-molecular weight asymmetric inducer. C.C.Kuenzle, M.H.Weibel, R.R.Pelloni and P.Hemmerich (Biochem. J. (1973) 133, 364) have verified by NMR that, under aprotic unpolar conditions, bilirubin assumes a skewed conformation, which is stabilized by strong hydrogen bridges between each propionate side chain and the outer cyclic amide group of the opposite pyrromethene half. An asymmetric inducer will select one enantiomeric conformer out of this racemic mixture. Any aggregation of bilirubin can be excluded under these conditions with high probability, but further experiments are under way in order to perfectly exclude this possibility. A specifically shaped hydrophobic region of the protein is, of course, prerequisite for the Blauer effect to be observed in a biopolymer system.

Blauer: According to our calculations and models, such internally hydrogen-bonded structures which had been proposed by us earlier for unbound bilirubin (see above, ref. 4), should not exhibit the large optical activity observed in the visible region, since the splitting would be too small in such conformations. However, the bound bilirubin could assume a variety of different conformations on the protein, which would most likely be stabilized mainly by hydrophobic interactions between the protein and the bile pigment. With suitable dihedral angles between the two dipyrromethene chromophores, the observed splitting and large optical activity could be accounted for.

Steinhardt: I seem to recall a Japanese paper on dimerization of bilirubin in slightly acid aqueous solutions.

Our 1966 measurements of n-alcohol binding constants to bovine serum albumin showed that they were quite high (about 5000 for octanol, over 100,000 for dodecanol). Pentanol might therefore compete for the bilirubin binding site and prevent the optical activity due to binding. Incidentally, alcohol binding produces large difference spectra at 280 and 232 nm. Both are red shifts. The 232 nm region has a very large effect.

Woody: Have you carried out experiments on alcohol effects at pH 4 as well as at neutral pH?

Blauer: We have also carried out such measurements at pH 4. I cannot present them now.

von Hippel: Is it possible that the increases in intrinsic viscosity you see with lower alcohols at the "iso-CD" form is due to increasing aggregation in these more alcoholic solutions?

Blauer: This should be checked. It is noteworthy, however, that in the presence of 0.03 M 1-pentanol or 0.3 M 1-butanol, under the conditions described, the intrinsic viscosity remained practically that measured in aqueous medium in the absence of these alcohols.

Section VI. Detergents and Electrolytes

Chairman: *R. Jaenicke*

The Nature of Specific and Non-Specific Interactions of Detergents with Proteins: Complexing and Unfolding

Jacinto Steinhardt

The interactions of proteins with ionic detergents differ from their interactions with other ligands in two important respects: 1) Ionic detergents having hydrocarbon-chain lengths of 12 or more carbon atoms are the most potent protein denaturants known (1). 2) Unlike all other classes of ligand except hydrogen-ion, they combine with most native proteins in multiple equilibria, i.e. many equivalents per mole of protein. They combine in still larger quantities with proteins that have been unfolded. In fact, they unfold proteins precisely because the difference in binding equivalence exists.

Thus, the two unique interactions of ionic detergents with proteins are closely interrelated. The observation by Anson in 1939 that a commercial sample of sodium dodecylsulfate (SDS) was a potent protein denaturant directed the attention of protein chemists to detergent interactions. Urea (8 M) has had its day as a protein denaturant, and guanidine hydrochloride (GuCl) at about 6 M has been the denaturant of choice since Tanford showed that in solutions of the latter reduced proteins lose all traces of secondary, tertiary, and quaternary structure, and behave like randomly coiled long-chain synthetic polymers. Dodecylsulfate, however, produces a maximum unfolding effect, especially of reduced proteins, at concentrations of less than 10^{-2} M, which may actually contain only 10^{-3} M or less of the detergent in the uncombined form. In 1% solutions barely more than 15 moles of detergent per mole of protein are required to produce unfolding of some of the protein molecules. The ratio for GuCl under identical conditions is about 40,000.

Anson's observation was followed by a spate of reports on the destructive effect of various detergents on a variety of proteins and viruses. The present review is far from exhaustive and omits, because of space limitations, references to many important contributions not only at this point but throughout the text.

A different kind of destructive effect, analogous to that of proteolytic enzymes, was investigated by Steinhardt and Fugitt (2), who demonstrated a powerful catalytic effect of dilute detergents on the acid hydrolysis of amide and peptide bonds in protein fibers and in dilute solutions of proteins. The proportion of SDS to amide content determined whether only amide, or both amide and peptide was attacked. These effects, however, appear to be unrelated to the denaturing effects already described - they are shared with other anions of high affinity, which do not denature as effectively as the long straight-chain ionic detergents.

Studies on the binding of anionic detergents to proteins quickly followed the work on denaturation. Electrophoresis revealed that binding had some of the aspects of an all-or-none process, i.e. after a few equivalents were bound by ovalbumin two boundaries appeared, one which travelled at the same speed, or only slightly faster than, the uncombined protein; and one which travelled considerably faster. The fraction of protein in the latter grew as the detergent concentration was increased, at the expense of the former (3). A third component appeared to bind double the number of equivalents displayed by the second boundary (4).

Numerous studies investigated the stoicheometry of precipitation of proteins by detergents. It was soon discovered that low concentrations of dodecylsulfate acted similarly to fatty acid anions or other organic anions in protecting serum albumin from denaturation by urea, guanidine hydrochloride, or heat coagulation (5). Investigation of this paradoxical effect led to the earliest conclusion, since confirmed, that a stable complex containing about eight dodecylsulfate ions or nine caprylate ions (6) is formed by serum albumins. A review by Putman (7) covers in much more detail all aspects of these early studies of protein-detergent interaction.

The first effort to determine the equilibrium binding isotherm of a protein (BSA) with a series of homologous sulfate half-esters was made by Karush and Sonnenberg (8). Their data do not permit distinguishing between binding to native and to unfolded protein. The latter purpose was achieved by Yang and Foster (9). They measured the binding of dodecylbenzenesulfonate to both BSA and ovalbumin, making use of electrophoretic patterns and mobilities. Since more definitive studies of the dodecylbenzenesulfonate -

serum albumin isotherm were later published by Decker and Foster (10), we give here the principal conclusions of the latter. Many of them were foreshadowed in the earlier paper. About 12 equivalents of the C_{12} compound are bound very tightly ($K > 10^6$) at neutral pH. Octylbenzenesulfonate has a decisively lower affinity (1.6×10^5). At higher pH the affinity of the C_{12} ion is lower but still very high (1.6×10^5 at pH 9.5).

The larger ion (C_{12}) still binds strongly at $\bar{v} \gg 12$; 60 to 70 equivalents are bound at a free concentration of 10^{-5} M. The C_8 compound binds less than 10 at this equilibrium concentration.

Electrophoresis schlieren pictures showed that the proportion of native protein commences to diminish at $\bar{v}$ over 12, and that a complex which binds 38 ions appears and increases until it represents 82% of the protein at $\bar{v} = 38$, at which point only 10% of the protein is still in the native state. At the same value of $\bar{v}$, a small amount of distinguishably still faster moving complex (PA_{76}) appears, and when $\bar{v}$ reaches 86, the new component contains all of the protein. A change in the Moffitt ORD parameter b_0 was seen to accompany the transition from the 12-ligand complex to the 38- and 76-ligand complexes. 25% of the helix content present in the native protein is lost in this transition.

Analysis of the mobilities of the three components and of the overall isotherm lead to $n_1 = 11 \pm 1$, $n_2 = 38 \pm 1$, and $n_3 = 76 \pm 2$. At higher pH values, PA_{38} is diminished because of disproportionation between PA_{38} and PA_{11} and PA_{76}.

The exposure of new sets of sites by unfolding at $\bar{v} > 11$ or 12 would not result in saturation of the new sites (an appearance of "cooperativity") unless the affinity of the new sites was great enough to combine ligands practically quantitatively at the equilibrium concentrations involved in unfolding. Sets of equations were formulated, a separate equation for the $\bar{v}$ value of each component. The total binding is not the sum of all three values of $\bar{v}$, since how much protein is present as A, B, C, is not specified. Foster's micellar alternative model is not described here since Decker and Foster showed that the two models could not be distinguished by the data.

Decker and Foster postulated that three forms of the protein are in equilibrium, which may be written

$$A \underset{}{\overset{K_{AB}}{\rightleftharpoons}} B \underset{}{\overset{K_{BC}}{\rightleftharpoons}} C$$

where A is uncombined native protein, B is uncombined unfolded protein, and C is another form of uncombined unfolded protein. They set up the equilibrium between $\sum A$ and $\sum B$ (all degrees of ligand binding) in terms of apparent equilibrium contants K'_{AB} and derived similar expressions for the second stage K'_{BC}. The important result was obtained that the relative amounts of native and unfolded protein depend on the equilibrium ligand concentration, when K_{AB}, the numbers of sites, and the various binding constants are given. The protein thus unfolds because there are more combining sites in the unfolded form having binding constants near enough to those native protein to combine with ligand, thus shifting the equilibrium toward the unfolded form. These relationships were made more explicit by Reynolds et al.(11).

Thus, the affinity of unfolded protein for detergent is only slightly lower than that of the native protein. Decker and Foster tentatively identified K_{AB} as a pH-dependent parameter responsible for the $N \rightleftharpoons F$ transformation near pH 4 (12). The fact that the $N \rightleftharpoons F$ equilibrium at a given pH is shifted toward N by the presence of small amounts of detergent indicates "protection" of the protein by small $\bar{v}$. The model described predicts such stabilization only if n strong sites are replaced by a larger number m of sites on unfolding, and if K_{AB} is a function of pH.

These ideas have the great merit that they relate binding and unfolding explicitly. It is a thermodynamic necessity for the protein to unfold when enough detergent is present if the unfolded protein makes available many more binding sites which are not very much weaker than the preexisting ones. The model predicts the observed stabilization when only small amounts of ligand ($\bar{v} < n$) are bound. It has the further virtue of isolating the intrinsic isomerization constant K_{AB} from the other parameters involved in unfolding (ligand concentration, binding strengths, numbers of sites on folded and unfolded protein).

A detailed investigation of the binding of both alkylsulfate half-esters (such as SDS), and the corresponding sulfonates, having chains of from 8 to

14 carbon atoms, was made by Reynolds et al. (11). With the sulfates, 12 or more carbon atoms were required to unfold BSA; the corresponding sulfonate did not unfold. Subsequent work showed that a 14 carbon sulfonate unfolded. (See also Tanford [15]). Other work has shown that much higher concentrations of decylsulfate than were used in this study also unfold. All of the detergents at very low concentrations showed normal statistical binding with from 6 to 11 high-affinity sites.

Reynolds et al. succeeded in fitting their experimental data for dodecylsulfate with an equation which embodied the same ideas as those of Decker and Foster, but which determined not only $\bar{v}$, but also the proportions of folded and unfolded protein in terms of the explicit parameters, including an intrinsic unfolding equilibrium constant, U. Only familiar statistical binding equilibria were involved; the "cooperativity" came out of the binding-induced unfolding. Their easily derived equation was:

$$\bar{v} = \frac{[A]}{1 + F[A]} \left[\frac{nK}{1 + K[A]} + \frac{F([A])\ mJ}{1 + J[A]} \right] \quad \text{(Eq.1)}$$

where $F([A]) = U\ (1 + J[A])^m/(1 + K[A])^n$

The latter expression differs from Foster's K_{AB} only in containing U, the conformational transition constant. K and J are binding constants for native and unfolded proteins, respectively, [A] is the concentration of free detergent, and n and m are the numbers of binding sites on native and unfolded proteins, respectively. Unfolding occurs when $mJ > nK$. The success of this concept in fitting the SDS-BSA data is shown in Fig. 1 . Equation 1 has been expanded to describe unfolding which occurs in two successive stages.

As early as 1958, Goodman (16) measured very high constants for the binding of the fatty acids to HSA. Stearate had the highest affinity of those tested: 1.1×10^8 for two of its 7 high-affinity sites. The other 5 had a K of 4×10^6. Reynolds et al. (17) found, with more gently purified protein, essentially the same binding constants for BSA as Goodman had found in HSA.[1] The unexpected conclusion was reached that the binding sites for

1. A number of recent papers by Spector and his collaborators (32) have attempted to obtain separate binding constants of fatty acid anions, instead of assuming grouped identical intrinsic constants as have other investigators.

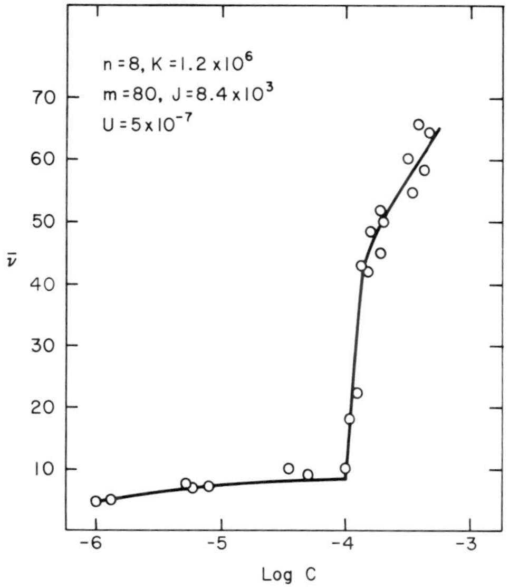

Fig. 1. The binding isotherm of BSA-dodecylsulfate at 2°, fitted by Eq.1 (11).

fatty acids and for sulfate detergents were different. Binding of long-chain examples of the latter causes a complex series of changes in the two ultra-violet absorption bands of both tryptophan and tyrosine (18); while the difference spectra initiated by binding fatty acids show effects on the tyrosine absorption only. Similar observations on the effects of binding fatty acids were also reported by Zakrzanski and Goch (19).

Few studies have been made with cationic detergents (see [20] for a review, and [21]). Enough has been done to show that the affinity for proteins of cationic detergents at neutral pH is substantially less than the affinity of anions of the same chain length. With insulin, the cations precipitate when present in stoichiometric equivalence to a single carboxylate group on the protein, but resolubilize gradually after addition beyond equivalence to all six carboxylates.[2] A non-ionic detergent Triton-X solubilizes

2. The large difference in affinity of anions and cations of the same chain length may be due to the large difference in electrostriction of water brought about by the deprotonation of carboxyl groups, and the deprotonation of RNH_3+ groups (33). An extra energy-requiring process appears to be involved when negative charges on proteins are neutralized than when positive charges are neutralized (34).

membrane proteins, but no unfolding occurs. Limited solubility prevents the attainment of unfolding concentrations of monomers (22).

A number of detailed studies of the effects of binding detergents on such optical properties as light absorption, fluorescence, and optical rotatory dispersion have yielded considerable information as to the nature of the environment of the binding sites, and have revealed remarkable differences between bovine and human serum albumins (12, 23, and 24).[3] Space does not permit an account of the phenomena or of their interpretation; we mention only that polarization of fluorescence data indicate that tryptophan is not free to rotate about its bond to the peptide backbone; this fact supports other indications that detergent-disorganized serum albumins retain considerable structure as compared with those unfolded by strong acid, urea, or GuCl. The dependence of the binding-induced fluorescence blue shift on $\bar{\bar{v}}$ leads to conclusions as to the distances between tryptophan and some of the binding sites for the first 15-20 equivalents bound; and provides compelling evidence that bound ligand diffuses rapidly (in less than 10^{-8} seconds) from occupied to unoccupied sites.

Kinetic studies of the unfolding and refolding of HSA at extremes of pH (25) which result in unmasking and remasking some of the tyrosines in this protein, have also shown that SDS unmasks these groups, which then ionize at a suitable pH. At high pH, however, small amounts bound partially suppress not only the unfolding which would otherwise occur at such pH values, but also the ionization itself (the pK is raised). This protection is given against unfolding by acid as well as against unfolding by base. An analogous stabilizing effect on HSA against the equilibrium unfolding action of urea was demonstrated by Markus, et al, (26) who showed that only the detergent bound by the native protein stabilized, and then only if the

3. The binding of detergents has also demonstrated differences between bovine and human serum albumins in the following respects, among others a) differences in binding isotherms, especially for tetradecylsulfate; b) unfolding of HSA in a single process as compared with a two-stage process in BSA (at pH 5.6 - both proteins unfold in a single stage at pH 6.9); c) a small effect of pH on the fluorescence of HSA-SDS complexes than on BSA-SDS complexes; and d) greater tryptophan tyrosine in HSA than in BSA complexes, the greatest efficiencies being obtained with the unfolding detergents.

ε-amino groups of the lysines were not blocked by acetylation or by formaldehyde (the effects of such blocking were manifested on the protection rather than on the SDS binding). The role of the cationic lysine groups led Markus et al. (26) to conclude that cross-linking (two-contact binding) might be involved.

Reynolds and Tanford have recently shown (27) that proteins of widely different compositions, examined after reduction of all disulfide bonds and complete disorganization in GuCl, all bind identical quantities of SDS on a gram per gram basis, i.e. the stoichiometry is remarkably independent of the molecular weight. Detergent ions thus appear to bind to completely unfolded proteins via the peptide bonds or monotonous structures composed of peptide bonds. Some of the experiments leading to this conclusion had been anticipated by Pitt-Rivers and Impiombato (28) who found 1.4 grams SDS/gram reduced protein; when the proteins contained intact disulfide bonds smaller amounts are bound (0.90-0.93 gms/gm). Since no GuCl was used in the experiments of Pitt-Rivers and Impiombato it appears that the disruption brought about in reduced proteins by SDS suffices to produce maximum binding.

In some experiments Reynolds varied SDS monomer concentrations indirectly by changing the ionic strength in such a way as to alter the CMC (the latter is equal to the monomer concentration). Thus, ionic strength was not held constant and the data given (based on numerous proteins) must not be interpreted as conventional binding isotherms. The observations just described provide the basis for the often used determination of molecular weights by measurements of electrophoretic mobility. Since the SDS-reduced protein complexes all form, at saturation, rod-shaped particles of uniform cross section and their charges are so high as to eliminate the effects of differences in composition, the mobilities are inversely proportional to the length of the rods, and thus of the molecular weights (29).

The constancy of the figure 1.4 grams SDS/gram protein has been questioned experimentally by Nelson (30) and by Allen (31). The former finds a spread in the saturation binding ratios with a number of proteins. The latter, using neither reducing agents, nor GuCl, and making use of gel filtration with S^{35} dodecylsulfate, gets a range of higher binding ratios with BSA than Reynolds and Tanford reported.

Our own most recent work with protein-detergent interactions (undertaken with K. S. Birdi of the Institute of Physical Chemistry of the Danish Technical University) is concerned with how large numbers of equivalents of detergents are held by the protein. Detergents above the CMC, which depends on the environment in solution, form micelles. There is thus a practical limit to the concentration of detergent monomers. These micelles act as ultramicroscopic droplets of hydrocarbon solvent, dispersed in an aqueous medium, which dissolve water-insoluble dyes. The absorption of the aqueous dispersion can be used as a direct measure of the amount of micelles present.

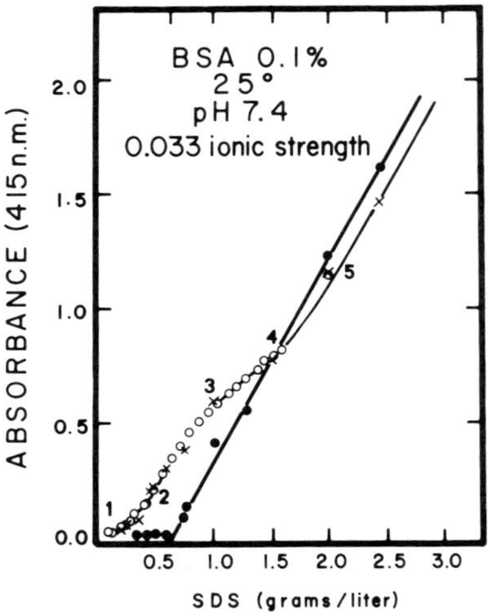

Fig. 2. The effect of 0.1% bovine serum albumin (BSA) on the solubilization of DMAB by SDS at 25°. O and X Protein present, ● Protein absent.

When protein is present, dye is dispersed at concentrations far below the CMC. The situation in Fig. 2 is typical: more dye is dispersed as the detergent concentration is increased. The protein curve and the no-protein or "micellar" curve eventually cross. At higher concentrations both curves are parallel, but now less dye is dispersed in the presence of protein. When the absorbance is plotted against the amount of SDS bound (as determined from dialysis isotherms) instead of against the amounts present, it is seen that the amounts of dye dispersed are proportional to the amounts of detergent bound. Furthermore, the amount dissolved per gram of detergent bound is nearly the same as the amount dissolved per gram of detergent micelles. Although Reynolds and Tanford have shown that micelles do not bind as such, nevertheless the protein mobilizes bound detergent, even below the CMC, so that it solubilizes as well as an equivalent weight of micelles. The maximum amount of SDS bound per HSA molecule is about 4 times as great as the amount in a single micelle. The ratio dye dissolved/SDS bound is preserved; thus about 4 dye molecules are dispersed per protein molecule.

In spite of this equal efficiency at moderate amounts bound, some binding must occur which does not solubilize, or which solubilizes "inefficiently", since at high concentration of detergent less dye is solubilized with protein present than with protein absent. Thus, the dispersion of dye by micelles in the parallel regions of Fig. 2 is reduced because there is less free detergent. These concepts have been tested by varying protein concentration, pH, and ionic strength. The proportionality of dye dissolved to detergent bound over the region $30 < \bar{v} < 160$ is preserved. When the protein is unfolded by reduction of its disulfide bonds in 6 M GuCl the dye dispersed is increased only at the lower concentrations of detergent - the unfolding caused by high concentrations of detergent almost suffices to match the effects of the unfolding action of GuCl and reductant. The maximum amounts which can be bound are found only when both protein concentration and ionic strength are low. Under these conditions, the intersection of the micellar curve with the protein curve is shifted far enough toward high concentrations so that the behavior of the latter can be observed up to where it levels off. Otherwise, a less clear end point must be estimated: the detergent concentration at which the protein curve becomes exactly parallel to and below the micellar curve at the right of the figure.

The same technique demonstrated that decylsulfate, which has practically the same affinity to native HSA as SDS, unfolds HSA only at very much higher concentrations than does SDS. Thus decylsulfate must have a much lower affinity to unfolded HSA than does SDS. With octylsulfate no attainable concentration produces unfolding.

We are now using the solubilization technique with SDS to rapidly screen the binding and solubilizing behavior of a large number of proteins. Some behave like the serum albumins described here, with relatively little SDS binding which does not solubilize dyes as effectively as SDS micelles (examples: ovalbumin, aldolase). Others bind equally well, but solubilize little if at all (examples: pepsin, apoferritin, IgG) and others are intermediate (β-lactoglobulin, conalbumin). In still other cases (e.g. lysozyme), precipitation by SDS occurs. No correlation between these classes and their amino-acid compositions or sequences has been found. Obviously, SDS is bound differently, at least in part, on the proteins of these three different classes. It is a challenge to make use of this fact to gain new information about the structures of the members of these classes.

Dye-dispersal experiments have further supported the concept that proteins in solutions of detergents exist in a limited number of states, in which the available binding sites in individual molecules are nearly completely filled, although the fraction of filled (unfolded) molecules varies with SDS concentration.

All of the experiments referred to above were done with dimethylaminoazobenzene (DMAB, Dimethyl-Yellow). My collaborator, Dr. Kulbir S. Birdi, has made some surprising observations with another dye, Orange OT. This dye is dispersed by SDS micelles precisely as is DMAB, but is not solubilized at all by detergent bound to protein. This remarkable and puzzling fact raises some important questions, and leads to some fascinating possibilities. How is the bound SDS organized on the protein so that it discriminates against one dye (Orange OT), but freely admits another (DMAB), while simple SDS micelles admit both of them.

One may quickly and easily determine the maximum detergent binding capacity of any protein by using Orange OT, by comparing the apparent CMC of SDS by solubilizing Orange OT both in the absence and the presence of the protein. The difference in grams SDS per liter is readily equated to grams per gram of protein, or equivalents per mole, if the concentration of the protein is known. Thus, for example, if the CMC at a fixed ionic strength with 0.1% protein present is 1.4 gms per liter and in its absence is 0.6 gm/liter then the bound SDS is 0.8 gm per gram of protein. Results of approximately this size have been obtained with BSA and with myoglobin at pH 7.4 at 0.033 ionic strength. Higher results are expected at lower ionic strengths (at which the SDS monomer concentration is higher) or when the disulfide bonds are reduced.

The interactions of detergents with proteins are worthy of further study. No other class of ligands combines so high a degree of multiple binding with both the uniformity and diversity sampled above. We believe the combined traits will lead to new information about the structure of individual proteins - or protein fragments (a subject we intend to pursue) - and to generalizations as to how structure affects binding properties.

1. Anson, M.L. (1939). J.Gen.Physiol. 23, 239.
2. Steinhardt, J. and Fugitt, C.H. (1942). J.Res.Natl.Bur.Std. 29, 315.
3. Lundgren, H.P., Elam, R.W., and O'Connell, R.A. (1943). J.Biol.Chem. 140, 183.
4. Putnam, F.W., and Neurath, H. (1945). J.Biol.Chem. 159, 195.
5. Boyer, P.D., Lum, F.G., Ballou, G.A., Luck, J.M., and Rice, R.G. (1946). J.Biol.Chem. 162, 181.
6. Duggan, E.L., and Luck, J.M. (1948). J.Biol.Chem. 172, 205.
7. Putnam, F.W. (1948). Advan. Protein Chem. 4, 79.
8. Karush, F., and Sonenberg, M. (1949). J.Am.Chem. Soc. 71, 1369.
9. Yang, J.T., and Foster, J.F. (1953). J.Am.Chem.Soc. 75, 5560.
10. Decker, R.V., and Foster, J.F. (1966). Biochemistry, 5, 1242; (1967). J.Biol.Chem. 242, 1526.
11. Reynolds, J.A., Herbert, S., Polet, H., and Steinhardt, J. (1967) Biochemistry 6, 937.
12. Leonard, W.J., and Foster, J.F. (1961). J.Biol.Chem. 236, 73.
13. Foster, J.F., and Aoki, K. (1958). J.Am.Chem.Soc. 80, 5215.
14. Foster, J.F(1960) in "The Plama Proteins" (F.W. Putnam, Ed.) Academic Press, New York.
15. Tanford, C. (1973). "The Hydrophobic Effect". John Wiley & Sons, New York.
16. Goodman, D.S. (1958). J.Am.Chem.Soc. 80, 3892.
17. Reynolds, J.A. Herbert, S., and Steinhardt, J. (1968). Biochemistry 7, 2606.
18. Polet, H., and Steinhardt, J. (1968). Biochemistry, 7, 1348.
19. Zakrzanski, K., and Goch, H. (1968). Biochemistry 7, 1835.
20. Steinhardt, J., and Reynolds, J.A. (1969). "Multiple Equilibria in Proteins" Academic Press, New York.
21. Birdi, K.S. (1973). Biochem. J. 135, 253.
22. Makino, S., Reynolds, J.A., and Tanford, C.(1973) J.Biol.Chem. 248, 4926.
23. Steinhardt, J., Krijn, J., and Leidy, J.G. (1971). Biochemistry 10, 4065.
24. Steinhardt, J., Leidy, J.G., and Mooney, J.P. (1972). Biochemistry 11, 1809.

25. Steinhardt, J., and Stocker, N. (1973). Biochemistry 12, 1789; 12, 2798.
26. Markus, G., Love, R.L., and Wissler, F.C. (1964). J.Biol.Chem. 239, 3687.
27. Reynolds, J.A., and Tanford, C. (1970). Proc.Nat.Acad.Sci.(U.S.) 60, 1002; J.Biol.Chem. 245, 5161.
28. Pitt-Rivers, R., and Impiombato, F.S.A. (1968). Biochem.J. 109, 825.
29. Fish, W.W., Reynolds, J.A., and Tanford, C. (1970). J.Biol.Chem. 245, 5166.
30. Nelson,C.A. (1971). J.Biol.Chem. 246, 3895.
31. Allen, G. (1974). Biochem.J., 137, 575.
32. Spector, A.A., Fletcher, J.E., and Ashbrook, J.D. (1971). Biochemistry, 10, 3229.
33. Rasper, J., and Kauzmann, W. (1962). J.Am.Chem.Soc. 84, 1771.
34. Kauzmann, W., Bodanszky, A., and Rasper, J. (1962). J.Am.Chem.Soc. 84, 1777.

Received July 1, 1974.

DISCUSSION

Maass: For a detailed understanding of the molecular basis of the interactions of detergents with proteins and of the detergents themselves when bound to the protein you should extend your measurements to a study of these interactions in unfolded oligopeptides as a function of chainlength and composition. This can give you, among others, particular information, for example, on the binding constants and the degree of cooperativity.

Steinhardt: This is our plan.

McNamee: Has the effect observed with orange-OT been seen with more than one protein?

Steinhardt: Yes. BSA, HSA, and myoglobin.

Blauer: You have stated that cationic detergents interact with proteins with less affinity than anionic detergents. Were these experiments conducted close to the isoionic points of the respective proteins?

Steinhardt: Yes, mostly at pH 5.6; although a total range of 3.8 to over 7.5 was examined.

Blauer: With regard to the rare case of protein-micelle interactions mentioned by you I would like to add that Dr. Yonath is going to present such a system in the afternoon session, namely the formation of complexes of ferrimyoglobin with laurylpyridinium halides above the critical micelle concentrations.

Sund: The binding capacity of protein molecules for SDS is one SDS molecule per two peptide bonds. Does this mean that to 50 % of the peptide bonds one SDS molecule is bound or could it also be possible that an associated SDS complex (dimer, trimer, etc) interacts with the peptide bonds?

Steinhardt: We do not know.

Veeger: A few years ago we observed that the binding of SDS to lipoamide dehydrogenase proceeds through three phases, an initial binding phase in which hardly anything occurs, up to about 50 moles/mole. During the second phase which is strongly pH and ionic strength dependent, the environment of the non-covalently bound flavin becomes more polar. The catalytic activity declines, but the flavin remains bound. Upon lowering the temperature the activity returns. In the third phase total dissociation of the flavin occurs. We have interpreted the data in terms of hydrophobic interaction in the first phase and electrostatic interaction in the second phase. Do you agree with such an interpretation? Furthermore, can you agree with the conclusion that at lower temperature a phase transition occurs in SDS micelles, leading to stronger mutual interaction and thus less efficient binding to the protein?

Steinhardt: I would have to study the papers before answering these questions.

Barrantes: I should like to make a comment which refers to both the problem of the structure of SDS micelles and its temperature dependence, arising from the questions of Drs. Veeger and Sund. Luzzati has shown some years ago that SDS micelles can exist in several conformations. His small-angle X-ray diffraction studies demonstrated the existence of hexagonally-packed tubular micelles besides the spherical ones, and, furthermore, the temperature and concentration dependence of such phases.

Steinhardt: I am not familiar with this work. I know only of the elliptically symmetrical micelles.

von Hippel: Does the hydrolysis of peptide bonds by SDS depend on impurities (perhaps proteases)? Does it proceed at rates which would be troublesome in SDS gel electrophoresis? I was not aware that this raised problems in practice anyway.

Steinhardt: The hydrolysis of peptide bond is very slow at pH values above 3.0 and at temperatures below 37°.

von Hippel: Some types of micelles do show temperature-dependent CMC's. (e.g., fatty acid amide micelles).

Wagner: Is there any evidence that the binding of SDS by the polypeptide chain is to some extent influenced by the amino acid side chain? To my knowledge histones behave in SDS gel electrophoresis differently from normal globular proteins.

Steinhardt: I believe one histone has been studied in Tanford's laboratory, and is like the others.

Sund: In the literature intrinsic viscosities for proteins in the presence of SDS are reported which are rather low, indicating globular, folded state. Is it correct to assume that in these cases the SDS concentration was above the CMC and, therefore, SDS was not bound to the protein, yielding a mixture of globular protein and globular SDS with low intrinsic viscosity?

Steinhardt: The intrinsic viscosity of protein-SDS complexes unfolded in guanidine-HCl and reduced is not low. The protein particles are very asymmetric.

Interaction of Detergents with Proteins: Effect of Counter Ions on the System Ferrimyoglobin-Laurylpyridinium Halides in Aqueous Medium

J. Yonath and *G. Blauer*

SUMMARY

Interactions of sperm whale ferrimyoglobin with laurylpyridinium halides in aqueous medium (pH 7.1; 25°C) were followed by spectral changes in the Soret-band absorption of the heme. These changes depend on the counter-ion of the cationic detergent, the effectiveness increasing with concentration and with the nature of the anion: $Cl^- < Br^- < I^- < SCN^-$. A similar dependence was observed for the decrease in the critical micelle concentration of the detergent.

From the transition curves, standard free-energy changes and the number of detergent molecules reacting with the protein were estimated, assuming an all-or-none interaction.

Since it appears that, in some cases, the protein-detergent interaction increases even above the critical micelle concentration of the detergent, the possibility of formation of micelle-like structures of detergent on the protein is being considered.

INTRODUCTION

In a previous communication [1] some of the properties of the myoglobin-laurylpyridinium chloride system in aqueous medium were analyzed thermodynamically. A number of observations pointed to similarities between detergent-detergent interactions, *i.e.*, micelle formation and protein-detergent interactions, which in the present case result in formation of a "green complex" [2,3]. The possible structure of such a complex has been discussed [3,1].

In both detergent-detergent and protein-detergent systems, the interaction is enhanced by increasing the hydrophobic chain length of the detergent. The standard free-energy change for these processes is found to decrease with the addition of $-CH_2$-residues [1,4-6]. Increasing the ionic strength diminishes the critical micelle concentration and increases the micelle size for many ionic detergents [6], including alkylpyridinium halides [7-9]. Similarly, the addition of NaCl decreases the detergent concentration needed for green complex formation and increases the number of detergent molecules attached to the protein [1].

The finding that the standard enthalpy change ΔH^0 is positive for the myoglobin-LPC system and that the main contribution to the standard free-energy change ΔG^0 is entropic ($\Delta S^0 > 0$) [1] also suggests the importance of hydrophobic interactions, as in the case of micelle formation. However, other interactions as well as configurational changes in the system may also contribute significantly to these thermodynamic parameters. An important feature, reminiscent of micelle formation, is the cooperativity of the process in which LPC reacts with myoglobin. The transition curves are sigmoidal and occur over a narrow detergent concentration range in this denaturation process.

In the present work, the effect of counter-ions on the protein-detergent interaction in aqueous medium is reported and is compared with the effect of counter-ions on micelle formation.

It has been reported that ions bearing the same charge as the polar detergent head have no effect on micelle formation [10]. As to the effect of counter-ions, their ability to decrease the CMC and to increase micelle molecular weight is well known (*e.g.*, [10,6]). However, the effect of the nature of the counter-ion, using either the sodium halides or the alkali metal chloride series, had been underestimated for a long time, probably because of the favored use of SDS and other anionic detergents. In this

---

Abbreviations. LPC, laurylpyridinium chloride; LPB, laurylpyridinium bromide; LPI, laurylpyridinium iodide; LPT, laurylpyridinium thiocyanate; FP, ferriprotoporphyrin IX; SDS, sodium dodecylsulfate; CMC, critical micelle concentration.

case, only very exact measurements may reveal the differences between Li^+, Na^+ and K^+ in their effects on the micellization of SDS [11,12]. Other investigators [7,9] could demonstrate much stronger effects of halide anions on the micellization of cationic detergents.

Decreasing the polarizability of the counter-ion in the series $Cs^+ > K^+ > Na^+ > Li^+$ or $SCN^- > I^- > Br^- > Cl^-$ lessens the effectiveness in decreasing the CMC and increasing the micellar size.

As will be shown below, the interaction of myoglobin with laurylpyridinium halides is also enhanced by changing the counter-ion in the same series: $Cl^- < Br^- < I^- < SCN^-$. Under certain conditions, a state may be reached where micellization and protein-detergent interaction occur simultaneously. Under the conditions of the previous work [1], carried out only with the chloride counter-ion, all protein-detergent interactions occurred below the CMC of the detergent.

The effects of counter-ions may be related to lipoprotein structure, as well as to membrane structure and function.

MATERIALS AND METHODS

Materials. Myoglobin (ferri), from sperm whale, cryst., 0.3% Fe, salt-free, lyophilized and pure, was from Koch-Light, Colnbrook, England. Laurylpyridinium chloride (LPC) (Dehyquart C), cryst., (94% pure + 6% water) was from Henkel, Düsseldorf, Germany. Laurylpyridinium bromide (LPB), iodide (LPI) and thiocyanate (LPT) were prepared from LPC solutions by two precipitations with concentrated sodium salt solutions [7]. The products were recrystallized twice from water, and their purity was checked by elementary analysis. All other chemicals used were reagent grade.

Preparation and Measurement of Solutions (see also [1]). Protein stock solutions were prepared in water and kept in the dark at 0° to 5°C for periods not exceeding ten days. Concentrations were based on the weight of the protein. The pH of the solutions was kept at 7.15 ± 0.10, and the temperature of measurements was 25 ± 0.5°C. Sodium phosphate buffer concentrations were less than 4 percent of the other counter-ions. In calculating total counter-ion concentrations, 1 mM buffer was taken as

1.7 mM, which is the amount of negative charge contributed at this pH. Aqueous detergent solutions were usually added as the last component of the mixture. The order of addition of these components has no influence on the final results, as verified in several experiments.

Myoglobin-detergent interactions were followed by measuring the absorbance changes in the Soret band (395-410 nm) using a Cary Model 14 recording spectrophotometer. Green-complex formation was checked spectrally in the range 450 to 650 nm. The final green-complex spectrum was independent of the anion used. Absorption spectra were taken only for clear solutions, within 20 minutes to 3 hours after their preparation. No significant spectral changes occurred during this period of time. Distilled water was used as a reference solvent, since all the components, except the heme, did not show significant absorption in the Soret region. Extinction coefficients (ε_{mM}) are given in units of $mM^{-1} \cdot cm^{-1}$ based on total heme.

Precipitation was observed with LPI and LPT at high salt concentrations, and therefore the experimental range of investigation in these anions was limited. In the presence of NaBr concentrations larger than 0.25 M, the titrations were made with LPC. Under these conditions, the Br^-/Cl^- molar ratio is larger than 200, and the results were taken as those of LPB. Even in the presence of only 50 mM NaBr, no difference was observed in the titration curve obtained with either LPC or LPB (3 mM).

Reversibility of the myoglobin-LPC system has been demonstrated previously by applying dilution tests [1].

Critical micelle concentrations (CMC) of all four detergents were determined by conductometric titrations at low salt concentrations, and were extended into the high-concentration range of added salt by surface tension measurements, using the Wilhelmy plate method [13]. Comparison with published data [7] was satisfactory.

RESULTS AND DISCUSSION

Typical transition curves are shown in Fig. 1, which compares the effects of LPT, LPI and LPB in the absence of added electrolyte, and those of LPB and LPC in the presence of 48 mM NaBr and NaCl, respectively. All these

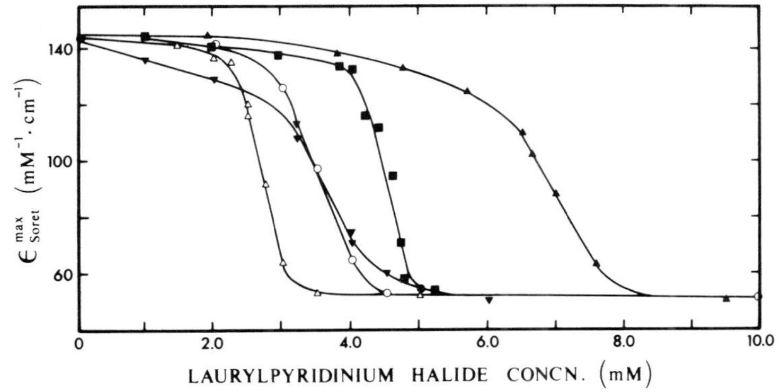

Fig. 1. Spectrophotometric titrations of ferrimyoglobin with laurylpyridinium halides below the CMC. The abscissa refers to the total detergent concentration. Total myoglobin, 1.0×10^{-5} M; pH 7.10; 25°C. (▼) LPT; (■) LPI; (▲) LPB; all without added electrolyte except 0.05 mM sodium phosphate; (Δ) LPB, with 48 mM NaBr; (o) LPC, with 48 mM NaCl.

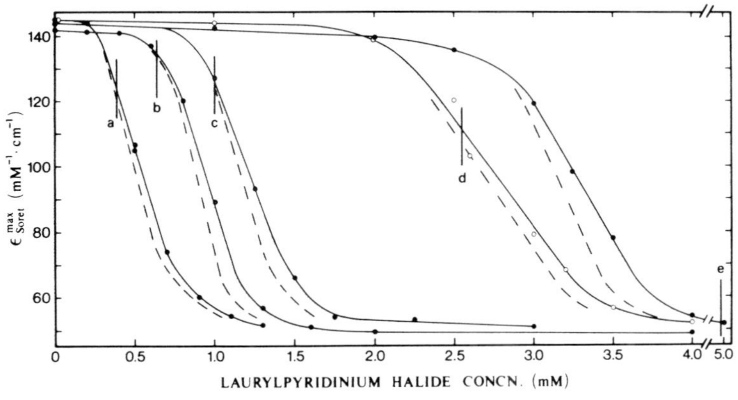

Fig. 2. Spectrophotometric titrations of ferrimyoglobin with LPB (●) and LPI (o). Total myoglobin, 1.0×10^{-5} M; pH 7.10; 25°C. The solid and dashed curves represent ε^{max}(Soret) *vs.* $[L]_t$ and $[L]_f$, respectively, according to Equation 3. The vertical bars show the CMC under the same conditions. (a) LPB, with 1.00 M NaBr and 20 mM sodium phosphate; (b) LPB, with 0.50 M NaBr and 20 mM sodium phosphate; (c) LPB, with 0.25 M NaBr and 10 mM sodium phosphate; (d) LPI, with 0.010 M NaI and 0.10 mM sodium phosphate; (e) LPB, with 0.030 M NaBr and 0.05 mM sodium phosphate.

transitions occur under conditions in which the detergent concentration is below its critical micelle concentration. Some other transition curves, which occur above or simultaneously with micelle formation, are shown in Fig. 2. The results were analyzed according to the following considerations, based on the assumption of an all-or-none process for the formation of the "green complex" (see [1]), which is formed by dissociation from the protein:

$$(1) \quad \text{Mb} + \nu\,\text{L} \rightleftharpoons \text{ApoMb-L}_{\tau} + \text{FP-L}_{(\nu-\tau)}$$

[Mb, ferrimyoglobin; L, laurylpyridinium halide; ApoMb-L_{τ}, complex between apomyoglobin and τ detergent molecules; FP-$L_{(\nu-\tau)}$, complex between ferriprotoporphyrin IX and $(\nu-\tau)$ detergent molecules; ν, apparent stoichiometric coefficient]. Application of the mass-action law to this equation (assuming unity for activity coefficients) results in:

$$(2) \quad [\text{L}]_f = \left[\frac{1}{K}\right]^{1/\nu} \cdot \left[\frac{(1-\alpha)^2}{\alpha}\right]^{1/\nu} \cdot [\text{Mb}]_t^{\,1/\nu}$$

where

$$(3) \quad [\text{L}]_f = [\text{L}]_t - \nu(1-\alpha)[\text{Mb}]_t$$

$[L]_f$ is the free detergent concentration; $[Mb]_t$ and $[L]_t$ are the total myoglobin and detergent concentrations, respectively; K is the apparent association constant, and α is the fraction of total protein in its unreacted form. α is the fractional change along the transition curves, whereby $\alpha = 1$ in the absence of detergent and $\alpha = 0$ when the transition is completed.

The average standard free-energy change of the reaction (ΔG^0) per mole detergent is:

$$(4) \quad \Delta G^0 = RT\ \ln\left[\frac{1}{K}\right]^{1/\nu} = RT\ \ln\{[\text{L}]_t(\alpha{=}0.38) - 0.62\nu[\text{Mb}]_t\} - \frac{RT}{\nu}\ \ln[\text{Mb}]_t$$

(Note that $\frac{(1-\alpha)^2}{\alpha} = 1$ when $\alpha = 0.38$). ν, which is necessary for the calculation of ΔG^0, has previously been derived from the dependence of the transition curves on the total myoglobin concentration, using a graphical linearization method which was described in a previous paper [1]. In the present study, all transition curves were measured at a single concentration, and ν was therefore derived from Equations 2 and 3 by applying them to two points on the curves, at $\alpha = 0.25$ and $\alpha = 0.75$. The result is:

$$(5) \quad \ln\left[\frac{[L]_t(\alpha{=}0.25) - 0.75\nu[Mb]_t}{[L]_t(\alpha{=}0.75) - 0.25\nu[Mb]_t}\right] = \frac{1}{\nu}\ln\left[\frac{(1-0.25)^2/0.25}{(1-0.75)^2/0.75}\right] = \frac{3.295}{\nu}$$

Denoting the left-hand side of Equation 5 by $\phi_1(\nu)$ and the right-hand side by $\phi_2(\nu)$, ν was solved graphically by plotting ϕ_1 and ϕ_2 as a function of ν and observing the point of intersection of the two curves. The results correspond well with those obtained by the previous method [1] for LPC: with 20 mM phosphate buffer, $\nu = 15$ (previously 14), and by addition of 500 mM NaCl, $\nu = 22$ (previously 25).

With ν values obtained in this manner, the free detergent concentration could be evaluated according to Equation 3 at each point on the transition curves. In Fig. 2, each transition is plotted also against $[L]_f$, which represents the concentration of detergent which is not bound to the protein. It is seen that in some cases $[L]_f$ exceeds the CMC at very early stages of green-complex formation. This excess detergent must therefore be in the form of micelles, as the concentration of free detergent molecules in solution cannot be appreciably greater than the CMC [10]. This means that the interaction of the detergent with the protein is a result of increasing the concentration of micelles, which could then be the interacting species.

The variation of ν and ΔG^0 with counter-ion concentration is shown in Fig. 3. The two parameters, ΔG^0 and ν, which characterize each transition curve, are related to its position and maximal slope along the detergent concentration axis. Under the present experimental conditions, the amount of bound detergent is small, $[L]_f$ is close to $[L]_t$, and ν measures the sharpness of the transition or its cooperativity, while ΔG^0 is mainly dependent on the detergent concentration needed to reach about half transition, namely $\alpha = 0.38$.

Fig. 4 summarizes all transition curves by plotting the values of $[L]_f$ at $\alpha = 0.50$ as a function of the total counter-ion concentration (see Materials and Methods). It also includes CMC values obtained under identical conditions, both in the presence and absence of protein. A general feature of Fig. 4 is that for each counter-ion, both the CMC and the

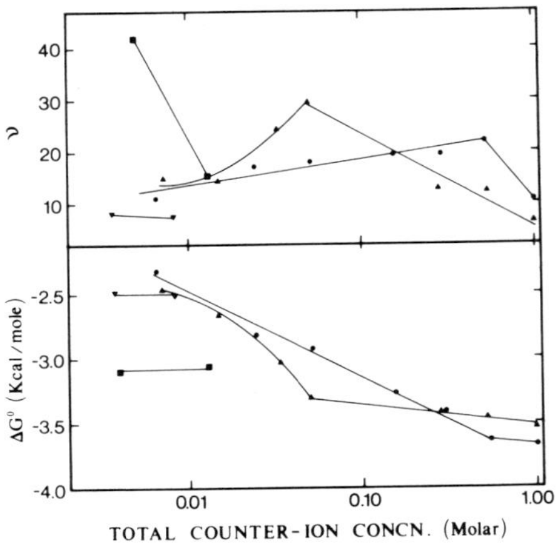

Fig. 3. ν and ΔG^0 for various counter-ions, *vs.* log total counter-ion concentration. (●) Cl^-; (▲) Br^-; (■) I^-; (▼) SCN^-.

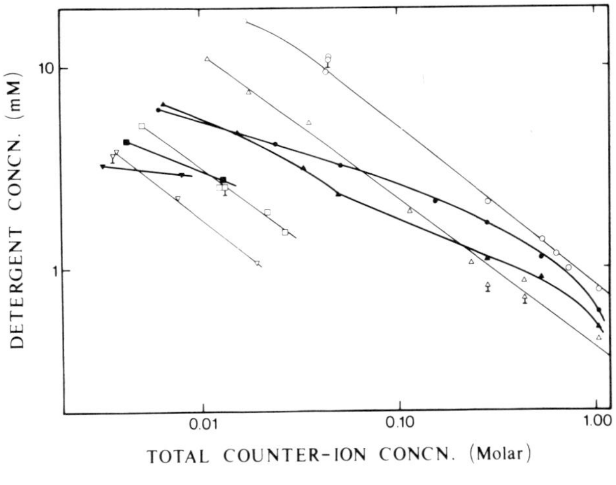

Fig. 4. Log-log plot of free detergent concentration at $\alpha = 0.5$ and CMC values *vs.* total counter-ion concentration. CMC values in the absence of protein: (○) Cl^-; (△) Br^-; (□) I^-; (▽) SCN^-. CMC values observed in the presence of 1.0×10^{-5} M myoglobin: (♀) Cl^-; (♀) Br^-; (♀) I^-; (♀) SCN^-. Free detergent concentrations at $\alpha = 0.5$: (●) Cl^-; (▲) Br^-; (■) I^-; (▼) SCN^-.

detergent concentration needed for green-complex formation are lowered by increasing the sodium halide concentration, though at different slopes. At a fixed counter-ion concentration, both are lowered in the series Cl^-, Br^-, I^-, SCN^-, again the magnitude being different for the two processes.

From Figs. 3 and 4 it may be concluded that for Cl^- concentrations up to about 500 mM, the protein-detergent interaction occurs before the free detergent concentration reaches the CMC, the sharpness of the transition (ν) increases and ΔG^0 decreases with increasing Cl^- concentration. From 500 to 1000 mM, the transition occurs close to the CMC, ν decreases and ΔG^0 seems to reach a constant value. Analogous phenomena are more pronounced in the case of Br^- for which, from about 50 mM on, the transition becomes less sharp, and the free detergent concentration present approaches and then exceeds the CMC.

For I^- and SCN^- the concentration range was limited experimentally. Even so, parts of the same phenomena are observable also for these ions. The effect of increasing ν with simultaneous decrease in ΔG^0 at the low concentration range of Cl^- and Br^- (Fig. 3) is analogous to that observed with micelles (*e.g.*, [7]). These effects are understandable if one takes into account the formation of micelles:

$$(6) \quad n\,L \rightleftharpoons Mc$$

where Mc is the micelle and n is the micelle number.

Consider a hypothetical case where protein interaction with monomeric detergent may be negligible; then equilibria (1) and (6) give:

$$(7) \quad Mb + \frac{\nu}{n}\,Mc \rightleftharpoons ApoMb\text{-}L_{\tau} + FP\text{-}L_{(\nu-\tau)}$$

which describes the interaction between detergent micelles and protein. Using Equation 7 instead of Equation 1, we obtain $\frac{\nu}{n}$, instead of ν in Equation 5. This could explain the decrease in the apparent ν shown in Fig. 3. (If one subtracts the CMC values in order to get the free micelle concentration for the use of Equation 5, still lower values are obtained, *e.g.*, $\frac{\nu}{n} = 2$, for 500 mM NaBr). Similarly, ΔG^0, calculated from Equation 4 is now related to the difference between the standard free-energy change of the micellization and that of the protein-detergent interaction. Due to the similar effects of added electrolyte on both processes, this difference is expected to remain approximately constant, as shown for ΔG^0 in Fig. 3.

A full analysis of the system requires, of course, a simultaneous solution of the equilibria (1) and (6). Such a solution may then be applied to experiments in which the micelle concentration may be measured simultaneously with the green-complex formation.

According to Fig. 3, ν appears to be a continuous function of the total counter-ion concentration. This may explain the absence of a noticeable change in the slope of the transition curves near the CMC (Fig. 2). Irrespective of the nature of the interaction products between the protein and the detergent above the CMC, these complexes may be formed kinetically either *via* the free detergent or the detergent micelles.

Micelle formation at concentrations lower than or near those required for protein-detergent complex formation was previously considered (*e.g.*, [5,14]). Recently, the dye-solubilization method was applied to the system cytochrome *c*-SDS [15] and the results showed that SDS micelles may be formed before or after the protein-detergent interaction. In other studies with SDS [16], dye solubilization indicated that SDS is attached to various proteins in the form of micelles. On the other hand, SDS binding to a number of proteins, including myoglobin, was considered to be dependent on the monomer detergent concentration [17], assuming that micelles do not bind to the proteins. It was concluded that the only effect exhibited by the ionic strength is to decrease the CMC, and hence the SDS monomer concentration.

The slopes of the CMC curves in Fig. 4 are proportional to the degree of association of counter-ions with the micelle. This effect is expected from equilibrium considerations [18]. Application of the same reasoning to the protein-detergent complex indicates that below the CMC a smaller fraction of the counter-ions is attached to each detergent ion associated with the protein, as compared with pure micelles.

The effect of counter-ions on micelle formation is in screening the electrostatic repulsive forces between charged groups on the micelle surface. The higher the counter-ion concentration and the smaller its distance of closest approach (*i.e.*, the larger are the polarizability and the unhydrated radius), the larger will be the effect on CMC lowering and increasing the micelle size [7,10-12,19].

If we assume that the positively-charged detergent molecules are attached to the protein by both hydrophobic and electrostatic interactions, then part of their charge will be neutralized by negative protein groups, and the effect of the counter-ions will be much less prominent than in the case of a micelle. At higher salt concentrations, additional interaction with detergent molecules may occur, leading to formation of micellar structures on the protein which may be similar to free micelles with respect to counter-ions. This consideration could explain the fact (see Fig. 4) that the difference between Cl^- and Br^- with respect to the protein-detergent interaction is observed only at high salt concentrations.

Further evidence will have to be presented for the possible existence of protein-detergent micelle complexes. This should include a quantitative comparison between protein-detergent interaction and detergent micelle formation, as well as an evaluation of possible conformational differences between the reactants and the products of these interactions.

ACKNOWLEDGEMENTS

The skillful technical assistance of Mr. P. Yanai is appreciated. This investigation was supported by a grant (No. 11 1772) from the Stiftung Volkswagenwerk.

REFERENCES

1. Yonath, J., Blauer, G.: Protein-detergent interactions. Properties and thermodynamic analysis of the system ferrimyoglobin-laurylpyridinium chloride. *Eur. J. Biochem. 41*, 163-170 (1974).
2. Niedick, B.: Über grüne Invertseifen-Chromoproteid-Komplexe. *Naturwissenschaften 45*, 163-164 (1958).
3. Blauer, G., Zvilichovsky, B.: Ultracentrifugation studies on the aggregation of ferriprotoporphyrin IX by electrolytes in aqueous alkaline medium. *Arch. Biochem. Biophys. 127*, 749-755 (1968).

4. Nakaya, K., Yamada, K., Onozawa, M., Nakamura, Y.: Interaction of cetylpyridinium chloride with oxymyoglobin. *Biochim. Biophys. Acta 251*, 7-13 (1971).

5. Tanford, C.: Hydrophobic free energy, micelle formation and the association of proteins with amphiphiles. *J. Mol. Biol. 67*, 59-74 (1972).

6. Osipow, L.I.: *Surface Chemistry*. Reinhold, New York, 1962, p. 167.

7. Ford, W.P.J., Ottewill, R.H., Parreira, H.C.: Light-scattering studies on dodecylpyridinium halides. *J. Colloid Interface Sci. 21*, 522-533 (1966).

8. Anacker, E.W.: Light scattering by cetylpyridinium chloride solutions. *J. Phys. Chem. 62*, 41-45 (1958).

9. Lange, H.: Über den Einfluss gewöhnlicher Elektrolyte auf die Mizellenbildung in Kolloidelektrolytlösungen. *Koll. Z. 121*, 66-71 (1951).

10. Debye, P.: Light scattering in soap solutions. *Ann. N.Y. Acad. Sci. 51*, 575-592 (1949).

11. Goddard, E.D., Harva, O., Jones, T.G.: The effect of univalent cations on the critical micelle concentration of sodium dodecyl sulphate. *Trans. Faraday Soc. 49*, 980-984 (1953).

12. Mukerjee, P., Mysels, K.J., Kapauan, P.: Counter ion specificity in the formation of ionic micelles-size, hydration and hydrophobic bonding effects. *J. Phys. Chem. 71*, 4166-4175 (1967).

13. Padday, J.F.: Surface tension. Part I. The theory of surface tension. In: Surface and Colloid Science (E. Matijević, Ed.). Wiley-Interscience, New York, Vol. 1, 1969, pp. 39-99.

14. Tanford, C.: Protein denaturation. *Advan. Protein Chem. 23*, 121-282 (1968).

15. Burkhard, R.K., Stolzenberg, G.E.: Interaction between sodium dodecyl sulfate and ferricytochrome *c*. *Biochemistry 11*, 1672-1677 (1972).

16. Pitt-Rivers, R., Impiombato, F.S.A.: The binding of sodium dodecyl sulfate to various proteins. *Biochem. J. 109*, 825-830 (1968).

17. Reynolds, J.A., Tanford, C.: Binding of dodecyl sulfate to proteins at high binding ratios. *Proc. Natl. Acad. Sci. U.S. 66*, 1002-1007 (1970).

18. Phillips, J.N.: The energetics of micelle formation. *Trans. Faraday Soc. 51*, 561-569 (1955).

19. Emerson, M.F., Holtzer, A.: On the ionic strength dependence of micelle number. *J. Phys. Chem. 71*, 1898-1907 (1961).

Received 12 July 1974

DISCUSSION

Steinhardt: Since the binding of these cations results in separation of the heme from the apoprotein, and the half-life of the apoprotein at room temperature is very short, the complexes studied may be with unfolded protein, just as with the anionic detergents.

Yonath: We do not know to what extent the protein in the complex is unfolded. In contrast to many studies with SDS we started with native protein and not with a reduced and guanidinium-treated protein.

Jaenicke: What is the size of the laurylpyridinium chloride micelle? Considering your ν values it would be interesting to know something about the stoichiometry of the Mb-detergent complex.

Yonath: Laurylpyridinium micelles vary in size from $\nu \approx 20$ at low Cl^- concentrations to $\nu \approx 100$ with I^-. For LPB, when about 30 molecules are attached to the protein, at the same concentration as the CMC, the micelle number is also about 30, which means that, on the average, only one micelle reacts with one protein molecule.

Jaenicke: If there are just one or two micelles per Mb molecule, do you expect the micelle to cluster on the surface of the molecule or could there be tubular structures as mentioned in the discussion of this morning's session?

Yonath: We do not have any evidence for tubular structures.

von Hippel: What is the micelle number when myoglobin is incorporated? Is it changed from the "without myoglobin" value? How is the myoglobin incorporated?

Yonath: ν values observed are in the same range as the micelle numbers. We do not have evidence of myoglobin incorporation into the micelle. We rather picture a spherical micelle or a part of it attached to the protein surface, and, in addition, possibly some detergent molecules attached at specific sites.

Hinz: I have some problems to understand the meaning of the ΔG^o values; one would have to denote them standard quantities, however, it did not become clear from your presentation to which standard state these ΔG^o values refer.

Yonath: ΔG^o was derived from the apparent equilibrium constant of Eq. 1., by using the relation $\Delta G^o = -\frac{RT}{\nu} \ln K$. The reference state is a free detergent molecule in a solution of unit activity.

Veeger: You mentioned in your discussion with Dr. Steinhardt, that in order to obtain the magic number of SDS binding one needs to denature the protein in guanidine-HCl plus reduce the S-S bridges. I disagree with this view. The experiments I described this morning in the discussion with Dr. Steinhardt were carried out with native protein containing an S-S bridge and still the magic number is almost reached. We therefore believe that it can be reached in the total denaturation of the protein by SDS. As far as I understand the main difference between SDS and laurylpyridinium binding is that in the latter case the detergent is bound in some micellar form.

Influence of Ions on the Kinetics of Cooperative Conformational Changes in Globular Proteins

Fritz M. Pohl

Changes of the tertiary or quaternary structure of proteins play an essential role in biological processes. In contrast to the importance of such events, relatively few thermodynamic and kinetic studies have been reported, where reversible conformational changes of globular proteins have been observed directly with methods which are sensitive to the spatial arrangement of the peptide chain in solution.

Studies of the unfolding and refolding reactions of simple enzymes, like chymotrypsin, trypsin, RNase, lysozyme etc., provide very convenient model systems for cooperative conformational changes of single subunit proteins. With a proper choice of conditions the unfolding reactions are completely reversible. This allows a detailed investigation of the thermodynamic and kinetic parameters as a function of external variables (1-5).

These "simple" proteins have the additional advantage that the folded conformations in crystals have been solved by X-ray diffraction. This offers the possibility of finding correlations between thermodynamic (and kinetic) properties of the protein in solution and the particular conformation of the folded peptide chain.

This report deals with the influence of different ions on the rate constants of the unfolding and refolding of such proteins and discusses a few aspects of the interaction of ions with different conformations of proteins.

EVALUATION OF RATE CONSTANTS

The kinetics of unfolding of proteins like chymotrypsin, trypsin, RNase, lysozyme etc. are observed most conveniently by the change of the absorbance of aromatic residues around 280 nm after a sudden perturbation of the temperature of the protein solution. A thermostated microcell allows to increase or decrease the temperature of the solution by a preselected amount within a few seconds (6). Most of the absorbance change after a temperature jump within the transition region follows an exponential decay curve, characterized by a single relaxation time τ . The amplitude of the relaxation curve contains the thermodynamic information.

Fig. 1 shows the basic experimental information obtained in this way. Fig. 1a gives the relative equilibrium concentration of unfolded molecules as a function of temperature, using the data for chymotrypsinogen B at acid pH as an example. The typical temperature dependence of the relaxation is shown in Fig. 1b. The over-all rate constants for the unfolding $\vec{k}$ and the refolding reaction $\overleftarrow{k}$ are calculated from

$$\vec{k} = \Theta/\tau \qquad \text{and} \qquad \overleftarrow{k} = (1 - \Theta)/\tau \qquad (1)$$

with Θ beeing the degree of transition. These expressions indicate that a measurement of <u>both</u> rate constants with reasonable accuracy is only possible within the transition region. Although many different elementary steps occur in such unfolding reactions the concentration of intermediate states appears to be very low. Therefore, the description by an "all-or-none" transition between two states of the protein is justified and the use of two 'steady-state' rate constants adequate. The rate constants depend on external variables as the temperature or the concentration of ions. Relatively simple expressions for the effect of molecules on the rate constants are obtained by using the following assumptions : the binding constant K_a of molecules in state A is the same for all n_a binding sites, but different to K_b, the binding constant in the unfolded state with n_b sites. With a slow rate of interconversion compared to the binding of ligands, the rate constants are given by

$$\vec{k} = \vec{k}_o \, (1 + E.\vec{d}/K_a)^{n_a} \, / \, (1 + E/K_a)^{n_a}$$
$$\overleftarrow{k} = \overleftarrow{k}_o \, (1 + E.\overleftarrow{d}/K_b)^{n_b} \, / \, (1 + E/K_b)^{n_b} \qquad (2)$$

(E: free effector concentration; $\vec{k}_o$ and $\overleftarrow{k}_o$: rate constants of interconversion in the absence of effector, and $\vec{d}$, $\overleftarrow{d}$: ratio of rate constants upon binding of one more ligand to state A or B.) The formalism of the allosteric model is obtained when n_a equals n_b.

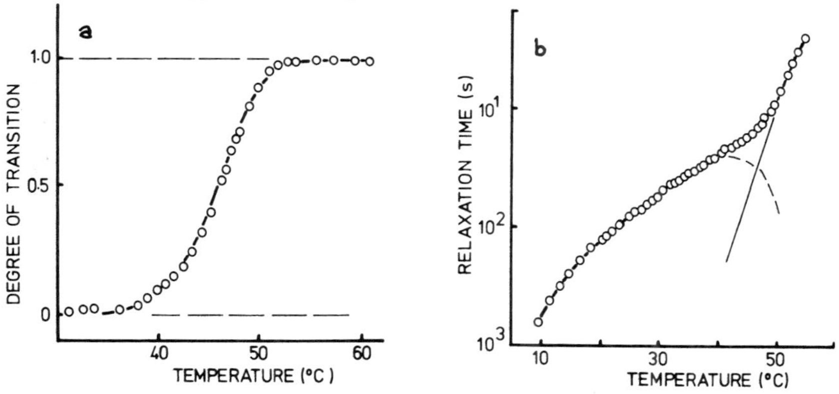

Fig. 1: Reversible unfolding of chymotrypsinogen B, kindly provided by Dr.L.B.Smillie, at pH = 2.0 as function of the temperature.
a) Degree of transition Θ, the relative concentration of unfolded molecules, determined from the amplitudes of relaxation curves after temperature-jumps of different size.
b) Relaxation time (ooo) and rate of unfolding (——) and rate of refolding (---) as function of the temperature.

EXPERIMENTAL RESULTS USING DIFFERENT IONS AS EFFECTORS

Equilibrium measurements allow no decision as to whether effector molecules influence the folded or the unfolded conformation of the protein. This question can be decided to some extent on the basis of kinetic measurements as indicated by equ. 2.

Protons

The unfolding of globular proteins at extreme pH-values is often assumed to be due to electrostatic effects in such a way that the high net charge of the protein leads to the repulsion between charged groups and thereby favours an extended conformation as compared to the folded one. Since no exact theory exists which allows to estimate the magnitude of the electrostatic interactions within protein molecules, one has to rely on experimental results.

Fig.2 shows the effect of increasing proton concentration on the rate constants $\vec{k}$ and $\overleftarrow{k}$. The unfolding rate increases with the third power of the proton concentration while the refolding rate shows only a very weak dependence. (A direct measurment of both rate constants is only possible within a relative narrow pH-range. The extrapolation according to equ. 2 is not very accurate and an unfolding rate of $10^{-5}s^{-1}$ at neutral pH can only be regarded as a very approximate one).

Experiments designed to evaluate the contribution of the increase of the net charge of the protein show that the electrostatic repulsion between the positive charges is only in a marginal way responsible for the observed behaviour. If 13 of the 16 carboxyl groups of chymotrypsinogen A are chemically modified the pH-dependence of the unfolding remains the same (Fig.2b). The protonation of three peculiar carboxyl groups is responsible for the destabilization of the protein by acids (8).

The analysis of the kinetics of unfolding according to equ.2 gives, with $n_a = n_b = 3$, a very low value of $pK_a = 1.7$, while pK_b for the unfolded state is about 3.7. Therefore one expects that at pH = 2 - 3 these particular carboxyl groups to be unprotonated in the folded state but protonated in the unfolded molecule. The unfolding of the protein should be

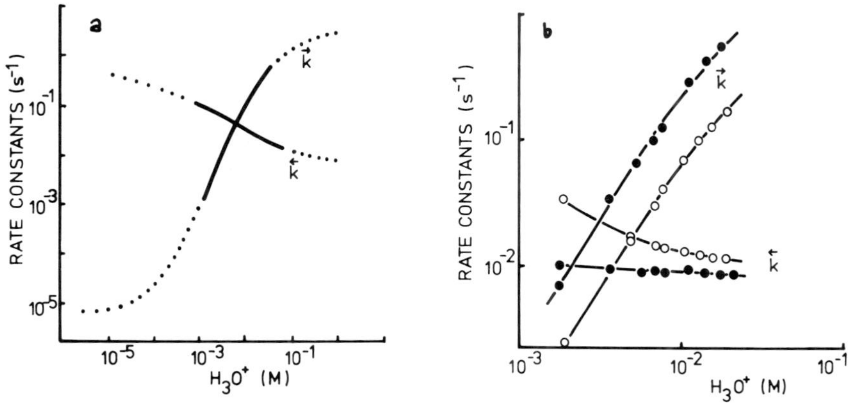

Fig. 2: pH-dependence of the rate of unfolding $\vec{k}$ and of refolding $\overleftarrow{k}$ of chymotrypsinogen A at 40°C.

a) Extrapolation of the experimental results (solid line) according to equ.2
b) Influence of the chemical modification of 13 carboxyl groups on the pH-dependence (●-●-●) as compared to the unmodified protein (o-o-o). The modified protein was kindly provided by Dr.M.Ladzdunski.

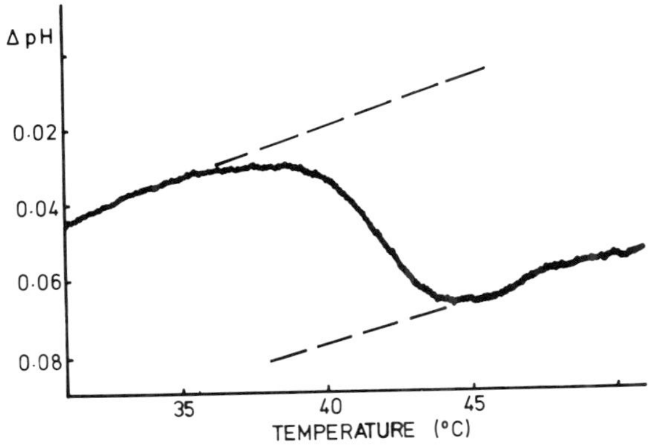

Fig. 3: Uptake of protons upon the unfolding of chymotrypsinogen A at pH=2.1.
The protein solution (10 mg/ml) was heated at a rate of 1°C/min and the pH measured with a combined microelectrode. The change of the pH in the transition region of the protein corresponds to the additional binding of about 2 protons to the unfolded form.

measurable by the uptake of two or three protons per protein molecule. Fig. 3 shows that this is indeed the case. Furthermore, the kinetics of the proton uptake or release show the same relaxation time as the change of absorbance after a temperature jump, in accordance with the model.

The interpretation that these peculiar carboxyl groups are involved in the formation of internal ionic bonds is supported by the X-ray structure. The free-energy change for one such ionic bond is about 3 kcal/mole and the protonation leads to an accelaration of the unfolding rate by a factor of 100 - 1000 per bond!

Salts

The stabilisation of the folded conformation by ionic bonds provides also an explanation for the peculiar salt dependence of the unfolding rate constant (Fig. 4a). The high positive netcharge of the protein should give rise to a monotonic decrease of k and/or an increase of k with increasing counterion concentration. Such a behaviour is observed for the refolding rate, but the unfolding rate shows a maximum around 0.2 M KCl. This indicates that two opposing effects of the ions are experienced by the folded molecule. At low ionic strength the counter ions appear to destabilize the ionic

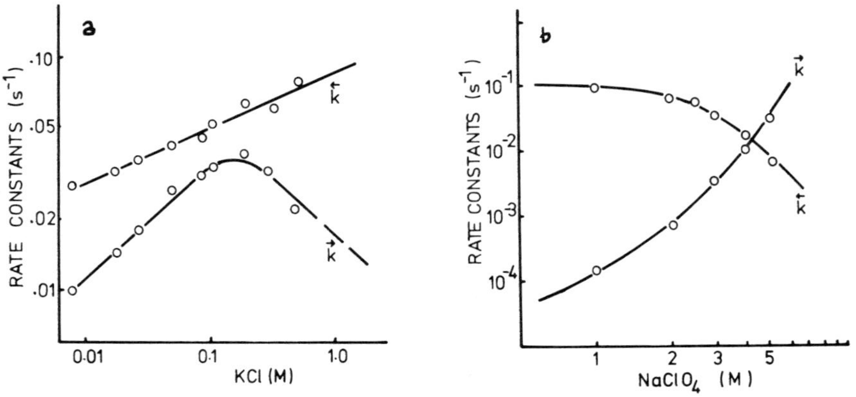

Fig. 4: Influence of salts on the rate constants of unfolding $\vec{k}$ and refolding at constant pH and temperature.

a) Variation of the kinetics of unfolding of trypsin at pH = 2.1 with increasing concentrations of KCl at 30°C.
b) Kinetics of the unfolding of pancreatic RNase as function of different sodium perchlorate concentrations at pH = 6.8 and 37°C.

bonds much more, overcoming the stabilisation of the folded molecule due to the shielding of the high net charge.

Different types of electrostatic interactions within a protein molecule may be deduced from such observations.

Sodium perchlorate

Very high concentrations of a number of neutral salts destabilize the native conformation of biopolymers (3). Fig. 4b shows that sodium perchlorate acts mainly via the acceleration of the unfolding rate. The rate of refolding is not very much affected, especially at low temperatures. The unfolding rate increases with about the 10th power of the perchlorate concentration and resembles rather closely the effect of urea. In both cases, the effector acts by accelerating mainly the unfolding rate, although no common structural basis can be seen. One possible explanation is that both molecules disturb the structure of the water around the folded protein in such a way that the mobility of the peptide chain increases considerably.

A reverse effect is observed with guanidine-HCl where the destabilisation is accompained by a decrease of the refolding rate (Fig.5).

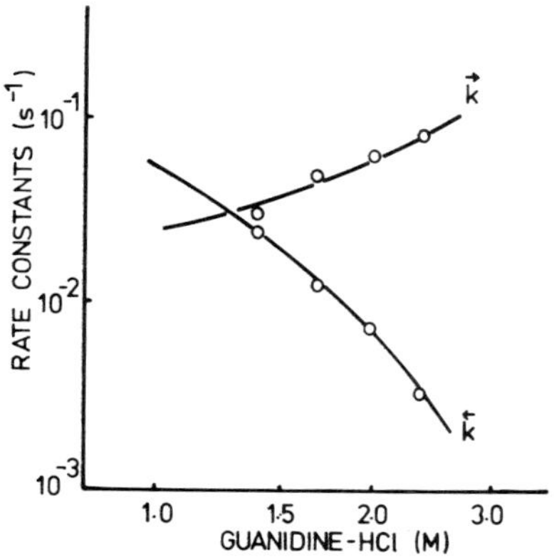

Fig. 5: Rate constants for the unfolding and refolding of lysozyme at 40°C as a function of the concentration of guanidine-HCl (pH = 6.8).

POSSIBLE IMPLICATIONS

Different ions or molecules like urea or detergents can be classified according to their effect on the rate constants for the unfolding or refolding reaction. Protons, perchlorate or urea, for example, accelerate in a cooperative way the unfolding rate. Other molecules, like guanidine-HCl (Fig. 5) give rise to a cooperative decrease of the refolding rate. (In such a case the unfolding can become a very slow process at low temperatures.)

The interpretation of the kinetics with respect to the model of differential binding to the two states of the protein, as given by equ. 2, is somewhat ambiguous. Even for such a simple model at least six parameters have to be evaluated and most of them can not be measured separately. Curve-fitting procedures give sets of parameters, but their interpretation in molecular terms of number of binding sites or binding constants of the ligands to different conformations is rather arbitrary. Only in the case of protons, as effectors of internal ionic bonds, a correlation between structural and thermodynamic or kinetic properties is possible.

The stabilisation of particular conformations of proteins by using ionic bonds may be an important mechanism in other systems as well. Ionic bonds in proteins can also involve acidic side chains and divalent cations, for example, Mg^{2+} or Ca^{2+}. It is known that the concentration of these ions plays a crucial role in the action of muscles or the nervous system.

Using such simple effector molecules offers the advantage that more than one binding site can be realized within a single peptide chain. This can lead to a non-linear, co-operative response of the protein conformation - and therefore its biological function - to a change of the ion concentration even by using single subunit proteins. (The free concentration of divalent ions may in turn be regulated via a complex formation with ATP, for example). The evolutionary restriction on the formation of binding sites for such simple effectors appears to be much less severe as compared to those of more complicated molecules (e.g. M.Bühner, this volume).

Other "ions", for example nucleic acids, are also able to change the conformation of a protein upon binding. A recent example is the conformational change of the gene-5 protein from bacteriophage fd upon the interaction with short oligonucleotides (9). This change results in a drastically enhanced association of the protein as deduced from cross-linking experiments. Such an enhanced association offers one way for bringing the ends of a single-stranded nucleic acid molecule in close proximity, in order to form a covalently closed circular molecule.

AKNOWLEDGEMENT

I thank Drs. L.B.Smillie and M.Ladzdunski for the gift of proteins and the Deutsche Forschungsgemeinschaft and the Fonds der Chemischen Industrie for support.

REFERENCES

1) Lumry,R., Biltonen,L.: Thermodynamic and kinetic aspects of protein conformation in relation to physiological function. Biological Macromolecules (ed. Timascheff,S.N., Fasman,G.D.;Marcel Dekker, New York) Vol. 2,65-212 (1969).

2) Brandts,J.F.: Conformational transitions of proteins in water and aqueous mixtures. ibid. 2, 213-290 (1969).

3) von Hippel,P.H., Schleich,T.: The effects of neutral salts on the structure and conformational stability of macromolecules in solution. ibid. 2, 417-574 (1969).

4) Tanford,C.: Protein denaturation. Advan. Protein Chem. 23, 121-282 (1969).

5) Pohl,F.M.: Cooperative conformational changes in globular proteins. Angew. Chem. Int. Ed. 11, 884-906 (1972).

6) Pohl,F.M.: Einfache Temperatursprung-Methode im Sekunden- bis Stundenbereich und die reversible Denaturierung von Chymotrypsin. Eur. J. Biochem. 4, 373-377 (1968).

7) Monod, J., Wyman, J., Changeux, J.-P.: On the nature of allosteric transitions: A plausible model. J. Mol. Biol. 12, 88-118 (1965).

8) Fersht, A.R., Sperling, J.: The charge relay system in chymotrypsin and chymotrypsinogen. J. Mol. Biol. 74, 137-149 (1973)

9) Rasched, I., Pohl, F.M.: Oligonucleotides and the quaternary structure of gene-5 protein from filamenteous bacteriophage. FEBS Letters 46, 115-118 (1974).

Received July 26, 1974

DISCUSSION

Steinhardt: I agree that the electrostatic repulsion model for protein denaturation is very limited. It has been demonstrated for some time that with heme proteins, acid denaturation occurs because there are more binding sites on the denatured than on the native protein. These binding sites have been identified - in myoglobin and both horse and human hemoglobin - as interior histidine groups. These proteins denature as H^+ ions are added for the same reason that serum albumin unfolds as SDS ions are added. Obviously, I agree with your views.

Pohl: I have indicated in equ. 2 a different number of binding sites for protons to the two states of the protein and if the pK_a of the groups in the folded conformation is much lower than the pH, where the measurements are done, this is an adequate description. But I prefer to use the formalism which involves equal number of binding sites, but differences in the pK for two reasons: One is that this model explains the kinetic results obtained at $pH \sim pK_a$ and the other, that these peculiar groups may become protonated without a corresponding unfolding reaction at low pH and temperature, and therefore, explain some of the optical changes observed.

Veeger: Have the data of your Fig. 1b been analyzed in terms of $\Delta S^{\ddagger}$ and $\Delta \overleftarrow{S}^{\ddagger}$, thus giving information about the magnitudes of these values and constancy?

Pohl: The entropy change will only be constant if there is no change in the specific heat of the reaction. These unfolding and refolding reactions involve a rather large change in the specific heat, as indicated also by non-linear Arrhenius plots. The unfolding reaction indicates a considerable increase in entropy, while the reverse reaction is governed by the loss of entropy. How far these values are due to the change in conformational freedom or to the change in the interaction with the solvent is difficult to estimate at present.

von Hippel: I would be very surprised if ion-pairs inside proteins play a major role in stabilizing protein structure. It can be shown that transfer of an ion-pair from high to low dielectric environment is unfavorable, and I believe X-ray results indicate internal ion-pairs are rare.

Pohl: Apparently, ion-pairs do play a major role in the stabilization of proteinases, like chymotrypsin or trypsin, where X-ray studies showed two ion-pairs in the interior of the protein. The formation of an ion-pair from the ions is certainly facilitated in a medium of low dielectric constant. One should, therefore, not only use the transfer data of an ion-pair itself, but also its formation from the hydrated ions. These processes involve rather high-energy contributions of different sign, and I would be very cautious in making predictions. But I agree that in other cases ion-pairs near the surface of the protein molecule may play the more important role.

Sund: If you get cross-linking of the gene-5-protein after binding to DNS, a closed packing and, therefore, some protein-protein interaction has to be assumed. Do you have information about the dissociation constants for the protein-DNS interaction and do they differ if you use a tetranucleotide instead of DNA?

Pohl: It appears from the titration data that a slightly higher protein concentration is necessary in the case of short oligomers as compared to polynucleotides, which may indicate a somewhat weaker interaction of the gene-5-protein with oligonucleotides. But no "hard" numbers have been obtained up to now with respect to your question.

Riesner: Do you have to assume a conformational transition in the gene-5-protein for polymerization in presence of the oligonucleotides? Charge neutralization due to the oligonucleotides should have the same effect.

Pohl: If the polymerization would be due only to the decrease in the electrostatic interaction, one expects to see an increased association of the protein alone with increasing salt concentration. But the reverse effect is observed. There may be other explanations for the increased association in the presence of the ligand, but a conformation change appears to be the simplest mechanism. It is also supported by changes of the circular-dichroism spectrum of the protein.

von Hippel: I was interested in your gene-5-polymerization mediated by oligomer binding. A student in my lab named Ray Kelly has found a very similar result with the "melting-protein" T_4 gene-32 protein.

Neutral Salt Effects on the Conformational Stability of Biological Macromolecules*

Peter H. von Hippel

In this lecture we will consider some aspects of the interactions involved in setting-up and maintaining the conformations of proteins and protein complexes, and the mechanisms by which these interactions are perturbed by various solvent additives, particularly, the neutral salts. Many of the experiments and ideas presented here have been summarized previously by our group (von Hippel & Schleich, 1969a,b; and especially von Hippel & Hamabata, 1973), and the reader is referred to these reviews, as well as the original literature, for further information, descriptions of the relevant results of others, and full experimental details.

I. CONFORMATIONAL STABILITY AND THE THERMODYNAMICS OF MACROMOLECULAR TRANSITIONS

Proteins consist basically of one or more linear (copolymer) polypeptide chains which range from ∿ 50 to 1000 residues in length, and consist of ∿ 20 chemically distinct types of monomers (amino acids) linked through peptide bonds in a genetically determined order. Obviously, within the limits set by amino acid sequence and polypeptide backbone stereochemistry, even the smallest chains can, in principle, adopt an enormous number of conformations. Nevertheless, in dilute aqueous salt solutions at neutral pH and moderate temperature, polypeptide chains of a particular sequence are generally arranged in space in a single structure (or, in some cases, a small number of closely related structures). Furthermore,

* These studies and the preparation of this manuscript were supported by USPHS Research Grants GM-15792 and GM-15423, and by a John Simon Guggenheim Memorial Fellowship. Most of the research discussed which derives from our laboratory was carried out in collaboration with Drs. Thomas Schleich, Alberto Hamabata and David Talbot, Mr. Kwok-Ying Wong and Gary Cook, and Mrs. Virginia Peticolas, Lisa Schack and Susan Chang, as detailed in the papers to which reference is made. Their major contributions to the ideas and results presented herein are gratefully acknowledged.

this structure almost certainly represents an equilibrium (free-energy minimum) state, since it has been shown for many single-chain (and some subunit) proteins that they can be reduced to random coils by heating or other types of solvent perturbation, and then "reannealed" to the native conformation by an appropriately programmed restoration of the original solvent environment.

This view of a specifically folded protein in aqueous solution as an equilibrium entity, with total free energy equal to the algebraic sum of the free energies of the various residue-residue and residue-solvent contacts, indicates that the positioning of the various chemically different residues within the molecular domain represents a balance of opposing forces. Obviously, such a structure requires a stable "inside", containing some (largely hydrophobic) residues which (thermodynamically) "prefer" residue-residue over residue-solvent contacts. Equally crucial is the existence of a stable "outside" surface of the protein, which must be largely populated by hydrophilic residues "preferring" residue-solvent contacts. A native protein in solution can thus be viewed as a microphase characterized by a stable inside and a stable outside. How stable is the "native" conformation of a protein; and what forces are involved in maintaining it?

As Kauzmann (1959) and Tanford (1962) pointed out some years ago, the major "force" destabilizing a specifically-folded native protein (relative to the random-coil conformation) is configurational entropy; the additional degrees of configurational freedom available to each peptide group in the random coil favor this form by $\sim$ 1.2 kcal/mole residue. Therefore more than 1.2 kcal of stabilizing interactions (per mole residue) must be present in the folded protein to offset the free-energy advantage of the unfolded state. Three possibly stabilizing types of interaction are generally considered: (i) Intramolecular charge-charge interactions (salt linkages): X-ray diffraction studies of globular proteins show these occur almost entirely on the surface of proteins and, indeed, theoretical calculations show that the transfer of charged groups (even as oppositely charged pairs) from a high (solvent) to a lower (protein interior) dielectric constant environment is thermodynamically unfavorable. (ii) Intramolecular hydrogen-bonding: recent calculations and measurements suggest that the net free energy gain involved in breaking hydrogen bonds with solvent and transferring paired donor-acceptor hydrogen bonding groups to the interior of a protein is small, if not zero.† Clearly, intra-chain hydrogen-bonding is primarily responsible for the conformational specificity of the protein interior, i.e., it is thermodynamically very unfavorable to transfer hydrogen bonding groups to the (anhydrous) inside in a non-hydrogen-bonded configuration, but the above reasoning and results suggest that the appropriately-paired intramolecular bonds contribute relatively little in the way of favorable interactions to the overall stability of the folded protein. (iii) Hydrophobic bonding: we define these "bonds" as the thermodynamic tendency of nonpolar groups to be removed from water. Both the "process of elimination" and many model experiments implicate these bonds as the main interactions stabilizing folded protein conformations.

For most proteins, the free energy of the destabilizing interactions, and thus also those of the stabilizing interactions, are quite large. However, the net free energy stabilizing the folded form is often very small. The delicacy of this balance of forces is easily demonstrated by the observation that small environmental changes are often capable of "pushing" the protein across a phase transition boundary, with the resultant formation of a totally different equilibrium structure. For example, the addition of ammonium sulfate or similar additives to the solvent increases the stability of residue-residue contacts relative to residue-solvent interactions, converting some previously stable surface residues into species for which the free energy balance now favors residue-residue interactions. Aggregation of the native protein results; this is the molecular basis of the "salting-out" procedures commonly used in isolating proteins. Other agents (urea, sodium perchlorate, etc.) have the opposite effect, stabilizing residue-solvent contacts relative to residue-residue interactions. As a consequence, the previously stable

---

† We are currently (Cook & von Hippel, in preparation) attempting to measure this free energy of transfer directly in model micellar systems containing "internal" hydrogen bonding groups.

inside residues now "prefer" outside positions and the protein is converted into a fully solvated random coil.

These aggregation or unfolding processes are highly cooperative, going to completion over a narrow range of additive concentration. However, these processes reflect progressive (generally linear in additive concentration) changes in the free energy of transfer of residues from the non-aqueous interior of the protein to the aqueous surround. This can be easily demonstrated by varying two environmental parameters simultaneously. For example, we often examine the stability of macromolecules as a function of temperature in the presence of increasing concentrations of solvent additive. Fig. 1 shows such a series of melting studies for native collagen molecules in the presence of various concentrations of the destabilizing neutral salt, $CaCl_2$ (von Hippel & Wong, 1963). [Here the transition is followed by changes in optical rotation and involves the conversion of the triple-helical native collagen molecule, characterized by an unsolvated interior, to three separate fully solvated random-coil chains of gelatin.] Obviously, the melting temperature (mid-point of the melting profile) of this macromolecule is depressed by increasing the concentration of $CaCl_2$ in the solvent, and indeed, a series of such experiments show that a linear (empirical) relationship exists between melting temperature and concentration of added salt:

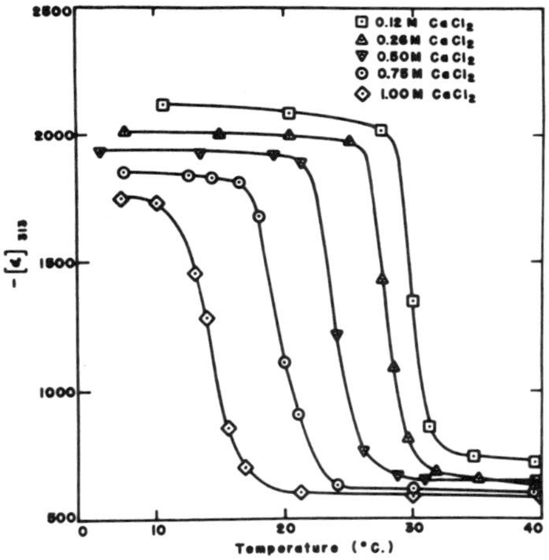

Figure 1

$$T_m - T_m^\circ = K_m c_s \tag{1}$$

where T_m° and T_m are the melting temperatures of the macromolecule in pure solvent, and in aqueous $CaCl_2$ solution of concentration c_s, respectively; and K_m (the slope of a plot of T_m versus c_s) represents the molar effectiveness of this salt in lowering T_m.

Such approximately linear relations between salt concentration and T_m apply to the effects of a number of salt solutions on the stability of collagen; some salts are characterized by positive values of K_m and thus are conformation stabilizers, while others show negative K_m values and thus are destabilizers. It can easily be shown (see von Hippel & Schleich, 1969a,b, for details) that the change in the T_m of the macromolecule is directly proportional to the change $[\Delta(\Delta G°_{conf})]$ in free energy of the native (relative to the random coil) conformation as a result of the transfer of the molecule from salt concentration $c_{s,1}$ to concentration $c_{s,2}$. That is:

$$\Delta(\Delta G°_{conf}) = A\, K_m\, (c_{s,2} - c_{s,1}) \qquad (2)$$

where A is a proportionality constant.

Studies of macromolecular systems as diverse as ribonuclease (von Hippel & Wong, 1965) and DNA (Hamaguchi & Geiduschek, 1962) showed that these effects are not unique to collagen. Investigation of the effects of a variety of salts on a number of macromolecular melting transitions has revealed that all obey eq. (1) quite closely, and yield values of K_m which vary widely from salt to salt, but, to a first approximation, are independent of the macromolecule under study. The effect of a series of neutral salts, ranked in order of molar effectiveness (K_m) as stabilizers or destabilizers of folding $\rightleftarrows$ unfolding transitions (as reflected in T_m alterations) for a number of different macromolecules, are summarized in Fig. 2. As has been described in detail elsewhere (von Hippel & Schleich, 1969a,b), this generality of the effects of neutral salts goes beyond intramolecular folding $\rightleftarrows$ unfolding transitions, applying equally to processes such as macromolecular association $\rightleftarrows$ dissociation and solubility equilibria. Analysis of the molecular details of these processes show that they all involve a net transfer of residues from an unsolvated (interior) to a solvated (exterior) environment.

Several other facets of these macromolecule-neutral salt effects are also brought out by implication in Fig. 2 (for details and documentation,

	← Helix / Native / Salting-out — Coil / Denatured / Salting-in →
Collagen – Gelatin	$SO_4^{=} < CH_3COO^- < Cl^- < Br^- < NO_3^- < ClO_4^- < I^- < CNS^-$
	$(CH_3)_4N^+ < NH_4^+ < Rb^+, K^+, Na^+, Cs^+ < Li^+ < Mg^{++} < Ca^{++} < Ba^{++}$
	$(CH_3)_4N^+ < (C_2H_5)_4N^+ \ll (C_3H_7)_4N^+, (C_4H_9)_4N^+$
Ribonuclease	$SO_4^{=} < CH_3COO^- < Cl^- < Br^- < ClO_4^- < CNS^-$
	$(CH_3)_4N^+, NH_4^+, K^+, Na^+ < Li^+ < Ca^{++}$
	$(CH_3)_4N^+ < (C_2H_5)_4N^+ < (C_3H_7)_4N^+ < (C_4H_9)_4N^+$
DNA	$Cl^-, Br^- < CH_3COO^- < I^- < ClO_4^- < CNS^-$
	$(CH_3)_4N^+ < K^+ < Na^+ < Li^+$
Polyvinylmethyl-oxazolidinone (Cloud Point)	$SO_4^{=} < CO_3^{=} < F^- < Cl^- < Br^- < ClO_4^- < SCN^-$
	$Na^+, K^+ < NH_4^+ < Li^+$

Figure 2

see von Hippel & Schleich, 1969a,b): (i) Cations and anions (and also uncharged perturbants) are additively effective in altering macromolecular stability. It is this fact which makes it possible to organize Fig. 2 into single-ion rankings. (ii) In order of effectiveness as salting-out agents (e.g., as reflected in Fig. 2 as changes in the solubility or "cloud-point" of the synthetic polymer polyvinylmethyloxazolidinone), the anions and cations both follow rather closely the classical series first demonstrated by Hofmeister for the salting-out of englobulins. These series have also been shown to apply to the salting-out of a number of simple nonelectrolytes from aqueous solution (see Long & McDevit, 1952). (iii) The tetraalkylammonium ions, and other perturbants containing hydrocarbon chains, increase in molar destabilizing effectiveness with increasing length of alkyl side-chains. These results, together with many other studies, demonstrate clearly that the various ions have specific and similar effects on the stability of a wide variety of macromolecular conformations and complexes. How then might the problem be simplified in an effort to get at molecular mechanisms?

II. MODEL TRANSFER EXPERIMENTS

In principle, we would like to know the unitary free energy for the transfer of all the various types of residues occurring in proteins from an anhydrous environment representative of the macromolecular interior to the solvent environment variously modified by stabilizing and destabilizing additives. For convenience, these model transfer experiments can be formally divided into two sequential (thermodynamic) steps: (1) transfer of the residue from an anhydrous environment to water; and (2) transfer of the residue from water to variously modified aqueous solutions. The equivalent of measurement (1) has been carried out by many workers (e.g., see Cohn & Edsall, 1943; Tanford, 1962, 1968, 1970); in general the interior of the protein is modeled by non-aqueous alcohol or other organic solvent. The equivalent of measurement (2) has also been made, generally by measuring changes in the solubility in aqueous solution of compounds related to the functional groups of proteins as a consequence of the addition of various concentrations of neutral salts (e.g., see Robinson & Jencks, 1965; Schrier & Schrier, 1967; Nandi & Robinson, 1972; Hamabata, Chang & von Hippel, 1973). These workers showed that the change in free energy of a model compound (containing various combinations of methyl, methylene and amide groups), on transfer from water to a particular salt solution, could be written:

$$\Delta(\Delta G^\circ_{tr,X}) = RT \ln (S_0/S) = 2.3\ RTK_s c_s \qquad (3)$$

where $\Delta(\Delta G^\circ_{tr,X})$ represents the unitary free energy of transfer of model compound X from water (in which its solubility is S_0) to a particular aqueous salt solution containing salt S at concentration c_s, at temperature T. K_s is termed the salting-out coefficient of the compound. The formal analogy between K_s and K_m is quite obvious on comparison of eqs. (2) and (3).

Two additional results emerged from the above studies: (i) It was shown that the salting-out coefficient (and thus the free energy of transfer) of a model compound could be approximately reduced to the algebraic sum of contributions from the various functional groups (e.g., the methyl, methylene and amide groups) involved; and (ii) that the "cross-over" point in the Hofmeister series in going from salts which decrease the solubility (positive values of K_s), to those which increase solubility (negative values of K_s), varies with the ratio of polar (amide) to nonpolar (methyl and methylene) groups in the model compound under test, the K_s value for a given salt increasing with decreasing polar to nonpolar ratio. Furthermore, this cross-over point occurs near Na^+ and Cl^- (Fig. 2) for most protein and nucleic acid transitions which have been studied, implying that the interior groups of these entities all have rather comparable average polar to nonpolar ratios.

III. BINDING CONSTANTS OF IONS TO POLYACRYLAMIDE AND POLYSTYRENE COLUMNS

The changes in the free energy of transfer of model compounds from non-aqueous to aqueous environments, reflected in the solubility studies cited above, can easily be interpreted in terms of preferential (positive or negative) binding of the perturbing ions to the model compounds in the dissolved state; or, by analogy to the macromolecular transitions, by preferential binding to macromolecular groups exposed in the transition. Since several molar concentrations of neutral salts are required to change appreciably the stability of macromolecular structures or the solubility of model compounds, it was apparent that these binding constants must be very small, and so we have utilized the amplification inherent in the method of recycling chromatography to measure these constants.

The most numerically significant groups present in the interior of native proteins are the nonpolar amino acid residues. Following the Kauzmann-Tanford view of hydrophobic bonding, most workers currently involved in this field attribute most of the stability of native protein conformations to the hydrophobic bonding resulting from this preferred location of non-polar side-chains in the protein interior. However, the creation of such a nonpolar interior also requires the "burial" of a certain number of polar (particularly peptide) groups, since every side-chain in a protein is, of necessity, attached to a peptide bond. Thus, as an initial model system, we carried out ion-binding studies to polyacrylamide columns, since this material (monomer formula: $—CH_2—CH—C(=O)NH_2$) contains both amide and nonpolar groups. And since we were primarily interested in changes in the free energy of transfer of functional groups on adding neutral salts to the aqueous environment (see Fig. 3), we determined the binding of ions relative to that of water, by measuring the volume by which a sample of salt injected into the column emerged behind, or ahead of, a tritiated water marker. [St. Pierre and Jencks (1969), also in the context of protein denaturation, have independently measured the relative affinity of a number of neutral salts for polyacrylamide columns.]

A typical experiment involving two salts (with a common cation) is illustrated in Fig. 3. Here we show that a stabilizing salt (NaF) emerges well ahead of the tritiated water marker, while a strongly denaturing salt (NaI) emerges well behind the water peak. This means that NaF shows "negative binding" relative to water, while NaI shows positive relative binding. From such data relative equilibrium constants ($K_{a,rel}$) for the binding of ions to acrylamide moieties can be calculated (von Hippel et al., 1973). A typical set of such binding constants, converted to single ion relative binding constants by attributing the binding of NaCl equally to Na^+ and Cl^-, is summarized in Table I.

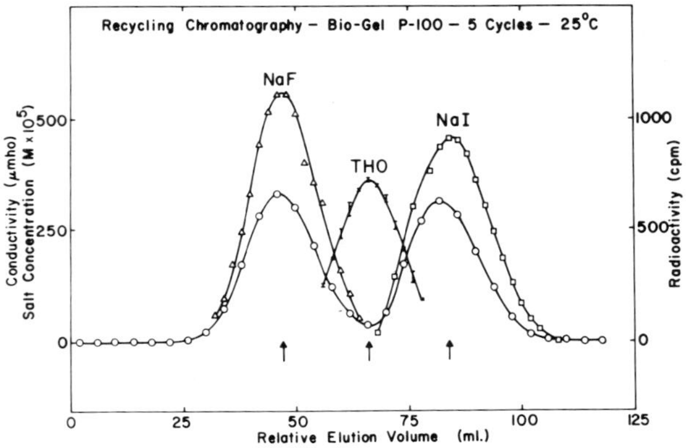

Figure 3

We can compare the anion and cation series in Table I with those in which these ions are arranged in order of effectiveness in altering the

TABLE I

"Single-ion" parameters for binding of ions to polyacrylamide at 25°[a]

Cation	$K_{a,rel}(M^{-1})$	$\Delta(\Delta G_{tr})_{res}$(cal/mole)[b]	Anion	$K_{a,rel}(M^{-1})$	$\Delta(\Delta G_{tr})_{res}$(cal/mole)[b]
K^+	-0.2×10^{-2}	+ 1.2	F^-	-3.1×10^{-2}	+18.0
Na^+	$+0.3 \times 10^{-2}$	- 1.8	$SO_4^=$	-1.4×10^{-2}	+ 8.3
Rb^+	$+0.5 \times 10^{-2}$	- 3.0	Cl^-	$+0.3 \times 10^{-2}$	- 1.8
Cs^+	$+1.0 \times 10^{-2}$	- 5.9	Br^-	$+2.9 \times 10^{-2}$	-17.0
Li^+	$+1.0 \times 10^{-2}$	- 5.9	NO_3^-	$+3.7 \times 10^{-2}$	-22.0
Mg^{++}	$+4.3 \times 10^{-2}$	-25.0	SCN^-	$+5.3 \times 10^{-2}$	-31.0
Ca^{++}	$+4.8 \times 10^{-2}$	-28.0	ClO_4^-	$+6.8 \times 10^{-2}$	-40.0
Ba^{++}	$+5.0 \times 10^{-2}$	-29.0			

[a]Based on assignment of the NaCl value of $K_{a,rel}$ equally to Na^+ and Cl^- (see text). The estimated uncertainty in $K_{a,rel} \sim \pm$ 10%. (Data from von Hippel et al., 1973).
[b]Calculated for 25°C and 1 M ion.

stability of protein conformations (Fig. 2). It is apparent not only that the order of ions in both rankings is the same, but also that the cross-over point in the ranking at which $K_{a,rel}$ goes from negative to positive values is the same as that at which the ions change from being stabilizers to being destabilizers of native protein conformations. Reasoning from the solubility results cited above, this suggests that the polar-nonpolar ratio in polyacrylamide is comparable to that of an average residue exposed in a typical protein unfolding transition, or that a typical residue exposed in such a transition might be represented by an acrylamide monomer containing an amide and two methylene groups. A variety of other evidence (see below) supports this conclusion.

IV. APPLICATIONS TO PROTEIN TRANSITIONS AND DISSOCIATIONS

Given values of $K_{a,rel}$, and the appropriateness of the acrylamide unit as a model for an average interior protein residue, we may use these data to calculate the difference between the free energy of transfer of such an average residue from the interior of a protein into a solution of stabilizing or destabilizing salt, and the free energy for the same transfer from the protein interior into pure water [$\Delta(\Delta G_{tr})_{res}$], using an equation modified from Schellman (1955):

$$\Delta G_{tr,\text{salt solution}} - \Delta G_{tr,H_2O} = \Delta(\Delta G_{tr})_{res} = -RT \sum_i \ln(1+K_{a,rel,i}[c_k]) \quad (3)$$

where $\Delta G_{tr,\text{salt solution}}$ and $\Delta G_{tr,H_2O}$ represent the change in the free energy of the acrylamide unit in going from an unsolvated environment to salt solution or water, respectively, $[c_i]$ is the concentration of ionic species i, and $K_{a,rel,i}$ is the relative equilibrium constant for salt binding to the amide moiety in polyacrylamide. Representative values of such single-ion parameters (for 1 M concentrations of ions) are also assembled in Table I. Such values can be calculated for any solution of neutral salts using eq. (3) and the concept of additivity of ion effects, and provide an estimate of $\Delta(\Delta G°_{tr})$ per average peptide unit exposed to the ionic solvent in the process of macromolecular unfolding (denaturation) or dissociation. The value of $\Delta(\Delta G°_{tr})$ per protein molecule can then be calculated directly for any transition in which the number of peptide groups exposed in a particular unfolding process is known. Alternatively, if $\Delta(\Delta G°_{tr})_{res}$ for transfer into the particular salt solution is known, one can link these parameters to changes in T_m using a melting-temperature-depression equation modified from Flory (1957):

$$\Delta T_m = T_m - T_m° = n\,[R(T_m°)^2 \Delta H_m°] \sum_i \ln(1+K_{a,rel,i}[c_i]) \quad (4)$$

where T_m and $T_m°$ are the melting temperature of the macromolecule in the denaturing (or stabilizing) neutral salt solution and in water, respectively, and $\Delta H_m°$ is the standard enthalpy change for the macromolecular unfolding process into the solvent under consideration. Thus, by measuring ΔT_m for a particular perturbant solution, one can, in principle, calculate n, the number of average residues exposed in the protein unfolding or dissociation process.

It is apparent from the notions outlined in Sections I and II that these concepts can be extended further. Thus, the change in the dissociation temperature ($\Delta T_{dissoc.}$) for an equilibrium protein subunit dissociation process as a function of perturbing salt concentration, or, alternatively, the change in perturbant concentration (Δc_s) needed to bring the dissociation transition to the experimental temperature in an isothermal measurement, can be formulated by equations similar to eqs. (3) and (4). For dissociation reactions involving the transfer of large numbers of residues from non-solvent-exposed to surface positions, the use of an average (acrylamide) residue to represent these groups is probably appropriate. However, for transitions in which only a few (and often "non-average") groups are exposed in the transition, new requirements and opportunities arise. To interpret these transitions in terms of the numbers and types of amino acid side chains present in the subunit-subunit "contact" surface, we will need to know the values of the free energies, enthalpies and entropies of transfer of the various side chains of proteins to various concentrations of the perturbing salts. Such model information could be gathered by recycling column procedures analogous to those described in Section III; with the appropriate values of $K_{a,rel,i}$ as a function of temperature in hand, it should, in principle, require only a few quick melting or salt-concentration-induced dissociation experiments on an unknown system to characterize the residues at the subunit interaction sites. By extending the "library" of model compound data available still further, similar approaches to the identification of stabilizing interactions in protein-lipid and protein-nucleic acid complexes should also be possible. We are currently extending the measurements of Robinson and Grant (1966) on nucleotide solubilities in aqueous salt solutions to provide such data for nucleic acid systems (von Hippel & Chang, in preparation).

V. SITES OF ION-BINDING

The above results show that relative binding constants for ions to acrylamide units can be obtained by recycling chromatographic methods, and that these constants can be used to obtain both thermodynamic and structural information about macromolecular transitions and association equilibria. This technique has also been used to obtain mechanistic information about the sites and mechanisms of ion-binding.

To this end, it was shown by varying the concentration of acrylamide units, that binding under conditions where amide groups are in local excess (relative to salt, in the column) is directly proportional to amide concentration. Also, recycling chromatographic experiments carried out on totally nonpolar (polystyrene) columns showed essentially no binding of ions to these columns, relative to the tritiated water marker. In addition to the insight they provide into the mechanisms of ion effects on macromolecules, these experiments also provide information relevant to the interpretation of hydrophobic bonding, since they show that nonpolar groups exposed to aqueous solutions are not surrounded by water (hydration) layers from which ions are measurably excluded, or into which they are measurably preferentially attracted [see von Hippel et al., (1973) for controls and experimental details; the thermodynamic analysis of these systems is described in Hamabata and von Hippel (1973) and von Hippel and Hamabata (1973)]. However, in the present context these experiments showed that the positive relative binding of ions to the acrylamide columns must be to the amide dipoles themselves, and that negative relative binding constants must arise primarily from the exclusion of stabilizing ions from the water of hydration of the amide dipole. [It can be shown by extrapolating to the totally excluded salt NaF, that this amounts to ∿ 2 water molecules per amide group, or ∿ 0.4 gm bound water per gm polyacrylamide, which is just about the level of water of hydration measured for proteins by a variety of techniques (e.g., see Kuntz et al., 1969).]

In summary, then, these polyacrylamide and polystyrene experiments have shown that the Hofmeister specificity of the neutral salts is fully developed in the relative binding of these salts to acrylamide units, that an acrylamide unit is an appropriate model of an average interior residue of a protein, and that binding is specifically to the amide dipole, and not to the nonpolar groups surrounding it.

VI. VICINAL GROUP EFFECTS IN ION BINDING TO AMIDES

Schrier and Schrier (1967) analyzed their solubility data on amide-containing model compounds in various neutral salts by dividing the measured salting-out coefficient (K_S) into a non-ion-specific part reflecting ion-binding to the amide dipole, and an ion-specific part reflecting specific salting-out of the vicinal methyl and methylene groups. This suggested that if one could measure relative ion-binding to amide residues surrounded by varying numbers of nonpolar residues, one might be able to test this hypothesis directly by extrapolating to zero vicinal nonpolar groups and determining whether, under these conditions, binding to the amide dipole becomes independent of ion type (for ions of a particular charge); i.e., whether for such binding to an "ideal" amide dipole the Hofmeister specificity of ion-binding is banished.

The measurements necessary to these experiments could not be made directly by recycling chromatographic techniques, since we were unable to obtain or prepare appropriately modified column packing materials. Nevertheless, the small magnitudes of the binding constants expected required the continued use of the amplification inherent in chromatographic procedures. Therefore, we inverted the previous procedure, using the various amides as the mobile (or eluant) phase and binding the ions to the stationary phase (column matrix) of ion-exchange resins. Pure cation or anion resins could not be used, since ion-ion interactions are much stronger than ion-dipole interactions, and thus ions can only be displaced from such resins by other ions and not by uncharged dipolar molecules. However, ions are primarily bound as ion-pairs to the paired acidic and basic sites in mixed cation-anion exchange resins (ion retardation resins). This permits the elution of ions from these resins by ion-dipole interactions; i.e., with an eluting phase containing only water or amide groups. Using this technique we have measured the relative (to water) binding constants of ions to variously substituted monomeric amides (for details see Hamabata & von Hippel, 1973).

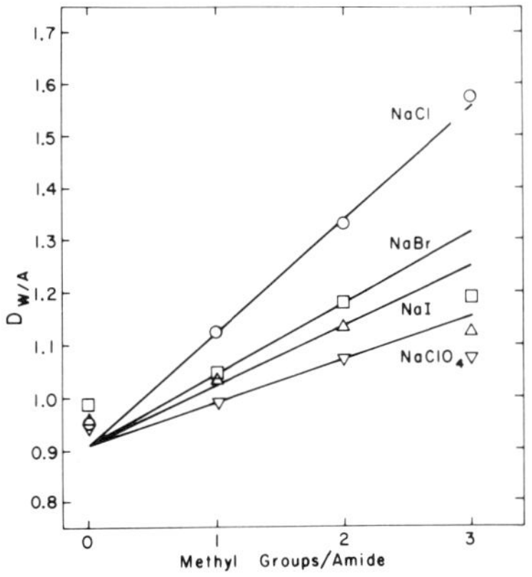

Figure 4

The central results are presented in four different salts in Fig. 4, in which the relative binding (distribution) constants ($D_{W/A}$) for several representative salts are plotted against the number of methyl groups per amide attached to the various amides tested. The amides examined include formamide ("zero methyl groups"), acetamide and N-methyl formamide (averaged for the "one methyl group" points), N-methyl acetamide and

N,N-dimethylformamide (averaged for the "two methyl groups" points) and N,N-dimethyl-acetamide (corresponding to the "three methyl group" data). Fig.4 shows that the (extrapolated) binding of mono-monovalent ions to the "ideal" amide dipole is indeed nonspecific, and that Hofmeister binding specificity develops as methyl groups are substituted preferentially around the amide dipole. Apparently, relative to water, all the ions treated bind preferentially (and with equal affinity) to the extrapolated "naked" amide dipole. As methyl groups are added, the binding diverges, becoming progressively less favorable to the amide-containing molecule. This progressive decrease in apparent relative binding affinity with added methyl groups is smallest for destabilizing anions such as ClO_4^- and I^-, and larger for such essentially inert ions as Cl^-. Thus, amide dipoles surrounded by two or three methyl groups appear to bind NaCl much less strongly than $NaClO_4$.

While these studies qualitatively support the original Schrier and Schrier hypothesis, detailed investigation shows that methyl groups attached to the amide group at different loci differ quantitatively in their effect on the development of the Hofmeister binding specificity. Thus methyl groups attached to the amide nitrogen were found to be less effective in decreasing binding to the amide group than methyl groups attached to the amide carbonyl carbon, etc. Furthermore, the results of Fig. 4 show that formamide itself retains some specificity of ion-binding; it approximates, but is not identical to, the non-specific amide dipole. Nevertheless, these results show conclusively that the specificity of ion-binding to amide groups is controlled (or "modulated") by the nonpolar surround of the amide dipole.

VII. CONCLUSIONS, EXTENSIONS AND EXTRAPOLATIONS

Studies of this sort can obviously be extended in a variety of directions in addition to those set out above:

(1) A direct thermodynamic connection has been established between the salting-out coefficient derived from solubility data [K_S; see eq. (2)] and the relative binding constant data (values of $D_{W/A}$) derived from column experiments (see Section VI). For details, see Hamabata and von Hippel

(1973) and von Hippel and Hamabata (1974).

(2) Direct solubility measurements have been made to determine the free energy of transfer, from water to various concentrations of the relevant neutral salt solutions, of methylene and methyl groups further removed from the polar amide group in model compounds (Nandi and Robinson, 1972; and Hamabata et al., 1973a). These and related data may be used to establish sets of values for the constituent groups of the amino acid side-chains of proteins, comparable to those listed in Table I for the acrylamide residue. Such values have been established for the acrylamide moiety by such an approach, and provide independent confirmation of the validity of this group as representative of the "average" residue in a protein "interior". Additional measurements will provide the necessary data to identify "contact residues" in protein subunit interactions, as outlined in Section IV.

(3) Finally, the results summarized in this lecture provide some, but by no means all, the information necessary to understand the molecular bases of the specificity of the Hofmeister series of ions. Thus we have shown that: (i) the Hofmeister specificity is fully developed for ion-binding (relative to water) to the acrylamide group of polyacrylamide; (ii) there is no significant preferential (relative to water) ion-binding to, or exclusion from, the nonpolar groups of polystyrene; (iii) binding is "non-salt-specific" to the ideal amide dipole, and becomes "salt-specific" with the insertion of vicinal methyl groups; and (iv) methylene groups far removed from the amide dipole, or isolated nonpolar groups, are salted-out of aqueous solution by ions with the characteristic Hofmeister specificity. Obviously, a two-part effect is involved: (a) an ion-dipole (salting-in) interaction at the amide locus; and (b) a "water-structure-modification" process manifested as salting-out of the nonpolar groups. Water-structure effects induced around the methyl groups directly adjacent to the hydrophilic amide dipole are modified by this proximity, so we can view these as "transitional" methylene groups relative to the "isolated" groups further removed from the amide group. The detailed nature of the "water-structure-modification effects" cited above is not known, and it is obvious that a final understanding of these "Hofmeister" effects still awaits a detailed unravelling of the equilibrium structures of water in

the presence of various ions, dipoles, and nonpolar groups, and an appreciation of the thermodynamic consequences of these structures.

REFERENCES

Cohn, E.J. and Edsall, J.T.: Proteins, Amino Acids and Peptides, Reinhold Publishing Corp. (1943).

Flory, P.J.: Crystallinity and dimensional changes in fibrous proteins. J. Cell. Comp. Physiol. 49 (suppl. 1), 175-183 (1957).

Hamabata, A., Chang, S. and von Hippel, P.H.: Solubility of fatty acid amides in ionic solutions. Biochem. 12, 1271-1278 (1973).

Hamabata, A. and von Hippel, P.H.: Effects of vicinal groups on the specificity of binding of ions to amide groups. Biochem. 12, 1264-1271 (1973).

Hamaguchi, K. and Geiduschek, E.P.: The effect of electrolytes on the stability of the DNA helix. J. Am. Chem. Soc. 84, 1329-1338 (1962).

Kauzmann, W.: Some factors in the interpretation of protein denaturation. Adv. Protein. Chem. 14, 1-63 (1959).

Kuntz, I.D., Brassfield, T.S., Law, G.D. and Purcell, G.V.: Hydration of macromolecules. Science 163, 1329-1332 (1969).

Long, F.A. and McDevit, W.F.: Activity of coefficients of nonelectrolyte solutes in aqueous salt solutions. Chem. Revs. 51, 119-169 (1952).

Nandi, P.K. and Robinson, D.R.: The effects of salts on the free energies of the peptide and nonpolar groups in model peptides. J. Am. Chem. Soc. 94, 1299-1315 (1972).

Robinson, D.R. and Grant, M.E.: The effects of aqueous salt solutions on the activity coefficients of purine and pyrimidine bases. J. Biol. Chem. 241, 4030-4042 (1966).

Robinson, D.R. and Jencks, W.P.: The effect of concentrated salt solutions on the activity coefficient of ATGEE. J. Am. Chem. Soc. 87, 2470-2479 (1965).

Schellman, J.A.: The stability of hydrogen-bonded peptide structures in aqueous solution. Compt. rend. trav. lab. Carlsberg, Ser. Chim. 29, 230-259 (1955).

Schrier, E.E. and Schrier, E.B.: The salting-out behavior of amides and its relation to the denaturation of proteins by salts. J. Phys. Chem. 71, 1851-1860 (1967).

St. Pierre, T. and Jencks, W.P.: Interaction of salts and denaturing agents with a polyacrylamide gel. Arch. Biochem. Biophys. 133, 99-104 (1969).

Tanford, C.: Contribution of hydrophobic interactions to the stability of the globular conformation of proteins. J. Am. Chem. Soc. 84, 4240-4247 (1962).

Tanford, C.: Protein denaturation. Adv. Protein Chem. 23, 121-282 (1968).

Tanford, C.: Protein denaturation. Adv. Protein Chem. 24, 1-97 (1970).

von Hippel, P.H. and Hamabata, A.: Model studies on the effects of neutral salts on the conformational stability of biological macromolecules. J. Mechanochem. Cell Motility 2, 127-138 (1973).

von Hippel, P.H., Peticolas, V., Schack, L. and Karlson, L.: Ion binding to polyacrylamide and polystyrene columns. Biochem. 12, 1256-1264 (1973).

von Hippel, P.H. and Schleich, T.: The effects of neutral salts on the structure and conformational stability of macromolecules in solution. In Biological Macromolecules, Vol. II, G. Fasman and S. Timasheff, eds. Marcel Dekker, New York, pp. 417-574 (1969a).

von Hippel, P.H. and Schleich, T.: Ion effects on the solution structure of biological macromolecules. Accts. Chem. Res. 2, 257-265 (1969b).

von Hippel, P.H. and Wong, K.-Y.: The collagen-gelatin phase transition. Biochem. 2, 1387-1398 (1963).

von Hippel, P.H. and Wong, K.-Y.: On the conformational stability of globular proteins; the ribonuclease transition. J. Biol. Chem. 240, 3909-3923 (1965).

Received, August 13, 1974

DISCUSSION

Blauer: How could you account for the smaller differences in binding within the cation series as compared with the anion series?

von Hippel: The differences between anions have always been known to be greater than the differences between cations in the Hofmeister series, in terms of effectiveness in various functions. There is no universally agreed upon explanation for this, but the smaller differences in cation affinity for amides shown here, obviously reflect this. Perhaps the lower charge densities of the anions make differences more prominent.

Blauer: How would other polar groups in proteins (hydroxyl, amino etc.) interact with the ions as compared with the amide groups investigated?

von Hippel: The small amount of data available suggests that binding to these (uncharged) moieties is not as great as to the more dipolar amides, but measurements have not been made in detail and would be required before a complete description of a real protein system could be given.

Pohl: The analysis of the static accessibility of different groups of proteins with known structure by Richards et al. show that most of the polar groups are in contact with the solvent, therefore, the contribution of these groups should be rather small compared with the overall change induced by the different salts in the case of unfolding.

von Hippel: Yes, but nevertheless those partially or totally unavailable, including peptide groups, would have to be taken into account in a complete analysis. As indicated in my lecture, the fact that for proteins which have been studied the "cross-over" point (see text) comes at approximately NaCl indicates that the average group actually exposed has a substantial polar component - i.e., it can be represented by one peptide group and two methylene groups.

Wagner: A nice application of your results and ideas about the interaction of salts with polyacrylamide gel was demonstrated by desalting of nucleotides. Taking NaCl, the nucleotide was eluted together with the salt, due to adsorptive effects of the nucleotide base towards the gel; however, desalting was successful with NaBr due to the larger affinity of the Br^- ion relative to Cl^- ions as you pointed out.

Steinhardt: Some years ago Kauzmann and his collaborators demonstrated that a positive charge at the protein surface has a very abnormal electrostriction effect on water. When separated from the surface it becomes more normal, finally becoming normal when the charge was removed by several chemical bonds. In thinking of ionic interactions in which positive charges are involved, the peculiarity of this surface water must be taken into account.

von Hippel: Yes, but to the extent that these groups are already at the surface, no differential effect would be observed. Only those formerly "inside" and transferred to "outside" by the transition (if any) would play a role here, but then their effects on the net free energy of transfer including reflections of water-structure effects, would have to be taken into account.

Wagner: Some time ago we measured very crudely the affinity of different anions towards poly-L-lysine. We found a dependence very like the Hofmeister series. We interpreted this as to reflect affinities of the anions directed towards the ε-amino groups of the lysines. This contribution may have some relevance in the lower concentration range of the salts.

Buehner: A very popular cation with crystallographers is ammonium. You have not shown a figure for it in your table. Does it go along with potassium? Some crystallographers have been trying to replace ammonium sulfate by lithium sulfate to trick some more cumbersome proteins into crystallizing, with not too much success thus far. This seems to be explained by your data which show a destabilizing effect of lithium. Could you please comment on this?

von Hippel: Yes, NH_4^+ falls somewhere near K^+ and Na^+, and is relatively inert. Li^+ would seem a bad choice for a system where maximum "salting-out" effectiveness is wanted, but in crystallization one often wants a balanced effect where only a few "aggregation nuclei" are created. Thus Li^+, coupled with a "salting-out" anion, might be appropriate. However, I understand from my crystallographic colleagues that protein crystallization is still more an "art" than a science.

Concluding Remarks

Peter H. von Hippel

Our friends and conference organizers, Horst and Gideon, have asked me to make some summarizing remarks in which they hoped I could put into perspective what we have learned, how it hangs together, and what the prospects for the field defined here, that of "Protein-Ligand Interactions" might be. I have listened as intently as possible throughout the week, learning about new developments in aspects of the field with which I was already partially familiar in some cases, and learning about the often fantastically promising developments in areas of which I knew little about before.

A comment about the <u>form</u> of the conference: I thought it was first-rate. Interactions between scientists, as well as between proteins and ligands were strong and usually favorable. I was astonished at the great coherence between scientists in rather distant disciplines, and how much we learned from one another at coffee, over dinner, etc., as well as in the formal sessions. It was much like the early Gordon Conferences, before they got so big. The preprints often helped in defining lectures in advance, and allowed for more intelligent discussions than is often the case.

But at the more mundane, scientific level, what did we learn? For me, at least, I felt the unifying impact of a somewhat different cross-section though the vast field of molecular biology, biochemistry, biophysics, physiology, etc., than is usually presented. Treating proteins as the functional units of life, their interactions with ligands were then indicated as crucial in discharging life's function. Thus proteins, which have repeatedly been defined in this conference as delicate, specifically balanced, pulsating entities, are portrayed as sitting in the midst of fields of ligands, to which evolution has accommodated them and to which they respond by changing specificities and functions. We see proteins as perched in free energy minima, separated from other potential minima by free energy differences of one or a few kcal/mole, but defined as specific entities by cooperative intra- and intermolecular interactions. Ligand binding shifts the relative free energy levels of the minima, and the proteins rapidly shift conformations in response to such binding.

The spectrum of the effects of ligands on proteins with which we have been presented ranges from unfolding and refolding phenomena under the influence of simple ions and detergents, which trigger conformational changes and influence rates by altering the relative stabilities of functional groups in these entities, through coenzymes and activators which bind to specific sites on enzymes and perturb subunit-subunit interactions, and thus enzyme rates and specificites, through agonists and antagonists of relatively simple structure which trigger receptors on membranes and induce the chain reactions involved in nerve conduction, and finally to proteins which bind specifically to sites on DNA and thus control the function and expression of the genetic apparatus.

In addition to this overall perspective, we have learned of powerful new and refurbished tools for studying these various systems: ranging from sophisticated uses of fluorescent probes from sources as esoteric as cobra venom, new uses of rapid techniques, developments in understanding of the interaction of chromophores in generating circular dichroism spectra, fancy new column chromatography techniques, and to the possibility, at least for some bacterial proteins, to use biological systems to produce families of proteins, for comparative study, which are modified at almost any amino acid locus one might wish to vary.

It is obvious that there will be many new developments to discuss at the next conference on this subject at Konstanz, to which we all hope to be invited.

Finally, one over-riding conclusion became even more apparent as the week progressed. Horst Sund and Gideon Blauer have arranged a delightful conference. When intellectual stimulation was needed we were served with excellent lectures; when sunshine was required to cheer our mid-week break it was provided; when food and drink were necessary to recharge our mental and physical batteries it appeared in copious quantity and highest quality; when problems arose they were solved smoothly and efficiently. In closing, I would like to request a rousing vote of thanks to Horst and Gideon, to their staff and to all who made our stay so pleasant. Many heartfelt thanks on behalf of us all!

Index of Contributors

* Symposium paper

Subject Index